C0-ARI-267

Contributions in Petroleum Geology and Engineering

Series Editor: George V. Chilingar, University of Southern California

Gulf Publishing Company
Houston, London, Paris, Zurich, Tokyo

Contributions in Petroleum Geology & Engineering 7

Hydrocarbon
Phase
Behavior

Tarek Ahmed

Contributions in Petroleum Geology and Engineering

Volume 7

Hydrocarbon Phase Behavior

Copyright © 1989 by Gulf Publishing Company, Houston, Texas. All rights reserved. Printed in the United States of America. This book, or parts thereof, may not be reproduced in any form without permission of the publisher.

Printed on acid-free paper (∞)

Library of Congress Cataloging-in-Publication Data

Ahmed, Tarek H., 1946–
Hydrocarbon phase behavior by Tarek Ahmed.
p. cm. — (Contributions in petroleum geology and engineering; v. 7)
Includes index.
1. Petroleum—Migration. 2. Phase rule and equilibrium. I. Title. II. Series: Contributions in petroleum geology & engineering; 7.
TN870.5.A35 1989   89-12003
622'.1828—dc20   CIP

ISBN 0-87201-589-0

ISBN 0-87201-066-X (Series)

Contents

Preface

This book explains the fundamentals of hydrocarbon phase behavior and their practical application in reservoir and production engineering. Although the book was developed from notes prepared for hydrocarbon phase behavior courses given to senior petroleum students, it should be useful as a reference book to practicing petroleum engineers.

Chapter 1 reviews the principles of hydrocarbon phase behavior and illustrates the use of phase diagrams in describing the volumetric behavior of single-component, two-component, and multi-component systems. Chapter 2 presents numerous mathematical and graphical correlations for estimating the physical and critical properties of the undefined petroleum fractions; and Chapter 3 deals with evaluation of properties of natural gases and introduces their applications in Darcy's equation and the material balance equation.

A complete and cohesive independent unit, Chapter 4 focuses on methods of determining the crude oil physical properties. Chapter 5 presents the concept and application of vapor-liquid phase equilibria. Chapter 6 reviews developments and advances in the field of empirical "cubic" equations of state and demonstrates their applications in petroleum engineering. Schemes of splitting and lumping petroleum fractions are illustrated in Chapter 7, and Chapter 8 discusses the simulation of laboratory PVT data by equations of state.

Much of the material on which this book is based was drawn from the publications of the Society of Petroleum Engineers of the American Institute of Mining, Metallurgical, and Petroleum Engineers; the American Gas Association; the Division of Production of the American Petroleum Institute; and the Gas Processors Suppliers Association. Tribute is due to these organizations and to the engineers, scientists, and authors who have made so many fine contributions to the field of hydrocarbon phase behavior.

I am indebted to my students at Montana Tech, whose enthusiasm for the subject has made teaching a pleasure. I would like to express my appreciation to all the people who have helped in the preparation of the book by technical comment and discussion and by giving permission to reproduce material. Special thanks to my colleagues: Professor Art Story, Dr. Herbert Warren, Dr. Gil Cody, Dr. Gene Collins, and Dr. Dan Bradley, for their encouragement and for making valuable suggestions for improvement of the book. I would also like to express my appreciation to the editorial staff of Gulf Publishing Company, especially Julia Starr. Thanks to Shanna for her patience and understanding, and believing that one day things would return to normal.

Tarek Ahmed

1
Basic Phase Behavior

A "phase" is defined as any homogeneous part of a system that is physically distinct and separated from other parts of the system by definite boundaries. For example, ice, liquid water, and water vapor constitute three separate phases of the pure substance H_2O because each is homogeneous and physically distinct from the others; moreover, each is clearly defined by the boundaries existing between them. Whether a substance exists in a solid, liquid, or gas phase is determined by the temperature and pressure acting on the substance. It is known that ice (solid phase) can be changed to water (liquid phase) by increasing its temperature and, by further increasing temperature, water changes to steam (vapor phase). This change in phases is termed Phase Behavior.

Hydrocarbon systems found in petroleum reservoirs are known to display multi-phase behavior over wide ranges of pressures and temperatures. The most important phases which occur in petroleum reservoirs are:

- Liquid phase, e.g., crude oils or condensates
- Gas phase, e.g., natural gases

The conditions under which these phases exist is a matter of considerable practical importance. The experimental or the mathematical determinations of these conditions are conveniently expressed in different types of diagrams, commonly called Phase Diagrams.

The objective of this chapter is to review the basic principles of hydrocarbon phase behavior and illustrate the use of phase diagrams in describing and characterizing the volumetric behavior of single-component, two-component, and multi-component systems.

SINGLE-COMPONENT SYSTEMS

The simplest type of hydrocarbon system to consider is that containing one component. The word "component" refers to the number of molecular

1

or atomic species present in the substance. A single-component system is composed entirely of one kind of atom or molecule. We often use the word "pure" to describe a single-component system.

The qualitative understanding of the relationship between temperature T, pressure p, and volume V of pure components can provide an excellent basis for understanding the phase behavior of complex petroleum mixtures. The foregoing relationship is conveniently introduced in terms of experimental measurements conducted on a pure component as the component is subjected to changes in pressure and volume at constant temperature. The effects of making these changes on the behavior of pure components are next discussed.

Suppose a fixed quantity of a pure component is placed in a cylinder fitted with a frictionless piston at a fixed temperature T_1. Furthermore, consider the initial pressure exerted on the system to be low enough that the entire system is in the vapor state. This initial condition is represented by point E on the pressure-volume phase diagram (p-V diagram) as shown in Figure 1-1. Consider the following sequential experimental steps taking place on the pure component:

Step 1. The pressure is increased isothermally by forcing the piston into the cylinder. Consequently, the gas volume decreases until it reaches point F on the diagram, where the liquid begins to condense. The corresponding pressure is known as the dew-point pressure p_d, and is defined as the pressure at which the first droplet of liquid is formed.

Step 2. The piston is moved further into the cylinder as more liquid condenses. This condensation process is characterized by a constant pressure and represented by the horizontal line FG. At point G, traces of gas remain and the corresponding pressure is called the bubble-point pressure p_b, and defined as the pressure at which the first sign of gas formation is detected.

A characteristic of a single-component system is that at a given temperature, the dew-point pressure and the bubble-point pressure are equal.

Step 3. As the piston is forced slightly into the cylinder, a sharp increase in the pressure (point H) is noted without an appreciable decrease in the liquid volume. That behavior evidently reflects the low compressibility of the liquid phase.

By repeating the above steps at progressively increasing temperatures, a family of curves of equal temperatures (isotherms) is constructed as shown in Figure 1-1. The dashed curve connecting the dew points is called the dew-point curve (line FC) and represents the states of the "saturated gas." The dashed curve connecting the bubble points is called the bubble-point

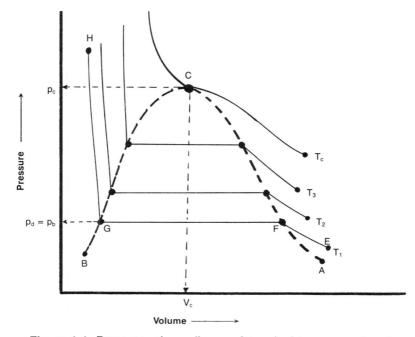

Figure 1-1. Pressure-volume diagram for a single component system.

curve (line GC) and similarly represents the "saturated liquid." These two curves meet at point C which is known as the critical point. The corresponding pressure and volume are called the critical pressure p_c and critical volume V_c, respectively. Notice that as the temperature increases, the length of the straight line portion of the isotherm decreases until it eventually vanishes, and the isotherm merely has a horizontal tangent and inflection point at the critical point. This isotherm temperature is called the critical temperature T_c of the single-component system. This observation can be expressed mathematically by the following relationship:

$$\left(\frac{\partial p}{\partial V}\right)_{Tc} = 0 \text{ (at the critical point)} \tag{1-1}$$

$$\left(\frac{\partial^2 p}{\partial V^2}\right)_{Tc} = 0 \text{ (at the critical point)} \tag{1-2}$$

Referring to Figure 1-1, the area enclosed by the phase envelope AFCGB is called the two-phase region. Within this defined region, vapor and liquid can coexist in equilibrium. Outside the phase envelope, only one phase can exist.

The critical point (point C) describes the critical state of the pure component and represents the limiting state for the existence of two phases, i.e., liquid and gas. In other words, for a single-component system, the critical point is defined as the highest value of pressure and temperature at which two phases can coexist. A more generalized definition of the critical point which is applicable to a single or multi-component system is: the critical point is the point at which all intensive properties of the gas and liquid phases are equal.

An intensive property is one that has the same value for any part of a homogeneous system as it does for the whole system, i.e., a property which is independent of the quantity of the system. Pressure, temperature, density, composition, and viscosity are examples of intensive properties.

Many characteristic properties of pure substances have been measured and compiled over the years. These properties provide vital information for calculating the thermodynamic properties of pure components as well as their mixtures. Those physical properties that are needed for hydrocarbon phase behavior calculations are presented in Table 1-1 for a number of hydrocarbon and non-hydrocarbon components.

Another means of presenting the results of this experiment is shown graphically in Figure 1-2 in which the pressure and temperature of the system are the independent parameters. Figure 1-2 shows a typical pressure-temperature diagram (p-T diagram) of a single-component system. The resulting curve, i.e., line AC, which terminates at the critical point (point C), can be thought of as being the dividing line between the area where liquid and vapor exists. The curve is commonly called the "vapor-pressure curve"

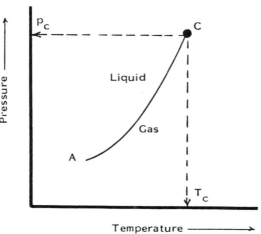

Figure 1-2. Pressure-temperature diagram for a pure component system.

or the "boiling-point curve." The corresponding pressure at any point on the curve is called the vapor pressure p_v.

Figure 1-2 shows that at the conditions of pressure and temperature specified by the vapor pressure curve, two phases can coexist in equilibrium. Systems represented by points located below the vapor-pressure curve are composed only of the vapor phase. Similarly, points above the curve represent systems that exist in the liquid phase. These remarks can be conveniently summarized by the following expressions:

if $p < p_v \rightarrow$ system is entirely in the vapor phase

$\quad p > p_v \rightarrow$ system is entirely in the liquid phase

$\quad p = p_v \rightarrow$ vapor and liquid coexist in equilibrium

where p is the pressure exerted on the pure substance. It should be pointed out that the above expressions are valid only if the system temperature is below the critical temperature T_c of the substance.

A method which is particularly convenient for plotting the vapor pressure as a function of temperature for pure substances is shown in Figure 1-3. The chart is known as the "Cox Charts." Notice that the vapor pressure scale is logarithmic, while the temperature scale is entirely arbitrary.

Example 1-1. A pure propane is held in a laboratory cell at 80°F and 200 psia. Determine the "existence state" (i.e., as a gas or liquid) of the substance.

Solution. From the Cox Charts, the vapor pressure $p_v = 150$ psi, and because the propane is under 200 psi (i.e., $p > p_v$), this means that the laboratory cell contains a liquefied propane.

The vapor pressure chart as presented in Figure 1-3 allows a quick estimation of the vapor pressure p_v of a pure substance at a specific temperature. For computer applications, however, an equation is more convenient. Lee and Kesler (1975) proposed the following generalized vapor pressure equation:

$$p_v = p_c \text{ EXP } (A + \omega B) \tag{1-3}$$

with

$$A = 5.92714 - \frac{6.09648}{T_r} - 1.2886 \ln (T_r) + 0.16934(T_r)^6 \tag{1-4}$$

$$B = 15.2518 - \frac{15.6875}{T_r} - 13.4721 \ln (T_r) + 0.4357(T_r)^6 \tag{1-5}$$

Table 1-1
Physical Properties for Pure Components

Physical Constants

Number	Compound	See Note No. -->	Formula	A. Molar mass (molecular weight)	B. Boiling point, °F 14.696 psia	Vapor pressure, psia 100 °F	C. Freezing point, °F 14.696 psia	D. Refractive index, n_D 60 °F	Critical constants Pressure, psia	Critical constants Temperature, °F	Critical constants Volume, ft³/lbm	Number
1	Methane		CH₄	16.043	-258.73	(5000)*	-296.44*	1.00042*	666.4	-116.67	0.0988	1
2	Ethane		C₂H₆	30.070	-127.49	(800)*	-297.04*	1.20971*	706.5	89.92	0.0783	2
3	Propane		C₃H₈	44.097	-43.75	188.64	-305.73*	1.29480*	616.0	206.06	0.0727	3
4	Isobutane		C₄H₁₀	58.123	10.78	72.581	-255.28	1.3245*	527.9	274.46	0.0714	4
5	n-Butane		C₄H₁₀	58.123	31.08	51.706	-217.05	1.33588*	550.8	305.62	0.0703	5
6	Isopentane		C₅H₁₂	72.150	82.12	20.445	-255.82	1.35631	490.4	369.10	0.0679	6
7	n-Pentane		C₅H₁₂	72.150	96.92	15.574	-201.51	1.35992	488.6	385.8	0.0675	7
8	Neopentane		C₅H₁₂	72.150	49.10	36.69	2.17	1.342*	464.0	321.13	0.0673	8
9	n-Hexane		C₆H₁₄	86.177	155.72	4.9597	-139.58	1.37708	436.9	453.6	0.0688	9
10	2-Methylpentane		C₆H₁₄	86.177	140.47	6.769	-244.62	1.37387	436.6	435.83	0.0682	10
11	3-Methylpentane		C₆H₁₄	86.177	145.89	6.103		1.37888	453.1	448.4	0.0682	11
12	Neohexane		C₆H₁₄	86.177	121.52	9.859	-147.72	1.37126	446.8	420.13	0.0667	12
13	2,3-Dimethylbutane		C₆H₁₄	86.177	136.36	7.406	-199.38	1.37730	453.5	440.29	0.0665	13
14	n-Heptane		C₇H₁₆	100.204	209.16	1.620	-131.05	1.38989	396.8	512.7	0.0691	14
15	2-Methylhexane		C₇H₁₆	100.204	194.09	2.272	-180.89	1.38714	396.5	495.00	0.0673	15
16	3-Methylhexane		C₇H₁₆	100.204	197.33	2.131		1.39091	408.1	503.80	0.0646	16
17	3-Ethylpentane		C₇H₁₆	100.204	200.25	2.013	-181.48	1.39566	419.3	513.39	0.0665	17
18	2,2-Dimethylpentane		C₇H₁₆	100.204	174.54	3.494	-190.86	1.38446	402.2	477.23	0.0665	18
19	2,4-Dimethylpentane		C₇H₁₆	100.204	176.89	3.293	-182.63	1.38379	396.9	475.95	0.0668	19
20	3,3-Dimethylpentane		C₇H₁₆	100.204	186.91	2.774	-210.01	1.38564	427.2	505.87	0.0662	20
21	Triptane		C₇H₁₆	100.204	177.58	3.375	-12.81	1.39168	428.4	496.44	0.0636	21
22	n-Octane		C₈H₁₈	114.231	258.21	0.53694	-70.18	1.39956	360.7	564.22	0.0690	22
23	Diisobutyl		C₈H₁₈	114.231	228.39	1.102	-132.11	1.39461	360.6	530.44	0.0676	23
24	Isooctane		C₈H₁₈	114.231	210.63	1.709	-161.27	1.38624	372.4	519.46	0.0656	24

No.	Compound	Formula	Mol. wt.	Boiling point °F	Vapor pressure 100°F, psia	Freezing point °F	Refractive index	Pc, psia	Tc, °F	Vc, ft³/lb	No.
25	n-Nonane	C9H20	128.258	303.47	0.17953	-64.28	1.40746	331.8	610.68	0.0684	25
26	n-Decane	C10H22	142.285	345.48	0.06088	-21.36	1.41385	305.2	652.0	0.0679	26
27	Cyclopentane	C5H10	70.134	120.65	9.915	-136.91	1.40896.	653.8	461.2	0.0594	27
28	Methylcyclopentane	C6H12	84.161	161.25	4.503	-224.40	1.41210	548.9	499.35	0.0607	28
29	Cyclohexane	C6H12	84.161	177.29	3.266	43.77	1.42862	590.8	536.6	0.0586	29
30	Methylcyclohexane	C7H14	98.188	213.68	1.609	-195.87	1.42538	503.5	570.27	0.0600	30
31	Ethene(Ethylene)	C2H4	28.054	-154.73	(1400)*	-272.47*	(1.228)*	731.0	48.54	0.0746	31
32	Propene(Propylene)	C3H6	42.081	-53.84	227.7	-301.45*	1.3130	668.6	197.17	0.0689	32
33	1-Butene(Butylene)	C4H8	56.108	20.79	62.10	-301.63*	1.3494*	583.5	295.48	0.0685	33
34	cis-2-Butene	C4H8	56.108	38.69	45.95	-218.06	1.3665*	612.1	324.37	0.0668	34
35	trans-2-Butene	C4H8	56.108	33.58	49.87	-157.96	1.3563*	587.4	311.86	0.0679	35
36	Isobutene	C4H8	56.108	19.59	63.02	-220.65	1.3512*	580.2	292.55	0.0682	36
37	1-Pentene	C5H10	70.134	85.93	19.12	-265.39	1.37426	511.8	376.93	0.0676	37
38	1,2-Butadiene	C4H6	54.092	51.53	36.53	-213.16		(653.)*	(340.)*	(0.065)*	38
39	1,3-Butadiene	C4H6	54.092	24.06	59.46	-164.02	1.3975*	627.5	305.	0.0654	39
40	Isoprene	C5H8	68.119	93.31	16.68	-230.73	1.42498	(558.)*	(412.)*	(0.065)*	40
41	Acetylene	C2H2	26.038	-120.49*		-114.5*		890.4	95.34	0.0695	41
42	Benzene	C6H6	78.114	176.18	3.225	41.95	1.50396	710.4	552.22	0.0531	42
43	Toluene	C7H8	92.141	231.13	1.033	-139.00	1.49942	595.5	605.57	0.0550	43
44	Ethylbenzene	C8H10	106.167	277.16	0.3716	-138.966	1.49826	523.0	651.29	0.0565	44
45	o-Xylene	C8H10	106.167	291.97	0.2643	-13.59	1.50767	541.6	674.92	0.0557	45
46	m-Xylene	C8H10	106.167	282.41	0.3265	-54.18	1.49951	512.9	651.02	0.0567	46
47	p-Xylene	C8H10	106.167	281.07	0.3424	55.83	1.49810	509.2	649.54	0.0570	47
48	Styrene	C8H8	104.152	293.25	0.2582	-23.10	1.54937	587.8	(703.)*	0.0534	48
49	Isopropylbenzene	C9H12	120.194	306.34	0.1884	-140.814	1.49372	465.4	676.3	0.0572	49
50	Methyl alcohol	CH4O	32.042	148.44	4.629	-143.79	1.33034	1174.	463.08	0.0590	50
51	Ethyl alcohol	C2H6O	46.069	172.90	2.312	-173.4	1.36346	890.1	465.39	0.0581	51
52	Carbon monoxide	CO	28.010	-312.68		-337.00*	1.00036*	507.5	-220.43	0.0532	52
53	Carbon dioxide	CO2	44.010	-109.257*		-69.83*	1.00048*	1071.	87.91	0.0344	53
54	Hydrogen sulfide	H2S	34.08	-76.497	394.59	-121.88*	1.00060*	1300.	212.45	0.0461	54
55	Sulfur dioxide	SO2	64.06	14.11	85.46	-103.86*	1.00062*	1143.	315.8	0.0305	55
56	Ammonia	NH3	17.0305	-27.99	211.9	-107.88*	1.00036*	1646.	270.2	0.0681	56
57	Air	N2+O2	28.9625	-317.8			1.00028*	546.9	-221.31	0.0517	57
58	Hydrogen	H2	2.0159	-422.955*		-435.26*	1.00013*	188.1	-399.9	0.5165	58
59	Oxygen	O2	31.9988	-297.332*		-361.820*	1.00027*	731.4	-181.43	0.0367	59
60	Nitrogen	N2	28.0134	-320.451		-346.00*	1.00028*	493.1	-232.51	0.0510	60
61	Chlorine	Cl2	70.906	-29.13	157.3	-149.73*	1.3878*	1157.	290.75	0.0280	61
62	Water	H2O	18.0153	212.000*	0.9501	32.00	1.33335	3198.8	705.16	0.0497s	62
63	Helium	He	4.0026	-452.09			1.00003*	32.99	-450.31	0.2300	63
64	Hydrogen chloride	HCl	36.461	-121.27	906.71	-173.52*	1.00042*	1205.	124.77	0.0356	64

NOTE: Numbers in this table do not have accuracies greater than 1 part in 1000; in some cases extra digits have been added to calculated values to achieve consistency or to permit recalculation of experimental values.

Courtesy of the Gas Processors Suppliers Association. Published in the GPSA Engineering Data Book, Tenth Edition, 1987.

(table continued on next page)

Table 1-1
Continued

Physical Constants

*See the Table of Notes and References.

Number	E. Density of liquid 14.696 psia, 60°F — Relative density (specific gravity) 60°F/60°F	E. lbm/gal.	E. gal./lb mole	F. Temperature coefficient of density, 1/°F	G. Acentric factor, ω	H. Compressibility factor of real gas, Z 14.696 psia, 60°F	I. Ideal gas 14.696 psia, 60°F — Relative density (specific gravity) Air = 1	I. ft³ gas/lbm	I. ft³ gas/gal. liquid	J. Specific Heat 60°F 14.696 psia Btu/(lbm·°F) — C_p, Ideal gas	J. C_p, Liquid	Number
1	(0.3)*	(2.5)*	(6.4172)*		0.0104	0.9980	0.5539	23.654	(59.135)*	0.52669		1
2	0.35619*	2.9696*	10.126*	-0.00162*	0.0979	0.9919	1.0382	12.620	37.476*	0.40782	0.97225	2
3	0.50699*	4.2268*	10.433*	-0.00119*	0.1522	0.9825	1.5226	8.6059	36.375*	0.38852	0.61996	3
4	0.56287*	4.6927*	12.386*	-0.00106*	0.1852	0.9711	2.0068	6.5291	30.639*	0.38669	0.57066	4
5	0.58401*	4.8690*	11.937*		0.1995	0.9667	2.0068	6.5291	31.790*	0.39499	0.57272	5
6	0.62470	5.2082	13.853	-0.00090	0.2280		2.4912	5.2596	27.393	0.38440	0.53331	6
7	0.63112	5.2617	13.712	-0.00086	0.2514		2.4912	5.2596	27.674	0.38825	0.54363	7
8	0.59666*	4.9744*	14.504*	-0.00106*	0.1963	0.9582	2.4912	5.2596	26.163*	0.39038	0.55021	8
9	0.66383	5.5344	15.571	-0.00075	0.2994		2.9755	4.4035	24.371	0.38628	0.53327	9
10	0.65785	5.4846	15.713	-0.00076	0.2780		2.9755	4.4035	24.152	0.38526	0.52732	10
11	0.66901	5.5776	15.451	-0.00076	0.2732		2.9755	4.4035	24.561	0.37902	0.51876	11
12	0.65385	5.4512	15.809	-0.00076	0.2326		2.9755	4.4035	24.005	0.38231	0.51367	12
13	0.66631	5.5551	15.513	-0.00076	0.2469		2.9755	4.4035	24.462	0.37762	0.51308	13
14	0.68820	5.7376	17.464	-0.00068	0.3494		3.4598	3.7872	21.729	0.38447	0.52802	14
15	0.68310	5.6951	17.595	-0.00070	0.3298		3.4598	3.7872	21.568	0.38041	0.52199	15
16	0.69165	5.7664	17.377	-0.00070	0.3232		3.4598	3.7872	21.838	0.37882	0.51019	16
17	0.70276	5.8590	17.103	-0.00069	0.3105		3.4598	3.7872	22.189	0.38646	0.51410	17
18	0.67829	5.6550	17.720	-0.00070	0.2871		3.4598	3.7872	21.416	0.38594	0.51678	18
19	0.67733	5.6470	17.745	-0.00073	0.3026		3.4598	3.7872	21.386	0.39414	0.52440	19
20	0.69772	5.8170	17.226	-0.00067	0.2674		3.4598	3.7872	22.030	0.38306	0.50138	20
21	0.69457	5.7907	17.304	-0.00068	0.2503		3.4598	3.7872	21.930	0.37724	0.49920	21
22	0.70696	5.8940	19.381	-0.00064	0.3977		3.9441	3.3220	19.580	0.38331	0.52406	22
23	0.69793	5.8187	19.632	-0.00067	0.3564		3.9441	3.3220	19.330	0.37571	0.51130	23
24	0.69624	5.8046	19.679	-0.00065	0.3035		3.9441	3.3220	19.283	0.38222	0.48951	24
25	0.72187	6.0183	21.311	-0.00061	0.4445		4.4284	2.9588	17.807	0.38246	0.52244	25
26	0.73421	6.1212	23.245	-0.00057	0.4898		4.9127	2.6671	16.326	0.38179	0.52103	26

12/12/86

No.												No.
27	0.42182	0.27199	33.856	5.4110	2.4215		0.1950	-0.00073	11.209	6.2570	0.75050	27
28	0.44126	0.30100	28.325	4.5090	2.9059		0.2302	-0.00069	13.397	6.2819	0.75349	28
29	0.43584	0.28817	29.452	4.5090	2.9059		0.2096	-0.00065	12.885	6.5319	0.78347	29
30	0.44012	0.31700	24.940	3.8649	3.3902		0.2358	-0.00062	15.216	6.4529	0.77400	30
31	0.57116	0.35697	39.167*	13.527	0.9686	0.9936	0.0865	-0.00173*	9.6889*	4.3432*	0.52095*	31
32	0.54533	0.35714	33.894*	9.0179	1.4529	0.9844	0.1356	-0.00112*	11.197*	5.0112*	0.60107*	32
33	0.52980	0.35446	35.366*	6.7636	1.9373	0.9699	0.1941	-0.00106*	10.731*	5.2288*	0.62717*	33
34	0.54215	0.33754	34.395*	6.7636	1.9373	0.9665	0.2029	-0.00117*	11.033*	5.0853*	0.60996*	34
35	0.54839	0.35574	33.574*	6.7636	1.9373	0.9667	0.2128		11.209*	5.0056*	0.60040*	35
36	0.51782	0.37690	33.856*	6.7636	1.9373	0.9700	0.1999	-0.00089	13.028	5.3834	0.64571	36
37	0.54029	0.36351	29.129	5.4110	2.4215		0.2333	-0.00101*	9.8605*	5.4857*	0.65799*	37
38	0.53447	0.34347	38.485*	7.0156	1.8677	(0.969)	0.2540		10.344*	5.2293*	0.62723*	38
39	0.51933	0.34120	36.687*	7.0156	1.8677	(0.965)	0.2007	-0.00110*	11.908	5.7205	0.68615	39
40		0.35072	31.869	5.5710	2.3520		0.1568	-0.00082				40
41		0.39754		14.574	0.8990	0.9930	0.1949		(7.473)	(3.4842)	(0.41796)	41
42	0.40989	0.24296	35.824	4.8581	2.6971		0.2093	-0.00067	10.593	7.3740	0.88448	42
43	0.40095	0.26570	29.937	4.1184	3.1814		0.2633	-0.00059	12.676	7.2691	0.87190	43
44	0.41139	0.27792	25.976	3.5744	3.6657		0.3027	-0.00056	14.609	7.2673	0.87168	44
45	0.41620	0.28964	26.363	3.5744	3.6657		0.3942	-0.00052	14.394	7.3756	0.88467	45
46	0.40545	0.27427	25.889	3.5744	3.6657		0.3257	-0.00053	14.658	7.2429	0.86875	46
47	0.40255	0.27471	25.800	3.5744	3.6657		0.3216	-0.00056	14.708	7.2181	0.86578	47
48	0.41220	0.27110	27.675	3.6435	3.5961		(0.2412)	-0.00053	13.712	7.5958	0.91108	48
49	0.42053	0.29170	22.804	3.1573	4.1500		0.3260	-0.00055	16.641	7.2228	0.86634	49
50	0.59187	0.32316	78.622	11.843	1.1063	0.9959	0.5649	-0.00066	4.8267	6.6385	0.79626	50
51	0.56610	0.33222	54.527	8.2372	1.5906	0.9943	0.6438	-0.00058	6.9595	6.6196	0.79399	51
52		0.24847	89.163*	13.548	0.9671	0.9846	0.0484		4.2561*	6.5812*	0.78939*	52
53		0.19911	58.807*	8.6229	1.5196	0.9802	0.2667	-0.00583*	4.4532*	6.8199*	0.81802*	53
54	0.50418	0.23827	74.401*	11.135	1.1767		0.0948		5.1005*	6.6817*	0.80144*	54
55	0.32460	0.14804	69.012*	5.9238	2.2118		0.2548	-0.00157*	5.4987*	11.650*	1.3974*	55
56	1.1209	0.49677	114.87*	22.283	0.5880	0.9877	0.2557		3.3037*	5.1550*	0.61832*	56
57		0.23988	95.557*	13.103	1.0000	1.0000			3.9713*	7.2930*	0.87476*	57
58		3.4038	111.54*	188.25	0.06960	1.0006	0.2202		3.4022*	9.59252*	0.071070*	58
59		0.21892	112.93*	11.859	1.1048	0.9992	0.0216		3.3605*	9.5221*	1.1421*	59
60		0.24828	91.413*	11.546	0.9672	0.9997	0.0372		4.1513*	6.7481*	0.80940*	60
61		0.11377	63.554*	5.3519	2.4482	(0.9875)	0.0878		5.9710*	11.875*	1.4244*	61
62	0.99974	0.44457	175.62	21.065	0.62202		0.3443	-0.00009	2.1609	8.33712	1.00000	62
63		1.2404	98.891*	94.814	0.1382	1.0006	0.		3.8376*	1.0430*	0.12510*	63
64		0.19086	73.869*	10.408	1.2589	0.9923	0.1259	-0.00300*	5.1373*	7.0973*	0.85129*	64

NOTE: Numbers in this table do not have accuracies greater than 1 part in 1000; in some cases extra digits have been added to calculated values to achieve consistency or to permit recalculation of experimental values.

Courtesy of the Gas Processors Suppliers Association. Published in the GPSA Engineering Data Book, Tenth Edition, 1987.

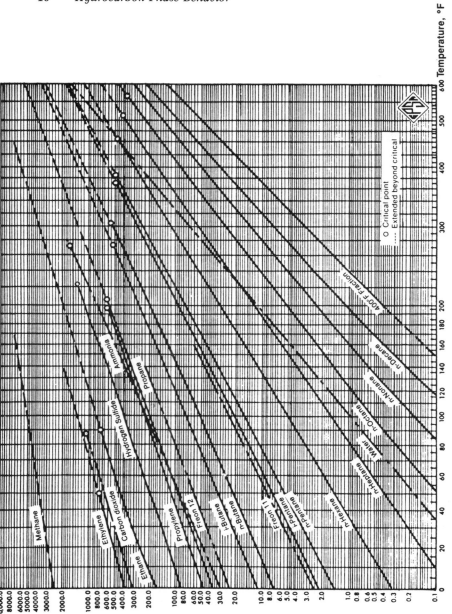

Figure 1-3. Vapor pressure chart for hydrocarbon components. Courtesy of the Gas Processors Suppliers Association. Published in the GPSA Engineering Data Book, Tenth Edition, 1987.

The term T_r is called the reduced temperature and is defined as the ratio of the absolute system temperature to the critical temperature of the fraction, or

$$T_r = \frac{T}{T_c}$$

where T_r = reduced temperature
$\quad\ T$ = substance temperature, °R
$\quad\ T_c$ = critical temperature of the substance, °R
$\quad\ p_c$ = critical pressure of the substance, psia
$\quad\ \omega$ = acentric factor* of the substance.

The acentric factor ω was introduced by Pitzer (1955) as a correlating parameter to characterize the acentricity or non-sphericity of a molecule, and is defined by the following expression:

$$\omega = -\text{Log}\left|\frac{p_v}{p_c}\right| - 1 \tag{1-6}$$

where p_v = vapor pressure of the substance at $T = 0.7\ T_c$, psia
$\quad\ p_c$ = critical pressure of the substance, psia

The acentric factor is frequently used as a third parameter in corresponding states and equation-of-state correlations. Values of the acentric factor for pure substances are tabulated in Table 1-1.

Example 1-2. Calculate the vapor pressure of propane at 80°F by using the Lee and Kesler correlation.

Solution.

• Obtain the critical properties and the acentric factor from Table 1-1.
$\quad\ T_c = 666.01$ °R
$\quad\ p_c = 616.3$ psia
$\quad\ \omega = 0.1522$

• Calculate the reduced temperature.

$$T_r = \frac{T}{T_c} = \frac{540}{666.01} = 0.81108$$

---

* When Equation 1-1 is employed to calculate p_v, it is recommended that Equation 2-13 from Chapter 2 be used to compute ω.

- Solve for the parameters A and B by applying Equations 1-4 and 1-5, respectively.

  A = − 1.27359
  B = − 1.147045

- Solve for pv by applying Equation 1-1.

  p_v = 616.3 EXP (− 1.27359 + 0.1572 (− 1.147045)) = 145 psia

The densities of the saturated phases of a pure component, i.e., densities of the coexisting liquid and vapor, may be plotted as a function of temperature, as shown in Figure 1-4. It should be noted that for increasing temperature, the density of the saturated liquid is decreasing, while the density of the saturated vapor increases. At the critical point C, the densities of vapor and liquid converge. At this critical temperature and pressure, all other properties of the phases become identical.

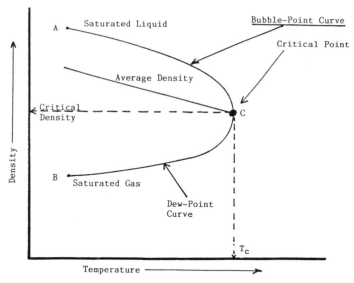

Figure 1-4. Typical density-temperature diagram for a pure component.

Figure 1-4 illustrates a useful observation, known as the Law of the Rectilinear Diameter, which states that the arithmetic average of the densities of the liquid and vapor phases is a linear function of the temperature. The straight line of average density versus temperature makes an easily defined intersection with the curved line of densities. This intersection then gives the critical temperature and density. Mathematically, this relationship is expressed as follows:

$$\frac{\rho_v + \rho_L}{2} = a + bT \tag{1-7}$$

where ρ_v = density of the saturated vapor, lb/ft^3
$\qquad \rho_L$ = density of the saturated liquid, lb/ft^3
$\qquad$ T = temperature, °R
$\qquad$ a, b = intercept and slope of the straight line.

At the critical point, Equation 1-7 can be expressed in terms of the critical density as follows:

$$\rho_c = a + b\ T_c \tag{1-8}$$

where ρ_c = critical density of the substance, lb/ft^3

Rackett (1970) proposed a simple generalized equation for predicting the saturated liquid density "ρ_L" of pure compounds. Rackett expressed the correlation in the following form:

$$\rho_L = \frac{(MW)\ p_c}{R\ T_c\ Z_c(1 + (1 - T_r)^{2/7})} \tag{1-9}$$

where ρ_L = saturated liquid density of the pure substance, lb/ft^3
$\qquad$ MW = molecular weight of the pure substance
$\qquad$ p_c = critical pressure of the substance, psia
$\qquad$ T_c = critical temperature of the substance, °R
$\qquad$ Z_c = critical gas compressibility factor
$\qquad$ R = gas constant, 10.73 ft^3 psia/lb-mole, °R
$\qquad$ $T_r = \dfrac{T}{T_c}$, reduced temperature
$\qquad$ T = temperature, °R

Spencer and Danner (1973) modified Rackett's correlation by replacing the critical compressibility factor Z_c in Equation 1-9 with the parameter Z_{RA} which is a unique constant for each compound. The authors proposed the following modification of the Rackett equation.

$$\rho_L = \frac{(MW)\ p_c}{R\ T_c\ Z_{RA}(1 + (1 - T_r)^{2/7})} \tag{1-10}$$

The values of Z_{RA} are given in Table 1-2 for selected components. If a value is not available, Yamada and Gunn (1973) suggested the following correlation to estimate Z_{RA}:

Table 1-2
Values of Z_{RA} for Selected Pure Components

Carbon dioxide	0.2722	n-Pentane	0.2684
Nitrogen	0.2900	n-Hexane	0.2635
Hydrogen sulfide	0.2855	n-Heptane	0.2604
Methane	0.2892	i-Octane	0.2684
Ethane	0.2808	n-Octane	0.2571
Propane	0.2766	n-Nonane	0.2543
i-Butane	0.2754	n-Decane	0.2507
n-Butane	0.2730	n-Undecane	0.2499
i-Pentane	0.2717		

$$Z_{RA} = 0.29056 - 0.08775 \, \omega \qquad (1\text{-}11)$$

where ω is the acentric factor of the compound.

Example 1-3. Calculate the saturated liquid density of propane at 160°F by using

a. The Rackett correlation
b. The modified Rackett equation

Solution.

- From Table 1-1
    T_c = 666.06 °R
    p_c = 616.0 psia
    MW = 44.097
    V_c = 0.0727 lb/ft^3
- Calculate Z_c by applying Equation 3-15 of Chapter 3.

$$Z_c = \frac{p_c \cdot V_c \cdot MW}{R \, T_c}$$

$$Z_c = \frac{(616.0) \, (0.0727) \, (44.097)}{(10.73) \, (666.06)} = 0.2763$$

$$T_r = \frac{160 + 460}{666.06} = 0.93085$$

a. The Rackett correlation. Solve for the saturated liquid density by applying the Rackett equation, i.e., Equation 1-9

$$\rho_L = \frac{(44.097)\ (616.0)}{(10.73)(666.06)(0.2763)^{1.4661}} = 25.05\ \text{lb/ft}^3$$

b. The modified Rackett equation. From Table 1-2, $Z_{RA} = 0.2766$

- Applying the modified Rackett equation, i.e., Equation 1-10

$$\rho_L = \frac{(44.097)(616)}{(10.73)(666.06)(0.2766)^{1.4661}} = 25.01\ \text{lb/ft}^3$$

TWO-COMPONENT SYSTEMS

A distinguishing feature of the single-component system is that, at a fixed temperature, two phases (vapor and liquid) can exist in equilibrium at only one pressure; this is the vapor pressure. For a binary system, two phases can exist in equilibrium at various pressures at the same temperature. The following discussion concerning the description of the phase behavior of a two-component system involves many concepts that apply to the more complex multi-component mixtures of oils and gases.

One of the important characteristics of the binary systems is the variation of their thermodynamic and physical properties with the composition. Therefore, it is necessary to specify the composition of the mixture in terms of mole or weight fractions. It is customary to designate one of the components as the more volatile component and the other the less volatile component, depending on their relative vapor pressure at a given temperature.

Suppose that the experiments previously described for a pure component are repeated, but this time we introduce into the cylinder a binary mixture of a known overall composition. Consider that the initial pressure p_1 exerted on the system, at a fixed temperature of T_1, is low enough that the entire system exists in the vapor state. This initial condition of pressure and temperature acting on the mixture is represented by point 1 on the p-V diagram of Figure 1-5. As the pressure is increased isothermally, it reaches point 2, at which an infinitesimal amount of liquid is condensed. The pressure at this point is called the dew-point pressure p_d of the mixture. It should be noted that at the dew-point pressure, the composition of the vapor phase is equal to the overall composition of the binary mixture. As the volume is decreased (by forcing the piston inside the cylinder), a noticeable increase in the pressure is observed as more and more liquid is condensed. This condensation process is continued until the pressure reaches point 3, at which traces of gas remain. At point 3, the corresponding pressure is called the bubble-point pressure p_b. Because at the bubble point the gas phase is only of infinitesimal volume, the composition of the liquid phase is therefore identical with that of the whole system. As the piston is forced further into the cylinder, the pressure rises steeply to point 4 with a corresponding decreasing volume.

Repeating the above experiments at progressively increasing temperatures, a complete set of isotherms is obtained as shown on the p-V diagram of Figure 1-6. The bubble-point curve, as represented by the line AC, represents the locus of the points of pressure and volume at which the first bubble of gas is formed. The dew-point curve (line BC) describes the locus of the points of pressure and volume at which the first droplet of liquid is formed. The two curves meet at the critical point (point C). The critical pressure, temperature, and volume are given by p_c, T_c, and V_c, respectively. Any point within the phase envelope (line ACB) represents a system consisting of two phases. Outside the phase envelope, only one phase can exist.

If the bubble-point pressure and dew-point pressure for the various isotherms on a p-V diagram are plotted as a function of temperature, a p-T diagram similar to that shown in Figure 1-7 is obtained. Figure 1-7 indicates that the pressure-temperature relationships can no longer be represented by a simple vapor pressure curve as in the case of a single-component system, but take on the form illustrated in Figure 1-7 by the phase envelope ACB. The dashed lines within the phase envelope are called the "quality lines"; they describe the pressure and temperature conditions of equal volumes of liquid. Obviously, the bubble-point curve and the dew-point curve represent 100% and 0% liquid, respectively.

Figure 1-8 demonstrates the effect of changing the composition of the binary system on the shape and location of the phase envelope. Two of the lines, as shown in Figure 1-8, represent the vapor pressure curves for meth-

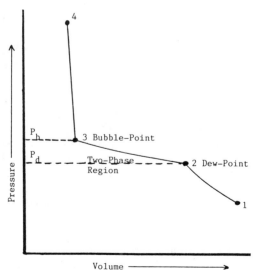

Figure 1-5. Pressure-volume diagram for a binary system.

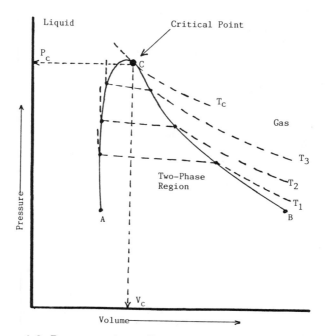

Figure 1-6. Pressure-volume diagram for a two-component system.

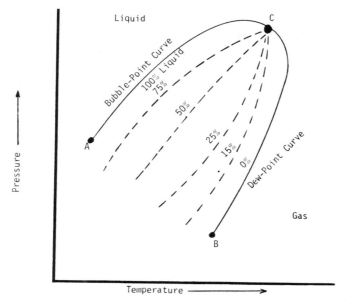

Figure 1-7. Typical pressure-temperature diagram for a binary system.

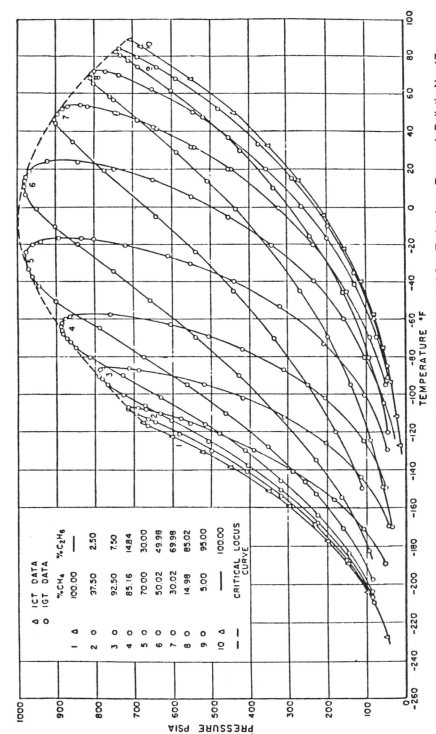

Figure 1-8. Phase diagram of methane-ethane mixture. Courtesy of the Institute of Gas Technology, Research Bulletin No. 17.

ane and ethane, which terminate at the critical point. Ten phase boundary curves (phase envelopes) for various mixtures of methane and ethane are also shown. These curves pass continuously from the vapor-pressure curve of the one pure component to that of the other as the composition is varied. The points labeled 1–10 represent the critical points of the mixtures as defined in the legend of Figure 1-8. The dashed curve illustrates the locus of critical points for the binary system.

It should be noted by examining Figure 1-8 that when one of the constituents becomes predominant, the binary mixture tends to exhibit a relatively narrow phase envelope and displays critical properties close to the predominant component. The size of the phase envelope enlarges noticeably as the composition of the mixture becomes evenly distributed between the two components.

Figure 1-9 shows the critical loci for a number of common binary systems. Obviously, the critical pressure of mixtures is considerably higher than the critical pressure of the components in the mixtures. The greater the difference in the boiling point of the two substances, the higher the critical pressure of the mixture will be.

The Phase Rule

It is appropriate at this stage to introduce and define the concept of the "Phase Rule." Gibbs (1876) derived a simple relationship between the number of phases in equilibrium, the number of components, and the number of independent variables that must be specified to describe the state of the system completely.

Gibbs proposed the following fundamental statement of the phase rule.

$$F = C - P + 2 \tag{1-12}$$

where F = number of variables required to determine the state of the system at equilibrium, or number of degrees of freedom
C = number of independent components
P = number of phases

A phase has been defined previously as a homogeneous system of uniform physical and chemical composition. In a system containing ice, liquid water, and water vapor in equilibrium there are three phases, i.e., P = 3. The number of independent components in the system is one, i.e., C = 1, since the system contains only H_2O. The degrees of freedom F for a system include the temperature, the pressure, and the composition (concentration) of phases. These independent variables must be specified to define the system completely.

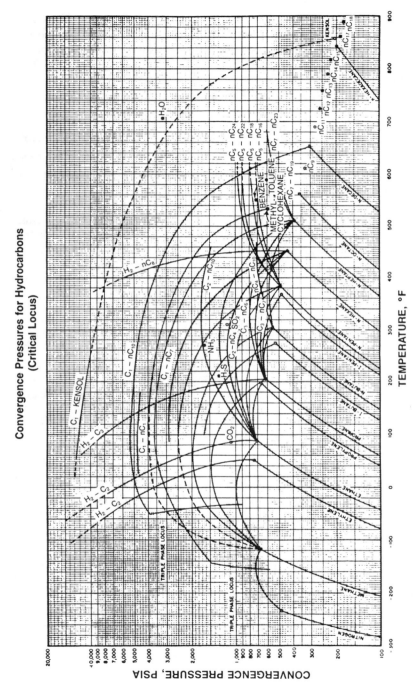

Figure 1-9. Critical loci of binary systems. Courtesy of the Gas Processors Suppliers Association. Published in the GPSA Engineering Data Book, Tenth Edition, 1987.

The phase rule as described by Equation 1-12 is useful in several ways. It indicates the maximum possible number of equilibrium phases that can co-exist and the number of components present. It should be pointed out that the phase rule does not determine the nature, the exact composition, or total quantity of the phases. Furthermore, it applies only to a system in stable equilibrium and does not determine the rate at which this equilibrium is attained.

The importance and the practical application of the phase rule are illustrated through the following examples:

Example 1-4. For a single-component system, determine the number of degrees of freedom required for the system to exist in the single-phase region.

Solution. Applying Equation 1-12, gives $F = 1 - 1 + 2 = 2$. There are two degrees of freedom that must be specified for the system to exist in the single phase region. These must be the pressure p and the temperature T.

Example 1-5. What are the degrees of freedom allowed for a two-component system in two phases?

Solution. Since $C = 2$ and $P = 2$, applying Equation 1-12 yields $F = 2 - 2 + 2 = 2$. The two degrees of freedom could be the system pressure and the system temperature, or p and the concentration (mole fraction), or some other combination of T, p, and composition.

Example 1-6. For a three-component system, determine the number of degrees of freedom that must be specified for the system to exist in the one-phase region.

Solution. Using the phase rule expression gives: $F = 3 - 1 + 2 = 4$. There are four independent variables that must be specified to fix the system. The variables could be the pressure, the temperature, and the mole fractions of two of the three components.

MULTI-COMPONENT SYSTEMS

The phase behavior of multi-component hydrocarbon systems in the liquid-vapor region is very similar to that of binary systems. However, as the system becomes more complex with a greater number of different components, the pressure and temperature ranges in which two phases exist increase significantly.

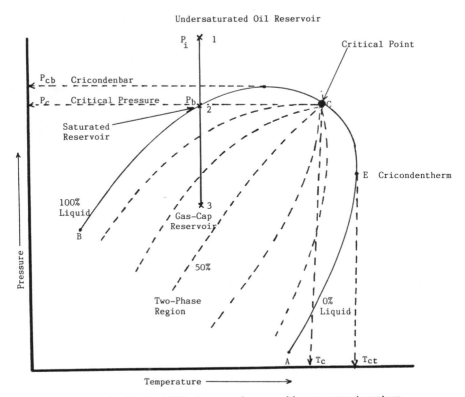

Figure 1-10. Typical P-T diagram for a multi-component system.

Figure 1-10 shows a typical pressure-temperature diagram of a multi-component system with a specific overall composition. Although a different hydrocarbon system would have a different phase diagram, the general configuration is similar.

These multi-component p-T diagrams are essentially used to

- Classify reservoirs
- Classify the naturally occurring hydrocarbon systems
- Describe the phase behavior of the reservoir fluid

To fully understand the significance of the p-T diagrams, it is necessary to identify and define the following key points on the p-T diagram:

- Cricondentherm (T_{ct})—The cricondentherm is defined as the maximum temperature above which liquid cannot be formed regardless of pressure (point E). The corresponding pressure is termed the cricondentherm pressure p_{ct}.

- Cricondenbar (p_{cb})—The cricondenbar is the maximum pressure above which no gas can be formed regardless of temperature (point D). The corresponding temperature is called the cricondenbar temperature T_{cb}.
- Critical point—The critical point for a multi-component mixture is referred to as the state of pressure and temperature at which all intensive properties of the gas and liquid phases are equal (point C). At the critical point, the corresponding pressure and temperature are called the critical pressure p_c and critical temperature T_c of the mixture.
- Phase envelope (two-phase region)—The region enclosed by the bubble-point curve and the dew-point curve (line BCA), wherein gas and liquid coexist in equilibrium, is identified as the phase envelope of the hydrocarbon system.
- Quality lines—The dashed lines within the phase diagram are called quality lines. They describe the pressure and temperature conditions for equal volumes of liquids. Note that the quality lines converge at the critical point (point C).
- Bubble-point curve—The bubble-point curve (line BC) is defined as the line separating the liquid phase region from the two-phase region.
- Dew-point curve—The dew-point curve (line AC) is defined as the line separating the vapor phase region from the two-phase region.

CLASSIFICATION OF RESERVOIRS AND RESERVOIR FLUIDS

Proper classification of a reservoir requires the knowledge of the thermodynamic behavior of the phases present in the reservoir and forces responsible for the production mechanism. In general, reservoirs are conveniently classified on the basis of the location of the point representing the initial reservoir pressure p_i and temperature T with respect to the p-T diagram of the reservoir fluid. Accordingly, reservoirs can be classified into essentially two types. These are:

- Oil reservoirs—If the reservoir temperature T is less than the critical temperature T_c of the reservoir fluid, the reservoir is classified as an oil reservoir.
- Gas reservoirs—If the reservoir temperature is greater than the critical temperature of the hydrocarbon fluid, the reservoir is considered a gas reservoir.

Oil Reservoirs

Depending upon initial reservoir pressure p_i, oil reservoirs can be subclassified into the following categories:

1. Undersaturated Oil Reservoir. If the initial reservoir pressure p_i (as represented by point 1 on Figure 1-10), is greater than the bubble-point pressure p_b of the reservoir fluid, the reservoir is labeled an undersaturated oil reservoir.

2. Saturated Oil Reservoir. When the initial reservoir pressure is equal to the bubble-point pressure of the reservoir fluid, as shown on Figure 1-10 by point 2, the reservoir is called a saturated oil reservoir.

3. Gas-cap Reservoir. If the initial reservoir pressure is below the bubble-point pressure of the reservoir fluid, as indicated by point 3 on Figure 1-10, the reservoir is termed a gas-cap or two-phase reservoir, in which the gas or vapor phase is underlain by an oil phase. The ratio of the gas-cap volume to reservoir oil volume is given by the appropriate quality line.

Crude oils cover a wide range in physical properties and chemical compositions, and it is often important to be able to group them into broad categories of related oils. In general, crude oils are commonly classified into the following types:

- Ordinary black oil
- Low-shrinkage crude oil
- High-shrinkage (volatile) crude oil
- Near-critical crude oil

The above classification is essentially based upon the properties exhibited by the crude oil, including:

- Physical properties
- Composition
- Gas-oil ratio
- Appearance
- Pressure-temperature phase diagram

Note that the reservoir temperature also plays a role in the classification of the crude oil.

1. Ordinary Black Oil. A typical p-T phase diagram for an ordinary black oil is shown in Figure 1-11. The phase diagram is characterized by quality lines that are approximately equally spaced. Following the pressure reduction path as indicated by the vertical line EF on Figure 1-11, the liquid shrinkage curve, as shown in Figure 1-12, is prepared by plotting the liquid volume percent as a function of pressure. The liquid shrinkage curve approximates a straight line except at very low pressures. When produced, ordinary black oils usually yield gas-oil ratios between 200–700 scf/STB and

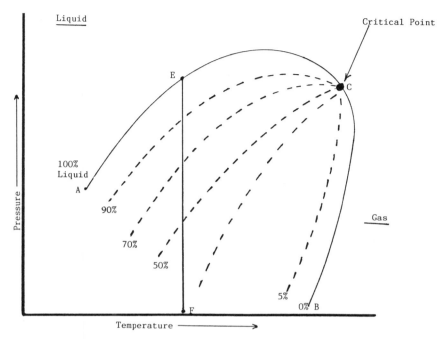

Figure 1-11. A typical P-T diagram for an ordinary black oil.

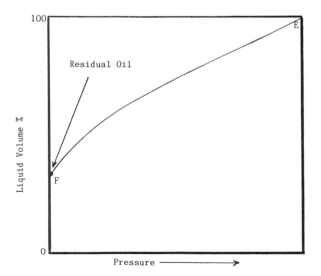

Figure 1-12. Liquid shrinkage curve for black oil.

oil gravities of 15 to 40°API. The stock-tank oil is usually brown to dark green in color.

2. Low-shrinkage Oil. A typical p-T phase diagram for a low shrinkage oil is shown in Figure 1-13. The diagram is characterized by quality lines that are closely spaced near the dew-point curve. The liquid shrinkage curve, as given in Figure 1-14, shows the shrinkage characteristics of this category of crude oils. The other associated properties of this type of crude oil are:

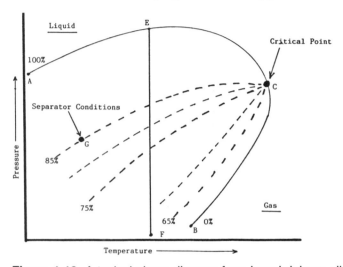

Figure 1-13. A typical phase diagram for a low shrinkage oil.

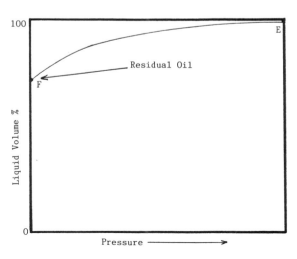

Figure 1-14. A typical liquid shrinkage curve for a low shrinkage oil.

- Gas-oil ratio less than 200 scf/STB
- Oil gravity less than 15°API
- Black or deeply colored
- Substantial liquid recovery at separator conditions as indicated by point G on the 85% quality line of Figure 1-13.

3. Volatile Crude Oil. The phase diagram for a volatile (high-shrinkage) crude oil is given in Figure 1-15. Note that the quality lines are close together near the bubble point and at lower pressures they are more widely spaced. This type of crude oil is commonly characterized by a high liquid shrinkage immediately below the bubble point as shown in Figure 1-16. The other characteristic properties of this oil include:

- Gas-oil ratios between 2,000–3,500 scf/STB
- Oil gravities between 45–55°API
- Lower liquid recovery of separator conditions as indicated by point G on Figure 1-15
- Greenish to orange in color

4. Near-critical Crude Oil. If the reservoir temperature T is near the critical temperature T_c of the hydrocarbon system, as shown in Figure 1-17, the hydrocarbon mixture is identified as a near-critical crude oil. Because all the

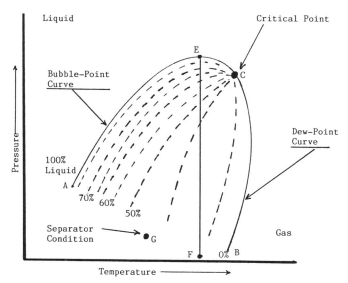

Figure 1-15. A typical P-T diagram for a volatile crude oil.

quality lines converge at the critical point, an isothermal pressure drop (as shown by the vertical line EF in Figure 1-17) may shrink the crude oil from 100% of the hydrocarbon pore volume at the bubble point to 55% or less at a pressure 10 to 50 psi below the bubble point. The shrinkage characteristic behavior of the near-critical crude oil is shown in Figure 1-18.

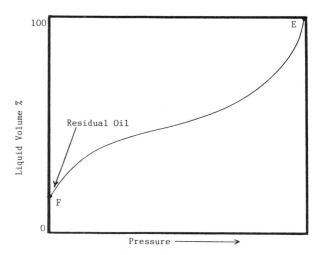

Figure 1-16. A typical liquid shrinkage curve for a volatile crude oil.

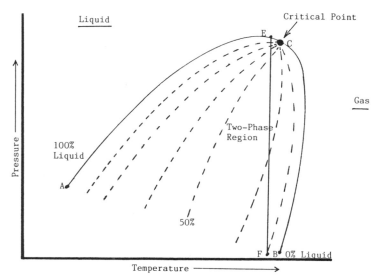

Figure 1-17. A schematic phase diagram for the near-critical crude oil.

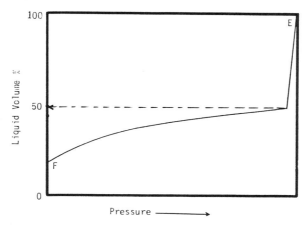

Figure 1-18. A typical liquid shrinkage curve for the near-critical crude oil.

Gas Reservoirs

In general, if the reservoir temperature is above the critical temperature of the hydrocarbon system, the reservoir is classified as a natural gas reservoir. Natural gases can be categorized on the basis of their phase diagram and the prevailing reservoir condition into four categories:

- Retrograde gas-condensate
- Near-critical gas-condensate
- Wet gas
- Dry gas

Retrograde Gas-condensate Reservoir. If the reservoir temperature T lies between the critical temperature T_c and cricondentherm T_{ct} of the reservoir fluid, the reservoir is classified as a retrograde gas-condensate reservoir. This category of gas reservoir is a unique type of hydrocarbon accumulation in that the special thermodynamic behavior of the reservoir fluid is the controlling factor in the development and the depletion process of the reservoir.

Consider that the initial condition of a retrograde gas reservoir is represented by point 1 on the pressure-temperature phase diagram of Figure 1-19. Because the reservoir pressure is above the upper dew-point pressure, the hydrocarbon system exists as a single phase (i.e., vapor phase) in the reservoir. As the reservoir pressure declines isothermally during production from the initial pressure (point 1) to the upper dew-point pressure (point 2), liquid begins to condense. As the pressure is further decreased, instead of expanding (if a gas) or vaporizing (if a liquid) as might be expected, the hydrocar-

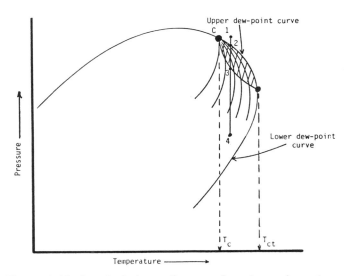

Figure 1-19. A typical phase diagram of a retrograde system.

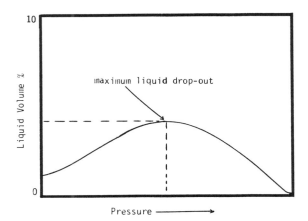

Figure 1-20. A typical liquid drop-out curve.

bon mixture tends to condense. This retrograde condensation process continues with decreasing pressure until the liquid drop-out reaches its maximum at point 3. However, at point 4, the dew-point curve must be crossed again. This means that all the liquid which formed must vaporize because the system is essentially all vapor at the lower dew-point.

The liquid shrinkage volume curve, commonly called the liquid drop-out curve, for a condensate system is shown in Figure 1-20. In most gas-condensate reservoirs, the condensed liquid volume seldom exceeds more than 10% of the pore volume. This liquid saturation is not large enough to allow any liquid flow. It should be recognized, however, that around the wellbore where the pressure drop is high, enough liquid drop-out might accumulate to give two-phase flow of gas and retrograde liquid.

The associated physical characteristics of this category are:

- Gas-oil ratios between 8,000 to 70,000 scf/STB. Generally, the gas-oil ratio for a condensate system increases with time due to the liquid drop-out and the loss of heavy components in the liquid.
- Condensate gravity above 50°API
- Stock-tank liquid is usually water-white or slightly colored.

Near-critical Gas-condensate Reservoir. If the reservoir temperature is near the critical temperature, as shown in Figure 1-21, the hydrocarbon mixture is classified as a near-critical gas-condensate. The volumetric behavior of this category of natural gas is described through the isothermal pressure declines as shown by the vertical line 1-3 in Figure 1-21 and also by the corresponding liquid drop-out curve of Figure 1-22. Because all the quality lines converge at the critical point, a rapid liquid build-up immediately below the dew-point will result (Figure 1-22) as the pressure is reduced to point 2.

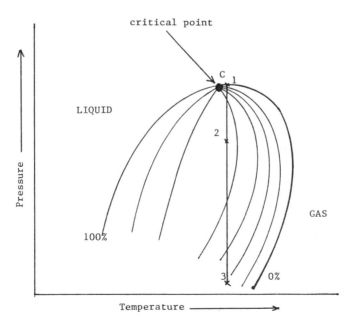

Figure 1-21. A typical phase diagram for a near critical-critical gas condensate reservoir.

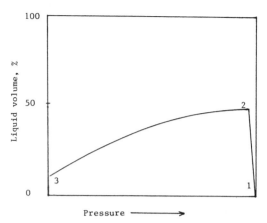

Figure 1-22. Liquid shrinkage curve for a near-critical gas-condensate system.

This behavior can be justified by the fact that several quality lines are crossed very rapidly by the isothermal reduction in pressure. At the point where the liquid ceases to build up and begins to shrink again, the reservoir goes from the retrograde region to a normal vaporization region.

Wet Gas Reservoir. A typical phase diagram of a wet gas is shown in Figure 1-23, where reservoir temperature is above the cricondentherm of the hydrocarbon mixture. Because the reservoir temperature exceeds the cricondentherm of the hydrocarbon system, the reservoir fluid will always remain in the vapor phase region as the reservoir is depleted isothermally, along the vertical line A-B. However, as the produced gas flows to the surface, the pressure and temperature of the gas will decline. If the gas enters the two-phase region, a liquid phase will condense out of the gas and be produced from the surface separators.

Wet gas reservoirs are characterized by the following properties:

- Gas-oil ratios between 60,000 to 100,000 scf/STB
- Stock-tank oil gravity above 60°API
- Liquid is water-white in color
- Separator conditions, i.e., separator pressure and temperature, lie within the two-phase region

Dry Gas Reservoir. The hydrocarbon mixture exists as a gas both in the reservoir and the surface facilities. The only liquid associated with the gas from a dry gas reservoir is water. A phase diagram of a dry gas reservoir is given in Figure 1-24. Usually a system having a gas-oil ratio greater than 100,000 scf/STB is considered to be a dry gas.

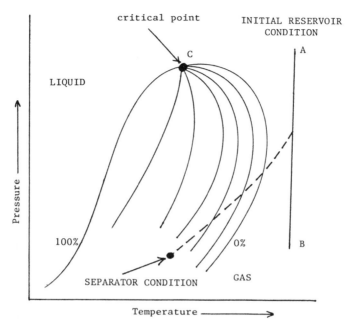

Figure 1-23. Pressure-temperature diagram for a wet gas reservoir.

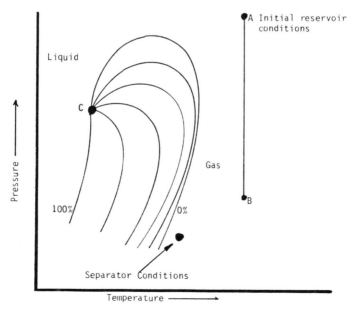

Figure 1-24. A typical pressure-temperature diagram for dry gas reservoir.

PROBLEMS

1. A pure component has the following vapor pressure:

T, °F	104	140	176	212
p_v, psi	46.09	135.04	345.19	773.75

   a. Plot the above data so as to obtain a nearly straight line
   b. Determine the boiling point at 200 psi
   c. Vapor pressure at 250°F

2. The critical temperature of a pure component is 260°F. The densities of the liquid and vapor phase at different temperatures are:

T, °F	86	122	158	212
ρ_L, lb/ft^3	40.28	38.16	35.79	30.89
ρ_v, lb/ft^3	0.886	1.691	2.402	5.054

   Determine the critical density of the substance.

3. Using the Lee and Kesler vapor correlation, calculate the vapor pressure of i-butane at 100°F. Compare the calculated vapor pressure with that obtained from the Cox charts.

4. Calculate the saturated liquid density of n-butane at 200°F by using:
   a. The Rackett correlation.
   b. The modified Rackett correlation.

5. What is the maximum number of phases that can be in equilibrium at constant temperature and pressure in one-, two-, and three-component systems?

6. For a seven-component system, determine the number of degrees of freedom that must be specified for the system to exist in the following regions:
   a. One-phase region
   b. Two-phase region

7. Figure 1-8 shows the phase diagrams of eight mixtures of methane and ethane along with the vapor pressure curves of the two components. Determine:
   a. Vapor pressure of methane at − 160°F
   b. Vapor pressure of ethane at 60°F
   c. Critical pressure and temperature of mixture 7
   d. Cricondenbar and cricondentherm of mixture 7

   e. Upper and lower dew-point pressure of mixture 6 at 20°F
   f. The bubble-point and dew-point pressures of mixture 8 at 60°F

8. Using Figure 1-8, prepare and identify the different phase regions of the pressure-composition diagram (commonly called the P-X diagram) for the following temperatures:
   a. – 120°F
   b. 20°F

9. Using Figure 1-8, prepare the temperature-composition diagram (commonly called the T-X diagram) for the following pressures:
   a. 300 psia
   b. 700 psia
   c. 800 psia

REFERENCES

1. Clark, N., "It Pays to Know Your Petroleum," *World Oil*, March 1953, Vol. 136, pp. 165–172.
2. Gibbs, J. W., *The Collected Works of J. Willard Gibbs*, Trans. Conn. Acad. Arts Sci., Vol. 1, Yale University Press, New Haven, reprinted 1948; original text published 1876.
3. Lee, B. I. and Kesler, M. G., "A Generalized Thermodynamics Correlation Based on Three-parameter Corresponding States," *AIChE Journal*, Vol. 21, No. 3, May 1975, pp. 510–527.
4. Pitzer, K. S., "The Volumetric and Thermodynamics Properties of Fluids," *J. Amer. Chem. Soc.*, Vol. 77, No. 13, July 1955, pp. 3427–3433.
5. Rackett, H. G., "Equation of State for Saturated Liquids," *J. Chem. Eng. Data*, Vol. 15, No. 4, 1970, pp. 514–517.
6. Spencer, F. F. and Danner, R. P., "Prediction of Bubble-point Density of Mixtures," *J. Chem. Eng. Data*, Vol. 18, No. 2, 1973, pp. 230–234.
7. Yamada, T. and Gunn, R., "Saturated Liquid Molar Volumes: The Rackett Equation," *J. Chem. Eng. Data*, Vol. 18, No. 2, 1973, pp. 234–236.

2
Pure Component Physical Properties and Characterizing Undefined Petroleum Fractions

Many of the physical properties of pure components have been measured and compiled over the years. These properties provide essential information for studying the volumetric behavior and determining the thermodynamic properties of pure components and their mixtures. The most important of these properties are:

- Critical pressure, p_c
- Critical temperature, T_c
- Critical volume, V_c
- Critical compressibility factor, Z_c
- Acentric factor, ω
- Molecular weight, MW

Petroleum engineers are usually interested in the behavior of hydrocarbon mixtures rather than pure components. However, the above characteristic constants of the pure component can be used with the independent state variables such as pressure, temperature, and composition to characterize and define the physical properties and the phase behavior of mixtures.

This chapter's primary objective is review. Several of the well-established physical property correlations are presented to illustrate how they can be used if no experimental data are available on the petroleum fraction.

GENERALIZED CORRELATIONS FOR ESTIMATING PHYSICAL PROPERTIES OF HYDROCARBON FRACTIONS

There are numerous correlations for estimating the physical properties of petroleum fractions. Most of these correlations use the specific gravity γ and the boiling point T_b as correlation parameters. Selecting proper values for the above parameters is very important because slight changes in these parameters can cause significant variations in the predicted results. Several of these correlations are presented below.

Riazi-Daubert Generalized Correlations

Riazi and Daubert (1980) developed a simple two-parameter equation for predicting the physical properties of pure compounds and undefined hydrocarbon mixtures. The proposed generalized empirical equation is based on the use of the normal boiling point and the specific gravity as correlating parameters. The basic equation is:

$$\Theta = a\, T_b^{\;b}\, \gamma^c \tag{2-1}$$

where Θ = any physical property
T_b = normal boiling point, °R
γ = specific gravity
a, b, c = correlation constants are given in Table 2-1 for each property

Table 2-1
Correlation Constants for Equation 2-1

Θ	a	b	c	Deviation % Average	Maximum
MW	4.5673×10^{-5}	2.1962	-1.0164	2.6	11.8
T_c, °R	24.2787	0.58848	0.3596	1.3	10.6
p_c, psia	3.12281×10^9	-2.3125	2.3201	3.1	9.3
V_c, ft³/lb	7.5214×10^{-3}	0.2896	-0.7666	2.3	9.1

The average error for each property is given in Table 2-1.

The prediction accuracy is reasonable over the boiling point range of 100–850°F.

Riazi and Daubert (1987), in their development of new correlations for improving the prediction of physical properties of petroleum fractions, considered various factors. These factors were accuracy, simplicity, generality, availability of input parameters, extrapolatability, and finally, comparability with similar correlations developed in recent years.

The authors proposed the following modification of Equation 2-1, which maintains the simplicity of the previous correlation while significantly improving its accuracy:

$$\Theta = a\ \Theta_1^b\ \Theta_2^c\ \text{EXP}\ [d\ \Theta_1 + e\ \Theta_2 + f\ \Theta_1\ \Theta_2]$$

where Θ = any physical property
         e–f = constants for each property

Riazi and Daubert stated that Θ_1 and Θ_2 can be any two parameters capable of characterizing molecular forces and molecular size of a compound. They identified (T_b, γ) and (MW, γ) as appropriate pairs of input parameters in the above equation. The authors finally proposed the following two forms of the generalized correlation:

Form 1. In this form, the boiling point T_b and the specific gravity γ of the petroleum fraction are used as correlating parameters.

$$\Theta = a\ T_b^{\ b}\ \gamma^c\ \text{EXP}\ [d\ T_b + e\ \gamma + f\ T_b\ \gamma] \qquad (2\text{-}2)$$

The constants a-f for each property Θ are given in Table 2-2.

Table 2-2
Correlation Constants for Equation 2-2

Θ	a	b	c
MW	581.96	0.97476	6.51274
T_c, °R	10.6443	0.81067	0.53691
p_c, psia	6.162×10^6	− 0.4844	4.0846
V_c, ft³/lb	6.233×10^{-4}	0.7506	− 1.2028

d	e	f
5.43076×10^{-4}	9.53384	1.11056×10^{-3}
$- 5.1747 \times 10^{-4}$	− 0.54444	3.5995×10^{-4}
$- 4.725 \times 10^{-3}$	− 4.8014	3.1939×10^{-3}
$- 1.4679 \times 10^{-3}$	− 0.26404	1.095×10^{-3}

Form 2. In this form, molecular weight MW and specific gravity γ of the component are used as correlating parameters.

$$\Theta = a\ (MW)^b\ \gamma^c\ \text{EXP}\ [d\ (MW) + e\ \gamma + f\ (MW)\ \gamma] \qquad (2\text{-}3)$$

where the correlation constants are given in Table 2-3.

In developing and obtaining the coefficients of the above two correlations, Riazi and Daubert used data on the properties of 38 pure hydrocarbons in the carbon number range 1–20, including paraffins, olefins,

Table 2-3
Correlation Constants for Equation 2-3

Θ	a	b	c
T_c, °R	544.4	0.2998	1.0555
p_c, psia	4.5203×10^4	− 0.8063	1.6015
V_c, ft³/lb	1.206×10^{-2}	0.20378	− 1.3036
T_b, °R	6.77857	0.401673	− 1.58262

d	e	f
$- 1.3478 \times 10^{-4}$	− 0.61641	0.0
$- 1.8078 \times 10^{-3}$	− 0.3084	0.0
$- 2.657 \times 10^{-3}$	0.5287	2.6012×10^{-3}
3.77409×10^{-3}	2.984036	$- 4.25288 \times 10^{-3}$

naphthenes, and aromatics in the molecular weight range 70–300 and the boiling point range 80–650°F.

Lin-Chao Generalized Correlation

Lin and Chao (1984) correlated the physical properties of hydrocarbon components with the molecular weight, specific gravity, and normal boiling point. The proposed correlation was developed by using perturbation theory; it contains 33 numerical constants for each physical property. The physical properties of C_1 to C_{20} n-alkanes (n-paraffins) were correlated with the molecular weight. Properties of other hydrocarbons and derivatives were expressed as perturbations from those of the n-paraffins with the boiling point and the specific gravity as the correlating parameters.

Lin and Chao expressed the physical properties of C_1 to C_{20} n-alkanes by the following generalized equation

$$\Theta_A = C_1 + C_2(MW) + C_3(MW)^2 + C_4(MW)^3 + C_5/(MW) \tag{2-4}$$

where Θ_A represents either T_c, $\ln(p_c)$, V_c, $(\omega \cdot T_c)$, or T_b of a n-alkane. The coefficients C_1 to C_5 are reported in Table 2-4.

The correlation produced an average absolute deviation of 0.15% for T_c, 1.0% for p_c (excluding methane), 1.2% for ω, 0.11% for T_b, and 0.07% for γ, when compared to the physical properties value, as listed in the American Petroleum Institute project 44 table values. Properties of the general hydrocarbons and derivatives were correlated as perturbations of those of n-alkanes according to the following equation:

Table 2-4
Coefficients for Θ_A in Equation 2-4

Θ_A	C_1	C_2
T_c, °R	490.8546	7.055982
$\ln(p_c)$, psi	6.753444	− 0.010182
$\omega \cdot T_c$	− 28.21536	2.209518
γ	0.66405	1.48130×10^{-3}
T_b, °R	240.8976	5.604282

C_3	C_4	C_5
− 0.02118708	2.676222×10^{-5}	− 4,100.202
2.51106×10^{-5}	$- 3.73775 \times 10^{-8}$	3.50737
17.943264×10^{-3}	$- 3.685356 \times 10^{-5}$	− 124.35894
$- 5.0702 \times 10^{-6}$	6.21414×10^{-9}	− 8.45218
− 0.012761604	13.84353×10^{-6}	− 2,029.158

$$\Theta = \Theta_A + A_1 \, \Delta\gamma + A_2 \, \Delta T_b + A_3 \, (\Delta\gamma)^2 + A_4 \, (\Delta\gamma)(\Delta T_b) + A_5 \, (\Delta T_b)^2$$
$$+ A_6 \, (\Delta\gamma)^3 + A_7 \, (\Delta\gamma)^2 (\Delta T_b) + A_8 \, (\Delta\gamma)(\Delta T_b)^2 + A_9 \, (\Delta T_b)^3 \quad (2\text{-}5)$$

with $\Delta\gamma = \gamma - \gamma_a$
$\Delta T_b = T_b - (T_b)_a$

where γ = specific gravity of the substance of interest
T_b = boiling point of the substance of interest, °R
$\gamma_a, (T_b)_a$ = specific gravity and boiling point of a hypothetical n-alkane with a molecular weight (MW) of the substance of interest (given by Equation 2-4).

The coefficients A_1–A_9 in Equation 2-5 are given by the following expression:

$$A_i = a_i + b_i \, MW \quad (2\text{-}6)$$

where values of the coefficients a_i and b_i are listed in Table 2-5.
The proposed correlation can be best explained through the following example:

Example 2-1. Estimate the critical temperature of a petroleum fraction with a boiling point of 731°R, specific gravity of 0.7515 and molecular weight of 127 by using the Lin and Chao method.

Solution.

Step 1. Using Equation 2-6, calculate the critical temperature $(T_c)_a$, specific gravity γ_a, and boiling point $(T_b)_a$ of a hypothetical n-alkane

Table 2-5
Coefficients for Equation 2-6

Coefficient	T_c, °R	$\ln(p_c)$, psia	$\omega \cdot T_c$
		θ	
a_1	2,844.45	9.71572	2.088792×10^3
a_2	-5.68509	$-1.844466667 \times 10^{-2}$	3.48210
a_3	-2.189862×10^4	-86.0375	5.009706×10^4
a_4	75.0653	0.3056211111	-2.05257×10^2
a_5	$-5.3688056 \times 10^{-2}$	$-2.777888889 \times 10^{-4}$	0.2532038889
a_6	3.908016×10^4	1.85927×10^2	-7.13722×10^4
a_7	-1.57999×10^2	-0.8395277778	5.08888×10^2
a_8	0.20029	$1.33582716 \times 10^{-3}$	-0.3390405556
a_9	$-0.851117284 \times 10^{-4}$	$0.6541941015 \times 10^{-6}$	$-0.5207160494 \times 10^{-3}$
b_1	-21.31776	-7.5037×10^{-2}	3.415698
b_2	5.77384×10^{-2}	$1.753983333 \times 10^{-2}$	2.41662×10^{-2}
b_3	1.992546×10^2	0.842854	-4.814316×10^2
b_4	-0.658450	$-2.897022222 \times 10^{-3}$	2.06071
b_5	$4.346166667 \times 10^{-4}$	$2.430015432 \times 10^{-6}$	$-2.900583333 \times 10^{-3}$
b_6	-367.641	-1.85430	7.66070×10^2
b_7	1.32064	$0.7558388889 \times 10^{-2}$	5.75141
b_8	$-1.26440556 \times 10^{-3}$	$-0.9997808642 \times 10^{-5}$	$4.814816667 \times 10^{-3}$
b_9	$2.698441358 \times 10^{-7}$	$0.3753412209 \times 10^{-8}$	$0.5407067901 \times 10^{-5}$

with a molecular weight of 127. These applications produce the
following results

$$(T_c)_a = 1,067.8°R$$
$$\gamma_a = 0.7977$$
$$(T_b)_a = 759.19°R$$

Step 2. Calculate the coefficients A_i by using Equation 2-6.

$A_1 = 137.09$ $A_2 = 1.6477$ $A_3 = 3,406.714$
$A_4 = 8.5579$ $A_5 = 0.001508$ $A_6 = -7,610.25$
$A_7 = 9.72228$ $A_8 = 0.03971$ $A_9 = -0.00005084$

Step 3. Solve for T_c by applying Equation 2-5.

$$\Delta\gamma = 0.7515 - 0.7977 = -0.0462$$
$$\Delta T_b = 731 - 759.19 = -28.19$$
$$\begin{aligned} T_c = {} & 1,067.8 + 137.09(-0.0462) + 1.6477(-28.19) \\ & + 3,406.714(-0.0462)^2 - 8.5579(-0.0462)(-28.19) \\ & + 0.001508(-29.19)^2 - 7,610.25(-0.0462)^3 \\ & + 9.72228 \ (-0.0462)^2 \ (-28.19) \\ & + 0.03971 \ (-0.0462)(-28.19)^2 \\ & - 0.00005084(-28.19)^3 = 1,012.2°R \end{aligned}$$

OTHER METHODS OF ESTIMATING PHYSICAL
PROPERTIES OF PETROLEUM FRACTIONS

Cavett's Correlations

Cavett (1962) proposed correlations for estimating the critical pressure and temperature of hydrocarbon fractions. The correlations have received a wide acceptance in the petroleum industry due to their reliability in extrapolating at conditions beyond those of the data used in developing the correlations. The proposed correlations were expressed analytically as functions of the normal boiling point T_b and API gravity. Cavett proposed the following expressions for estimating the critical temperature and pressure of petroleum fractions:

Critical Temperature

$$T_c = a_0 + a_1 T_b + a_2 T_b^2 + a_3 (API)(T_b) + a_4(T_b)^3$$
$$+ a_5(API)(T_b)^2 + a_6 (API)^2 (T_b)^2 \qquad (2\text{-}7)$$

Critical Pressure

$$\text{Log } (p_c) = b_0 + b_1 (T_b) + b_2 (T_b)^2 + b_3 (API) (T_b) + b_4 (T_b)^3$$
$$+ b_5 (API) (T_b)^2 + b_6 (API)^2 (T_b) + b_7 (API)^2 (T_b)^2 \qquad (2\text{-}8)$$

where T_c = critical temperature, °R
p_c = critical pressure, psia
T_b = normal boiling point, °F
API = API gravity of the fraction

The coefficients of Equations 2-7 and 2-8 are tabulated in Table 2-6. Cavett presented these correlations without reference to the type and source(s) of data used for their development.

Kesler-Lee Correlations

Kesler and Lee (1976) proposed a set of equations to estimate the critical temperature, critical pressure, acentric factor, and molecular weight of petroleum fractions. The equations, as expressed below, use specific gravity and boiling point in °R as input parameters.

Table 2-6
Coefficients of Equations 2-7 and 2-8

i	a_i	b_i
0	768.07121	2.8290406
1	1.7133693	$0.94120109 \times 10^{-3}$
2	-0.0010834003	$-0.30474749 \times 10^{-5}$
3	-0.0089212579	$-0.20876110 \times 10^{-4}$
4	$0.38890584 \times 10^{-6}$	$0.15184103 \times 10^{-8}$
5	$0.53094920 \times 10^{-5}$	$0.11047899 \times 10^{-7}$
6	$0.32711600 \times 10^{-7}$	$-0.48271599 \times 10^{-7}$
7	—	$0.13949619 \times 10^{-9}$

Critical Pressure

$$\ln (p_c) = 8.3634 - 0.0566/\gamma - (0.24244 + 2.2898/\gamma \\ + 0.11857/\gamma^2)10^{-3} T_b + (1.4685 + 3.648/\gamma \\ + 0.47227/\gamma^2)10^{-7} T_b^2 - (0.42019 + 1.6977/\gamma^2)10^{-10} T_b^3$$

$$(2\text{-}9)$$

Critical Temperature

$$T_c = 341.7 + 811.1 \gamma + (0.4244 + 0.1174\gamma) T_b \\ + (0.4669 - 3.26238\gamma) 10^5/T_b$$

$$(2\text{-}10)$$

Molecular Weight

$$MW = -12{,}272.6 + 9{,}486.4 \gamma + (4.6523 - 3.3287\gamma) T_b \\ + (1 - 0.77084\gamma - 0.02058\gamma^2)(1.3437 - 720.79/T_b)10^7/T_b \\ + (1 - 0.80882\gamma + 0.02226\gamma^2)(1.8828 - 181.98/T_b)10^{12}/T_b^3$$

$$(2\text{-}11)$$

The above equation was obtained by regression analysis using the available data on molecular weights ranging from 60 to 650.

Acentric Factor. Defining the Watson characterization factor K and the reduced boiling point Θ by the following relationships

$$K = \frac{(T_b)^{1/3}}{\gamma}$$

$$\Theta = \frac{T_b}{T_c}$$

where T_b = boiling point, °R

Kesler and Lee proposed the following two expressions for calculating the acentric factor

For $\Theta > 0.8$:

$$\omega = -7.904 + 0.1352K - 0.007465K^2 + 8.359\ \Theta + (1.408 - 0.01063K)/\Theta \qquad (2\text{-}12)$$

For $\Theta < 0.8$:

$$\omega = \frac{-\ln(p_c/14.696) - 5.92714 + 6.09648/\Theta + 1.28862\ \ln(\Theta) - 0.169347\ \Theta^6}{15.2518 - 15.6875/\Theta - 13.4721\ \ln(\Theta) + 0.43577\Theta^6}$$

$$(2\text{-}13)$$

where
$\quad p_c$ = critical pressure, psia
$\quad T_c$ = critical temperature, °R
$\quad T_b$ = boiling point, °R
$\quad \omega$ = acentric factor
$\quad MW$ = molecular weight
$\quad \gamma$ = specific gravity

Kesler and Lee stated that Equations 2-9 and 2-10 give values for p_c and T_c that are nearly identical with those from the API Data Book up to a boiling point of 1,200°F. Modifications were introduced to extend the correlations beyond the boiling point limit of 1,200°F. These extensions (extrapolations) were achieved by ensuring that the critical pressure approaches the atmospheric pressure as the boiling point approaches critical temperature.

Winn-Sim-Daubert Correlations

Sim and Daubert (1980) concluded that the Winn (1957) nomograph is the most accurate method for characterizing petroleum fractions. For this reason, Sim and Daubert represented the critical pressure, critical temperature, and molecular weight of the Winn nomograph analytically by the following equations:

$$p_c = 3.48242 \times 10^9\ T_b^{-2.3177}\ \gamma^{2.4853} \qquad (2\text{-}14)$$
$$T_c = EXP\ [3.9934718\ T_b^{0.08615}\ \gamma^{0.04614}] \qquad (2\text{-}15)$$
$$MW = 1.4350476 \times 10^{-5}\ T_b^{2.3776}\ \gamma^{-0.9371} \qquad (2\text{-}16)$$

where p_c = critical pressure, psia
T_c = critical temperature, °R
T_b = boiling point, °R

Watansiri-Owens-Starling Correlations

Watansiri, et. al. (1985) developed a set of correlations to estimate the critical properties and acentric factor of coal compounds and other hydrocarbons and their derivatives. The proposed correlations express the characterization parameters as functions of the normal boiling point, specific gravity, and molecular weight. These relationships have the following forms:

Critical Temperature

$$\ln(T_c) = -0.0650504 - 0.0005217\ T_b + 0.03095\ \ln(MW) \\ + 1.11067\ \ln(T_b) + MW\ [0.078154\ \gamma^{1/2} \\ - 0.061061\ \gamma^{1/3} - 0.016943\ \gamma] \tag{2-17}$$

where T_c = critical temperature, °R

Critical Volume

$$\ln(V_c) = 76.313887 - 129.8038\ \gamma + 63.1750\ \gamma^2 - 13.175\ \gamma^3 \\ + 1.10108\ \ln(MW) + 42.1958\ \ln(\gamma) \tag{2-18}$$

where V_c = critical volume, ft³/lb-mole

Critical Pressure

$$\ln(p_c) = 6.6418853 + 0.01617283\ (T_c/V_c)^{0.8} - 8.712(MW/T_c) \\ - 0.08843889(T_b/MW) \tag{2-19}$$

where p_c = critical pressure, psia

Acentric Factor

$$\omega = [5.12316667 \times 10^{-4}\ T_b + 0.281826667(T_b/MW) + 382.904/MW \\ + 0.074691 \times 10^{-5}(T_b/\gamma)^2 - 0.12027778 \times 10^{-4}(T_b)(MW) \\ + 0.001261(\gamma)(MW) + 0.1265x10^{-4}\ (MW)^2 \\ + 0.2016x10^{-4}\ (\gamma)(MW)^2 - 66.29959(T_b)^{1/3}/MW \\ - 0.00255452\ T_b^{2/3}/\gamma^2](5\ T_b/(9\ MW)) \tag{2-20}$$

The proposed correlations produce an average absolute relative deviation of 1.2% for T_c, 3.8% for V_c, 5.2% for p_c, and 11.8% for ω.

Edmister's Correlation

Edmister (1958) proposed a correlation for estimating the acentric factor ω of pure fluids and petroleum fractions. The equation, widely used in the petroleum industry, requires boiling point, critical temperature, and critical pressure. The proposed expression is given by the following relationship:

$$\omega = \frac{3[\text{Log }(p_c/14.70]}{7[(T_c/T_b - 1]} - 1 \tag{2-21}$$

where ω = acentric factor
p_c = critical pressure, psia
T_c = critical temperature, °R
T_b = normal boiling point, °R

If the acentric factor is available from another correlation, the Edmister equation can be rearranged to solve for any of the three other properties (providing the other two are known).

CRITICAL COMPRESSIBILITY FACTORS

The critical compressibility factor is defined as the component compressibility factor calculated at its critical point. This property can be conveniently computed by the real gas equation-of-state at the critical point, or

$$Z_c = \frac{p_c\, V_c}{R\, T_c} \tag{2-22}$$

where R = universal gas constant, 10.73 psia.ft³/lb-mole.°R
V_c = critical volume, ft³/lb-mole

If the critical volume V_c is given in ft³/lb, Equation 2-22 is written as

$$Z_c = \frac{p_c\, V_c\, MW}{R\, T_c}$$

where MW = molecular weight
V_c = critical volume, ft³/lb

The accuracy of Equation 2-22 depends on the accuracy of the values of p_c, T_c, and V_c. Table 2-7 presents a summary of the critical compressibility estimation methods.

Table 2-7
Critical Compressibility Estimation Methods

Method	Year	Z_c	Equation No.
Haugen	1959	$Z_c = 1/(1.28\ \omega + 3.41)$	2-23
Reid-Prausnitz-Sherwood	1977	$Z_c = 0.291 - 0.080\ \omega$	2-24
Salerno et al.	1985	$Z_c = 0.291 - 0.800\ \omega - 0.016\ \omega^2$	2-25
Nath	1985	$Z_c = 0.2918 - 0.0928\ \omega$	2-26

Where ω is the acentric factor.

Example 2-3. Estimate the critical properties, molecular weight, and acentric factor of a petroleum fraction with a boiling point of 198°F and specific gravity of 0.7365, by using the following methods:

1. Riazi-Daubert (Equation 2-1)
2. Riazi-Daubert (Equation 2-2)
3. Cavett's
4. Kesler-Lee
5. Winn-Sim-Daubert
6. Watansiri-Owens-Starling

Solution.

1. **Riazi-Daubert:** Equation 2-1

- $MW = 4.5673 \times 10^{-5}\ (658)^{2.1962}\ (0.7365)^{-1.0164} = 96.4$
- $T_c = 24.2787\ (658)^{0.58848}\ (0.7365)^{0.3596} = 990.67°R$
- $p_c = 3.12281 \times 10^9\ (658)^{-2.3125}\ (0.7365)^{2.3201} = 466.9\ \text{psia}$
- $V_c = 7.5214 \times 10^{-3}\ (658)^{0.2896}\ (0.7365)^{-0.7666} = 0.06227\ \text{ft}^3/\text{lb}$
- Solve for Z_c by applying the above calculated properties in Equation 2-22:

$$Z_c = \frac{p_c\ V_c\ MW}{R\ T_c} = \frac{(466.9)(0.06227)(96.4)}{(10.73)\ (990.67)} = 0.26365$$

- Solve for ω by applying Equation 2-21:

$$\omega = \frac{3\ [\log\ (466.9/14.7)]}{7\ [(990.67/658) - 1]} - 1 = 0.2731$$

2. **Riazi-Daubert:** Equation 2-2
Applying Equation 2-2 and using the appropriate constants yields:

- MW = 96.911
- T_c = 986.7 °R
- p_c = 465.83 psia
- V_c = 0.06257 ft³/lb
- Solve for the acentric factor and the critical compressibility factor by applying Equations 2-21 and 2-22, respectively:

- ω = 0.2877
- Z_c = 0.2668

3. Cavett's Correlation:

- Solve for T_c by applying Equation 2-7:

    T_c = 978.1 °R

- Calculate p_c with Equation 2-8:

    p_c = 466.1 psia

- Solve for the acentric factor by applying the Edmister correlation, Equation 2-21:

$$\omega = \frac{3 \, [\text{Log} \, (4.66.1/14.7)]}{7 \, [(980/658) - 1]} - 1 = 0.3147$$

- Compute the critical compressibility by using Equation 2-25:

    $Z_c = 0.291 - (0.08)(0.3147) - 0.016 \, (0.3147)^2 = 0.2642$

- Estimate V_c from Equation 2-22:

$$V_c = \frac{Z_c \, R \, T_c}{p_c} = \frac{(0.2642)(10.731)(980)}{466.1} = 5.9495 \text{ ft}^3/\text{lb-mole}$$

Assume MW = 96

$$V_c = \frac{5.9495}{96} = 0.06197 \text{ ft}^3/\text{lb}$$

4. Kesler-Lee

- Calculate p_c from Equation 2-9:

    p_c = 470 psia

- Solve for T_c by using Equation 2-10:

    T_c = 980 °R

- Calculate the molecular weight MW by using Equation 2-11:

    MW = 98.7

- Compute the Watson characterization factor K and the parameter Θ:

$$K = \frac{(658)^{1/3}}{0.7365} = 11.8$$

$$\Theta = \frac{658}{980} = 0.671$$

- Solve for acentric factor by applying Equation 2-13:

$$\Theta = 0.306$$

- Estimate for the critical gas compressibility Z_c by using Equation 2-26:

$$Z_c = 0.2918 - (0.0928)(0.306) = 0.2634$$

- Solve for V_c by applying Equation 2-22:

$$V_c = \frac{Z_c \, R \, T_c}{p_c \, MW} \frac{(0.2634)(10.73)(980)}{(470)(98.7)} = 0.0597 \text{ lb/ft}^3$$

5. Winn-Sim-Daubert:

- Estimate p_c from Equation 2-16:

$$p_c = 478.6 \text{ psia}$$

- Solve for T_c by applying Equation 2-15:

$$T_c = 979.2 \text{ °R}$$

- Calculate MW from Equation 2-16:

$$MW = 95.93$$

- Solve for the acentric factor from Equation 2-21:

$$\omega = 0.3280$$

- Solve for Z_c by applying Equation 2-24:

$$Z_c = 0.291 - (0.08)(0.3280) = 0.2648$$

- Calculate the critical volume V_c from Equation 2-22:

$$V_c = \frac{(0.2648)(10.731)(979.2)}{(478.6)\,(95.93)} = 0.06059 \text{ ft}^3/\text{lb}$$

6. Watansiri-Owens-Starling:

- Because Equations 2-17 through 2-19 require the molecular weight, assume MW = 96

- Calculate T_c from Equation 2-17:

  $T_c = 980.0 \ ^\circ R$

- Determine the critical volume from Equation 2-18 to give

  $V_c = 0.06548 \ ft^3/lb$

- Solve for the critical pressure of the fraction by applying Equation 2-19 to produce

  $P_c = 426.5 \ psia$

- Calculate the acentric factor from Equation 2-20 to give

  $\omega = 0.2222$

- Compute the critical compressibility factor by applying Equation 2-26

  $Z_c = 0.27112$

Table 2-8 summarizes the results of Example 2-3.

Table 2-8
Summary of the Calculated Results of Example 2-3

Method	T_c °R	p_c psia	V_c ft³/lb-mole	MW	ω	Z_c
Riazi-Daubert No. 1	990.67	466.9	0.06227	96.4	0.2731	0.26365
Riazi-Daubert No. 2	986.7	465.83	0.06257	96.911	0.2877	0.668
Cavett	978.1	466.1	0.06197	—	0.3147	0.2642
Kesler-Lee	980	479	0.0597	98.7	0.3060	0.2634
Winn	979.2	478.6	0.06059	95.93	0.3280	0.2648
Watansiri	980	426.5	0.06548	—	0.2222	0.27112

CHARACTERIZING HYDROCARBON HEAVY FRACTIONS

Nearly all naturally occurring hydrocarbon systems contain a quantity of heavy fractions that are not well defined and are not mixtures of discretely identified components. These heavy fractions are often lumped together and identified as the plus fraction, e.g., C_{7+} fraction.

A proper description of the physical properties of the plus fractions and other undefined petroleum fractions in hydrocarbon mixtures is essential in performing reliable phase behavior calculations and compositional model-ing studies. Frequently a distillation analysis or a chromatographic analysis is available for this undefined fraction. Other physical properties such as molecular weight and specific gravity may also be measured for the entire fraction or various cuts of it.

To use any of the thermodynamic property-prediction models, e.g., equation of state, to predict the phase and volumetric behavior of complex hydrocarbon mixtures, one must be able to provide the acentric factor, critical temperature, and critical pressure for both the defined and undefined (heavy) fractions in the mixture. The problem of how to adequately characterize these undefined plus fractions in terms of their critical properties and acentric factors has been long recognized in the petroleum industry. Whitson (1984) presented an excellent documentation on the influence of various heptanes-plus (C_{7+}) characterization schemes on predicting the volumetric behavior of hydrocarbon mixtures by equations-of-state.

Numerous characterization procedures have been proposed over the years. Some of the most widely accepted and used procedures are reviewed below.

METHODS BASED ON THE "PNA" DETERMINATION

The vast number of hydrocarbon compounds making up naturally occurring crude oil have been grouped chemically into several series of compounds. Each series consists of those compounds similar in their molecular make-up and characteristics. Within a given series, there exist compounds ranging from extremely light, or chemically simple, to heavy, or chemically complex. In general, it is assumed that the heavy (undefined) hydrocarbon fractions are composed of three hydrocarbon groups, namely:

- Paraffins (P)
- Naphthenes (N)
- Aromatics (A)

The PNA content of the plus fraction of the undefined hydrocarbon fraction can be estimated experimentally from a distillation analysis and/or a chromatographic analysis. Both types of analysis provide information valuable for use in characterizing the plus fractions.

In the distillation process, the hydrocarbon plus fraction is subjected to a standardized analytical distillation, first at atmospheric pressure, and then in a vacuum at a pressure of 40 mm Hg. Usually the temperature is taken when the first droplet distills over. Ten fractions (cuts) are then distilled off, the first one at 50°C and each successive one with a boiling range of 25°C. For each distillation cut, the volume, specific gravity, and molecular weight, among other measurements, are determined. Cuts obtained in this manner are identified by the boiling-point ranges in which they were collected.

Generally, there are five different methods of defining the normal boiling point for petroleum fractions. These are:

1. Volume Average Boiling Point (VABP), which is defined mathematically by the following expression:

$$VABP = \sum_i v_i T_{bi}$$ (2-27)

   where T_{bi} = boiling point of the distillation cut i, °R
   v_i = volume fraction of the distillation cut i

2. Weight Average Boiling Point (WABP), defined by the following expression:

$$WABP = \sum_i w_i T_{bi}$$ (2-28)

   where w_i = weight fraction of the distillation cut i

3. Molar Average Boiling Point (MABP), given by the following relationship:

$$MABP = \sum_i x_i T_{bi}$$ (2-29)

   where x_i = mole fraction of the distillation cut i

4. Cubic Average Boiling Point (CABP) which is defined as

$$CABP = \left[\sum_i x_i T_{bi}^{1/3} \right]^3$$ (2-30)

5. Mean Average Boiling Point (MeABP):

$$MeABP = \frac{MABP + CABP}{2}$$ (2-31)

As indicated by Edmister and Lee (1984), these five expressions for calculating normal boiling points result in values that do not differ significantly from one another for narrow boiling petroleum fractions.

Figure 2-1 shows a typical graphical presentation of the molecular weight, specific gravity, and the true boiling point TBP as a function of the volume fraction of liquid vaporized. It should be pointed out that when a single boiling point is given for a plus fraction, it is given as its Volume Average Boiling Point.

Bergman et al. (1977) outlined the chromatographic analysis procedure by which distillation cuts are characterized by the density and molecular weight as well as Weight Average Boiling Point.

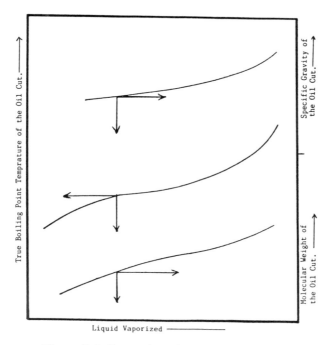

Figure 2-1. Properties of a crude oil fraction.

All three parameters (i.e., molecular weight, specific gravity, and VABP/ WABP) are employed, as discussed below, to estimate the PNA content of the heavy hydrocarbon fraction which in turn is used to predict the critical properties and acentric factor of the fraction. Hopke and Lin (1974), Erbar (1977), Bergman et al. (1977), and Robinson and Peng (1978) have used the PNA concept to characterize the undefined hydrocarbon fractions. As a representative of this characterization approach, the Robinson-Peng method and the Bergman method are discussed below.

Robinson-Peng Method

Robinson and Peng (1978) proposed a detailed procedure for characterizing heavy hydrocarbon fractions. The procedure is summarized in the following steps:

Step 1. Calculate the PNA content (X_p, X_N, X_A) of the undefined fraction by solving the following three rigorously defined equations:

$$\sum_{i = P,N,A} X_i = 1 \qquad (2\text{-}32)$$

$$\sum_{i = P,N,A} [MW_i \; T_{bi} \; X_i] = (MW)(WABP) \tag{2-33}$$

$$\sum_{i = P,N,A} [MW_i \; X_i] = MW \tag{2-34}$$

where $\quad X_p$ = mole fraction of the paraffinic group in the undefined fraction

$\quad X_N$ = mole fraction of the naphthenic group in the undefined fraction

$\quad X_A$ = mole fraction of the aromatic group in the undefined fraction

$\quad$ WABP = weight average boiling point of the undefined fraction, °R

$\quad$ MW = molecular weight of the undefined fraction

$\quad (MW_i)$ = average molecular weight of each cut, i.e., PNA

$\quad (T_b)_i$ = boiling point of each cut, °R

Equations 2-32 through 2-34 can be written in a matrix form as follows:

$$\begin{bmatrix} 1 & 1 & 1 \\ [MW \cdot T_b]_P & [MW \cdot T_b]_N & [MW \cdot T_b]_A \\ [MW]_P & [MW]_N & [MW]_A \end{bmatrix} \begin{bmatrix} X_P \\ X_N \\ X_A \end{bmatrix} = \begin{bmatrix} 1 \\ MW \cdot WABP \\ MW \end{bmatrix} \tag{2-35}$$

Robinson and Peng pointed out that it is possible to obtain negative values for the PNA contents. To prevent these negative values, the authors imposed the following constraints:

$$0 \leq X_P \leq 0.90$$
$$X_N \geq 0$$
$$X_A \geq 0$$

To solve Equation 2-35 for the PNA content requires the weight average boiling point and molecular weight for the cut of the undefined hydrocarbon fraction. If the experimental values of these cuts are not available, the following correlations proposed by Robinson and Peng can be used:

Determination of $(T_b)_P$, $(T_b)_N$, and $(T_b)_A$.

- Paraffinic group $\quad Ln(T_b) = Ln(1.8) + \sum_{i = 1}^{6} [a_i(n - 6)^{i-1}] \tag{2-36}$

- Naphthenic group $\quad Ln(T_b) = Ln(1.8) + \sum_{i = 1}^{6} [a_i(n - 7)^{i-1}] \tag{2-37}$

• Aromatic group $\mathrm{Ln(T_b)} = \mathrm{Ln}(1.8) + \sum_{i=1}^{6} [a_i(n-7)^{i-1}]$ (2-38)

where   n = number of carbon atoms in the undefined hydrocarbon fraction
       a_i = coefficients of the equations and are given in Table 2-9

Table 2-9
Coefficients a_i in Equations 2-36 through 2-38

Coefficient	Paraffin P	Naphthene N	Aromatic A
a_1	5.83451830	5.8579332	5.86717600
a_2	$0.84909035 \times 10^{-1}$	$0.79805995 \times 10^{-1}$	$0.80436947 \times 10^{-1}$
a_3	$-0.52635428 \times 10^{-2}$	$-0.43098101 \times 10^{-2}$	$-0.47136506 \times 10^{-2}$
a_4	$0.21252908 \times 10^{-3}$	$0.14783123 \times 10^{-3}$	$0.18233365 \times 10^{-3}$
a_5	$-0.44933363 \times 10^{-5}$	$-0.27095216 \times 10^{-5}$	$-0.38327239 \times 10^{-5}$
a_6	$0.37285365 \times 10^{-7}$	$0.19907794 \times 10^{-7}$	$0.32550576 \times 10^{-7}$

Determination of $(MW)_P$, $(MW)_N$, and $(MW)_A$.

• Paraffinic group $(MW)_P = 14.026\, n + 2.016$ (2-39)

• Naphthenic group $(MW)_N = 14.026\, n - 14.026$ (2-40)

• Aromatic group $(MW)_A = 14.026\, n - 20.074$ (2-41)

Step 2. Having obtained the PNA content of the undefined hydrocarbon fraction, as outlines in Step 1, calculate the critical pressure of the fraction by applying the following expression:

$$P_c = X_P(P_c)_P + X_N(P_c)_N + X_A(P_c)_A \qquad (2\text{-}42)$$

where P_c = critical pressure of the heavy hydrocarbon fraction, psia

The critical pressure for each cut of the heavy fraction is calculated according to the following equations:

• Paraffinic group $(P_c)_P = \dfrac{206.126096n + 29.67136}{(0.227n + 0.340)^2}$ (2-43)

• Naphthenic group $(P_c)_N = \dfrac{206.126096n - 206.126096}{(0.227n - 0.137)^2}$ (2-44)

- Aromatic group $(P_c)_A = \dfrac{206.126096n - 295.007504}{(0.227n - 0.325)^2}$ (2-45)

Step 3. Calculate the acentric factor of each cut of the undefined fraction by using the following expressions:

- Paraffinic group $(\omega)_P = 0.0432n + 0.0457$ (2-46)

- Naphthenic group $(\omega)_N = 0.0432n - 0.0880$ (2-47)

- Aromatic group $(\omega)_A = 0.0445n - 0.0995$ (2-48)

Step 4. Calculate the critical temperature of the fraction under consideration by using the following relationship:

$$T_c = X_P \, (T_c)_P + X_N \, (T_c)_N + X_A \, (T_c)_A \tag{2-49}$$

where T_c = critical temperature of the fraction, °R

The critical temperatures of the various cuts of the undefined fractions are calculated from the following expressions:

- Paraffinic group $(T_c)_P = S(T_b)_P \left[1 + \dfrac{3\mathrm{Log}[(P_c)_P] - 3.501952}{7\,[1 + (\omega)_P]} \right]$ (2-50)

- Naphthenic group $(T_c)_N = S_1(T_b)_N \left[1 + \dfrac{3\mathrm{Log}[(P_c)_N] - 3.501952}{7\,[1 + (\omega)_P]} \right]$ (2-51)

- Aromatic group $(T_c)_A = S_1(T_b)_A \left[1 + \dfrac{3\mathrm{Log}[(P_c)_A] - 3.501952}{7\,[1 + (\omega)_A]} \right]$ (2-52)

where the correction factors S and S_1 are defined by the following expressions:

$S = 0.996704 + 0.00043155n$
$S_1 = 0.99627245 + 0.00043155n$

Step 5. Calculate the acentric factor of the heavy hydrocarbon fraction by using the Edmister correlation (Equation 2-21) to give

$$\omega = \dfrac{3\,[\mathrm{Log}\,(P_c/14.7)]}{7\,[(T_c/T_b) - 1]} - 1 \tag{2-53}$$

where ω = acentric factor of the heavy fraction
      P_c = critical pressure of the heavy fraction, psia
      T_c = critical temperature of the heavy fraction, °R
      T_b = average-weight boiling point, °R

Example 2-4. Calculate the critical pressure, the critical temperature, and acentric factor of an undefined hydrocarbon fraction with a measured molecular weight of 94 and a weight average boiling point of 655°R. The number of carbon atoms of the component is 7.

Solution.

Step 1.   Calculate the boiling point of each cut by applying Equations 2-36 through 2-38 to give

$$(T_b)_P = 666.58°R, \; (T_b)_N = 630.0°R, \; (T_b)_A = 635.85°R$$

Step 2.   Compute the molecular weight of various cuts by using Equations 2-39 through 2-41 to yield

$$(MW)_P = 100.198, \; (MW)_N = 84.156, \; (MW)_A = 78.18$$

Step 3.   Solve Equation 2-35 for X_P, X_N, and X_A, to give

$$X_P = 0.6313, \; X_N = 0.3262, \; X_A = 0.0425$$

Step 4.   Calculate the critical pressure of each cut in the undefined fraction by applying Equations 2-43 and 2-45.

$$(P_c)_P = 395.7 \text{ psia}, \; (P_c)_N = 586.61 \text{ psia}, \; (P_c)_A = 718.46 \text{ psia}$$

Step 5.   Calculate the critical pressure of the heavy fraction from Equation 2-42 to give

$$P_c = 471.7 \text{ psia}$$

Step 6.   Compute the acentric factor for each cut in the fraction by using Equations 2-46 through 2-48 to yield

$$(\omega)_P = 0.3481, \; (\omega)_N = 0.2144, \; (\omega)_A = 0.212$$

Step 7.   Solve for $(T_c)_P$, $(T_c)_N$, and $(T_c)_A$ by using Equations 2-50 through 2-52 to give

$$(T_c)_P = 969.4°R, \; (T_c)_N = 947.3°R, \; (T_c)_A = 1{,}014.9°R$$

Step 8.   Solve for (T_c) of the undefined fraction from Equation 2-49.

$$T_c = 964.1°R$$

Step 9. Calculate the acentric factor from Equation 2-53, to give

$$\omega = 0.3680$$

Bergman's Method

Bergman et al. (1977) proposed a detailed procedure for characterizing the undefined hydrocarbon fractions based on calculating the PNA content of the fraction under consideration. The proposed procedure was originated from analyzing extensive experimental data on lean gases and condensate systems. The authors, in developing the correlation, assumed that the paraffinic, naphthenic, and aromatic groups have the same boiling point. The computational procedure is summarized in the following steps:

Step 1. Estimate the weight fraction of the aromatic content in the undefined fraction by applying the following expression:

$$w_A = 8.47 - K_w \tag{2-54}$$

where w_A = weight fraction of aromatics
  K_w = Watson characterization factor, defined mathematically by the following expression:

$$K_w = (T_b)^{1/3}/\gamma \tag{2-55}$$

where γ = specific gravity of the undefined fraction
  T_b = weight average boiling point, °R

Bergman et al. imposed the following constraint on the aromatic content:

$$0.03 \leq w_A \leq 0.35$$

Step 2. With the estimate of the aromatic content, the weight fractions of the paraffinic and naphthenic cuts are calculated by solving the following system of linear equations:

$$w_P + w_N = 1 - w_A \tag{2-56}$$

$$\frac{w_P}{\gamma_P} + \frac{w_N}{\gamma_N} = \frac{1}{\gamma} - \frac{w_A}{\gamma_A} \tag{2-57}$$

where W_P = weight fraction of the paraffin cut
  w_N = weight fraction of the naphthene cut

γ = specific gravity of the undefined fraction

$\gamma_P, \gamma_N, \gamma_A$ = specific gravity of the three groups at the weight average boiling point of the undefined fraction. These gravities are calculated from the following relationships:

$$\gamma_P = 0.582486 + 0.00069481 \, (T_b - 460)$$
$$- 0.7572818(10^{-6})(T_b - 460)^2$$
$$+ 0.3207736(10^{-9})(T_b - 460)^3 \qquad (2\text{-}58)$$

$$\gamma_N = 0.694208 + 0.0004909267(T_b - 460)$$
$$- 0.659746(10^{-6})(T_b - 460)^2$$
$$+ 0.330966(10^{-9})(T_b - 460)^3 \qquad (2\text{-}59)$$

$$\gamma_A = 0.916103 - 0.000250418(T_b - 460)$$
$$+ 0.357967(10^{-6})(T_b - 460)^2$$
$$- 0.166318(10^{-9})(T_b - 460)^3 \qquad (2\text{-}60)$$

A minimum paraffin content of 0.20 was set by Bergman et al. To insure that this minimum value is met, the estimated aromatic content that results in negative values of w_P is increased in increments of 0.03 up to a maximum of 15 times until the paraffin content exceeds 0.20. They pointed out that this procedure gives reasonable results for fractions up to C_{15}.

Step 3. Calculate the critical temperature, the critical pressure, and acentric factor of each cut from the following expressions:

• Paraffins:

$$(T_c)_P = 275.23 + 1.2061(T_b - 460)$$
$$- 0.00032984(T_b - 460)^2 \qquad (2\text{-}61)$$

$$(P_c)_P = 573.011 - 1.13707(T_b - 460) + 0.00131625(T_b - 460)^2$$
$$- 0.85103(10^{-6})(T_b - 460)^3 \qquad (2\text{-}62)$$

$$(\omega)_P = 0.14 + 0.0009(T_b - 460) + 0.233(10^{-6})(T_b - 460)^2 \qquad (2\text{-}63)$$

• Naphthenes:

$$(T_c)_N = 156.8906 + 2.6077(T_b - 460) - 0.003801(T_b - 460)^2$$
$$+ 0.2544(10^{-5})(T_b - 460)^3 \qquad (2\text{-}64)$$

$$(P_c)_N = 726.414 - 1.3275(T_b - 460) + 0.9846(10^{-3})(T_b - 460)^2$$
$$- 0.45169(10^{-6})(T_b - 460)^3 \qquad (2\text{-}65)$$

$$(\omega)_N = (\omega)_P - 0.075 \qquad (2\text{-}66)$$

Bergman et al. assigned the following special values of the acentric factor to the C_8, C_9, and C_{10} naphthenes.

C_8: $(\omega)_N = 0.26$
C_9: $(\omega)_N = 0.27$
C_{10}: $(\omega)_N = 0.35$

- Aromatics:

$$(T_c)_A = 289.535 + 1.7017(T_b - 460) - 0.0015843(T_b - 460)^2$$
$$+ 0.82358(10^{-6})(T_b - 460)^3 \tag{2-67}$$

$$(P_c)_A = 1{,}184.514 - 3.44681(T_b - 460) + 0.0045312(T_b - 460)^2$$
$$- 0.23416(10^{-5})(T_b - 460)^3 \tag{2-68}$$

$$(\omega)_A = (\omega)_P - 0.1 \tag{2-69}$$

Step 4. Calculate the critical pressure, the critical temperature, and the acentric factor of the undefined fraction from the following relationships:

$$P_c = X_P(P_c)_P + X_N(P_c)_N + X_A(P_c)_A \tag{2-70}$$

$$T_c = X_P(T_c)_P + X_N(T_c)_N + X_A(T_c)_A \tag{2-71}$$

$$\omega = X_P(\omega)_P + X_N(\omega)_N + X_A(\omega)_A \tag{2-72}$$

Whitson (1984) suggested that the Peng-Robinson and the Bergman PNA methods are not recommended for characterizing reservoir fluids containing fractions heavier than C_{20}.

OTHER METHODS OF CHARACTERIZING
THE HYDROCARBON HEAVY FRACTIONS

Rowe's Characterization Method

Rowe (1978) proposed a set of correlations for estimating the normal boiling point, the critical temperature, and the critical pressure of the heptanes-plus fraction, i.e., C_{7+}. The prediction of the C_{7+} properties is based on the assumption that the "lumped" fraction behaves as a normal paraffin hydrocarbon. Rowe used the number of carbon atoms n as the only correlating parameter. He proposed the following set of formulas for characterizing the C_{7+} fraction:

- Calculation of the critical temperature:

$$(T_c)_{C_{7+}} = 1.8 [961 - 10^a] \tag{2-73}$$

where $(T_c)_{C_{7+}}$ = critical temperature of C_{7+}, °R

a = coefficient of the equation and is given by the following expression:

$a = 2.95597 - 0.090597n^{2/3}$

where n is the number of carbon atoms and is calculated from the molecular weight of the C_{7+} fraction by the following relationship:

$$n = [MW_{C_{7+}} - 2.0]/14 \tag{2-74}$$

where $MW_{C_{7+}}$ is the molecular weight of the heptanes-plus fraction.

- Calculation of the critical pressure:

$$(P_c)_{C_{7+}} = 10^{(5 + Y)}/(T_c)_{C_{7+}} \tag{2-75}$$

with

$$Y = -0.0137726826 \, n + 0.6801481651$$

where $(P_c)_{C_{7+}}$ is the critical pressure of C_{7+} in psia.

- Calculation of the critical temperature:

$$(T_b)_{C_{7+}} = 0.0004347(T_c)^2_{C_{7+}} + 265 \tag{2-76}$$

where $(T_b)_{C_{7+}}$ is the normal boiling point in °R.

Standing's Method

Mathews, Roland, and Katz (1942) presented graphical correlations for determining the critical temperature and pressure of the heptanes-plus fraction. Standing (1977) expressed these graphical correlations more conveniently in mathematical forms as follows:

$$(T_c)_{C_{7+}} = 608 + 364 \, Log \, [(MW)_{C_{7+}} - 71.2] + [2,450 \, Log \, (MW)_{C_{7+}} - 3,800] \, Log \, (\gamma)_{C_{7+}} \tag{2-77}$$

$$(P_c)_{C_{7+}} = 1,188 - 431 \, Log \, [(MW)_{C_{7+}} - 61.1] + [2,319 - 852 \, Log \, [(MW)_{C_{7+}} - 53.7]] \, ((\gamma)_{C_{7+}} - 0.8) \tag{2-78}$$

where $(MW)_{C_{7+}}$ and $(\gamma)_{C_{7+}}$ are the molecular weight and specific gravity of the C_{7+}.

Example 2-5. If the molecular weight and specific gravity of the heptanes-plus fraction are 216 and 0.8605, respectively, calculate the critical temperature and pressure by using

1. Rowe's Correlations
2. Standing's Correlations

Solution.

1. Solution by using Rowe's Correlation

Step 1. Calculate the number of carbon atoms of C_{7+} from Equation 2-74 to give

$$n = 15.29$$

Step 2. Solve for the critical temperature from Equation 2-73 to yield

$$(T_c)_{C_{7+}} = 1,279.8°R$$

Step 3. Solve for the critical pressure from Equation 2-75 to give

$$(P_c)_{C_{7+}} = 230.4 \text{ psia}$$

2. Solution by using Standing's Correlations

Step 1. Solve for the critical temperature by using Equation 2-77 to give

$$(T_c)_{C_{7+}} = 1,269.3°R$$

Step 2. Calculate the critical pressure from Equation 2-78 to yield

$$(P_c)_{C_{7+}} = 270 \text{ psia}$$

Katz-Firoozabadi Method

Katz and Firoozabadi (1978) presented a generalized set of physical properties for the petroleum fractions C_6 through C_{45}. The tabulated properties include the average boiling point, specific gravity, and molecular weight. The authors' proposed tabulated properties are based on the analysis of the physical properties of 26 condensates and naturally occurring liquid hydrocarbons.

Whitson (1983) found inconsistency in the Katz and Firoozabadi tabulated molecular weight data after analyzing and comparing these data with sources from which they were developed. Whitson pointed out that this inconsistency was found in the hydrocarbon fractions C_{22} through C_{45}.

Whitson modified the original tabulated physical properties to make their use more consistent. The modification was accomplished by employing the Riazi and Daubert correlation form (Equation 2-1) to extrapolate the molecular weight data from C_{22} to C_{45}. The coefficients a, b, and c of Equation 2-1 were recalculated by using a nonlinear regression model to fit the molecular weight data of C_6 through C_{22}. The equation was then used to calculate molecular weights of C_{23} through C_{45}. The author also calculated the critical properties and acentric factors of C_6 through C_{45} in terms of their boiling point, specific gravity, and the modified values of the molecular weight. These generalized properties are given in Table 2-10.

Ahmed (1985) correlated Katz-Firoozabadi-Whitson tabulated physical properties with the number of carbon atoms of the fraction by using a regression model. The generalized equation has the following form:

$$\Theta = a_1 + a_2 \, n + a_3 \, n^2 + a_4 \, n^3 + (a_5/n) \tag{2-79}$$

where $\quad \Theta$ = any physical property
$\quad\quad\quad$ n = number of carbon atoms, i.e., 6, 7, . . ., 45
$\quad\quad\quad$ a_1–a_5 = coefficients of the equations and are given in Table 2-11.

Riazi-Daubert Method

Riazi and Daubert (1980 and 1987) suggested that their proposed correlations for calculating the physical properties of pure petroleum fractions (Equations 2-1 through 2-3) can be used to predict the physical properties of the undefined pertroleum fractions. The authors proposed that the following "average" boiling points be used for estimating the physical properties.

Physical Property Θ	Average Boiling Point
T_c	Molal Average Boiling Point (MABP)
P_c, MW, γ, V_c	Mean Average Boiling Point (MeABP)

Cavett's Method

With the appropriate boiling point, Cavett's correlations, as given by Equations 2-7 and 2-8, can be used to predict the critical temperature and the critical pressure of the heavy hydrocarbon fraction. Cavett proposed that the molal average boiling point (MABP) is used in Equation 2-7 to calculate T_c and the mean average boiling point (MeABP) is used in Equation 2-8 for calculating P_c.

Table 2-10
Generalized Physical Properties

Group	T_b (°R)	γ	K	M	T_c (°R)	P_c (psia)	ω	V_c (ft³/lb)	Group
C_6	607	0.690	12.27	84	923	483	0.250	0.06395	C_6
C_7	658	0.727	11.96	96	985	453	0.280	0.06289	C_7
C_8	702	0.749	11.87	107	1,036	419	0.312	0.06264	C_8
C_9	748	0.768	11.82	121	1,085	383	0.348	0.06258	C_9
C_{10}	791	0.782	11.83	134	1,128	351	0.385	0.06273	C_{10}
C_{11}	829	0.793	11.85	147	1,166	325	0.419	0.06291	C_{11}
C_{12}	867	0.804	11.86	161	1,203	302	0.454	0.06306	C_{12}
C_{13}	901	0.815	11.85	175	1,236	286	0.484	0.06311	C_{13}
C_{14}	936	0.826	11.84	190	1,270	270	0.516	0.06316	C_{14}
C_{15}	971	0.836	11.84	206	1,304	255	0.550	0.06325	C_{15}
C_{16}	1,002	0.843	11.87	222	1,332	241	0.582	0.06342	C_{16}
C_{17}	1,032	0.851	11.87	237	1,360	230	0.613	0.06350	C_{17}
C_{18}	1,055	0.856	11.89	251	1,380	222	0.638	0.06362	C_{18}
C_{19}	1,077	0.861	11.91	263	1,400	214	0.662	0.06372	C_{19}
C_{20}	1,101	0.866	11.92	275	1,421	207	0.690	0.06384	C_{20}
C_{21}	1,124	0.871	11.94	291	1,442	200	0.717	0.06394	C_{21}
C_{22}	1,146	0.876	11.95	300	1,461	193	0.743	0.06402	C_{22}
C_{23}	1,167	0.881	11.95	312	1,480	188	0.768	0.06408	C_{23}
C_{24}	1,187	0.885	11.96	324	1,497	182	0.793	0.06417	C_{24}
C_{25}	1,207	0.888	11.99	337	1,515	177	0.819	0.06431	C_{25}
C_{26}	1,226	0.892	12.00	349	1,531	173	0.844	0.06438	C_{26}
C_{27}	1,244	0.896	12.00	360	1,547	169	0.868	0.06443	C_{27}
C_{28}	1,262	0.899	12.02	372	1,562	165	0.894	0.06454	C_{28}
C_{29}	1,277	0.902	12.03	382	1,574	161	0.915	0.06459	C_{29}
C_{30}	1,294	0.905	12.04	394	1,589	158	0.941	0.06468	C_{30}
C_{31}	1,310	0.909	12.04	404	1,603	143	0.897	0.06469	C_{31}

C_{32}	1,326	12.05	0.912	415	1,616	138	0.909	0.06475
C_{33}	1,341	12.05	0.915	426	1,629	134	0.921	0.06480
C_{34}	1,355	12.07	0.917	437	1,640	130	0.932	0.06489
C_{35}	1,368	12.07	0.920	445	1,651	127	0.942	0.06490
C_{36}	1,382	12.08	0.922	456	1,662	124	0.954	0.06499
C_{37}	1,394	12.08	0.925	464	1,673	121	0.964	0.06499
C_{38}	1,407	12.09	0.927	475	1,683	118	0.975	0.06506
C_{39}	1,419	12.10	0.929	484	1,693	115	0.985	0.06511
C_{40}	1,432	12.11	0.931	495	1,703	112	0.997	0.06517
C_{41}	1,442	12.11	0.933	502	1,712	110	1.006	0.06520
C_{42}	1,453	12.13	0.934	512	1,720	108	1.016	0.06529
C_{43}	1,464	12.13	0.936	521	1,729	105	1.026	0.06532
C_{44}	1,477	12.14	0.938	531	1,739	103	1.038	0.06538
C_{45}	1,487	12.14	0.940	539	1,747	101	1.048	0.06540

Permission to publish by the Society of Petroleum Engineers of AIME. Copyright SPE-AIME.

Table 2-11
Coefficients of Equation 2-79

Θ	a_1	a_2	a_3	a_4	a_5	MAD* %	Max. Error %
MW	-131.11375	24.96156	-0.34079022	2.4941184×10^{-3}	468.32575	0.418	1.31
T_c, °R	915.53747	41.421337	-0.7586859	5.8675351×10^{-3}	-1.3028779×10^3	0.126	0.5
P_c, psia	275.56275	-12.522269	0.29926384	$-2.8452129 \times 10^{-3}$	1.7117226×10^3	2.071	5.3
T_b, °R	434.38878	50.125279	-0.9027283	7.0280657×10^{-3}	-601.85651	0.178	0.7
ω	-0.50862704	8.700211×10^{-2}	$-1.8484814 \times 10^{-3}$	1.4663890×10^{-5}	1.8518106	1.152	3.7
γ	0.86714949	3.4143408×10^{-3}	-2.839627×10^{-5}	2.4943308×10^{-8}	-1.1627984	0.103	0.47
V_c, ft³/lb	5.223458×10^{-2}	7.8709139×10^{-4}	$-1.9324432 \times 10^{-5}$	1.7547264×10^{-7}	4.4017952×10^{-2}	0.099	0.49

* MAD = Mean Average Deviation

Willman-Teja Method

Willman and Teja (1987) proposed correlations for determining the critical pressure and critical temperature of the n-alkane homologous series. The authors used the normal boiling point and the number of carbon atoms of the n-alkane as a correlating parameter. The applicability of the Willman and Teja proposed correlations can be extended to predict the critical temperature and pressure of the undefined petroleum fraction by recalculating the exponents of the original expressions. These exponents were recalculated by using a nonlinear regression model to best match the critical properties data of Berman et al. (1977) and Whitson (1980). The empirical formulas are given by

$$T_c = T_b [1 + (1.25127 + 0.137242n)^{-0.884540633}] \tag{2-80}$$

$$P_c = [339.0416805 + 1,184.157759n][0.873159 + 0.54285n]^{-1.9265669} \tag{2-81}$$

where n = number of carbon atoms
     T_b = average boiling point of the undefined fraction, °R

It should be noted that the Edmister acentric factor correlation (Equation 2-21) and the Kesler and Lee correlations (Equations 2-9 through 2-13) can also be used to characterize the undefined hydrocarbon fraction from their average boiling point and specific gravity.

Example 2-6. Calculate the critical properties and the acentric factor of C_{7+} with a measured molecular weight of 198.71 and specific gravity of 0.8527. Employ the following methods:

1. Rowe's correlation
2. Standing's correlation
3. Riazi-Daubert correlation

Solution.

1. Solution by using Rowe's Correlation.

Step 1. Calculate the number of carbon atoms of the fraction

     n = 14.0507

Step 2. Determine T_c from Equation 2-73 to give

     $T_c = 1,247.06°R$

Step 3. Compute P_c from Equation 2-75 to yield

P_c = 245.89 psia

Step 4. Determine T_b by applying Equation 2-76 to give

T_b = 941.03°R

Step 5. Solve for the acentric factor by applying Equation 2-21 to give

ω = 0.6123

2. **Solution by using Standing's Correlation.**

Step 1. Solve for the critical temperature of C_{7+} by using Equation 2-77 to give

$(T_c)_{C_{7+}}$ = 1,247.73°R

Step 2. Determine the critical pressure from Equation 2-78 to give

$(P_c)_{C_{7+}}$ = 291.41 psia

3. **Solution by using Riazi-Daubert Correlation.**

Step 1. Solve Equation 2-3 for T_c to give

T_c = 1,294.1°R

Step 2. Calculate P_c from Equation 2-3 to give

P_c = 263.67

Step 3. Determine T_b by applying Equation 2-3

T_b = 958.5°R

Step 4. Solve for the acentric factor from Equation 2-21 to give

ω = 0.5346

DETERMINATION OF THE PHYSICAL PROPERTIES OF THE HEAVY PETROLEUM FRACTION FROM GRAPHICAL CORRELATIONS

Several mathematical correlations for determining the physical and critical properties of petroleum fractions have been presented. These correlations are readily adapted to computer applications. However, it is important to present the properties in graphical forms for a better understanding of the behaviors and interrelationships of the properties.

Boiling Points

Numerous graphical correlations have been proposed over the years for determining the physical and critical properties of petroleum fractions. Most of these correlations use the normal boiling point as one of the correlation parameters. As stated previously, there are five different methods of defining the normal boiling point:

a. Volume Average Boiling Point (VABP)
b. Weight Average Boiling Point (WABP)
c. Molal Average Boiling Point (MABP)
d. Cubic Average Boiling Point (CABP)
e. Mean Average Boiling Point (MeABP)

Figure 2-2 shows the conversions between the VABP and other boiling points. The following steps summarize the use of Figure 2-2.

Step 1. On the basis of ASTM D-86 distillation data, calculate the volumetric average boiling point from the following expressions:

$$VABP = (t_{10} + t_{30} + t_{50} + t_{70} + t_{90})/5 \qquad (2\text{-}82)$$

where t is the temperature in °F and the subscripts 10, 30, 50, 70, and 90 refer to the volume percent recovered during the distillation.

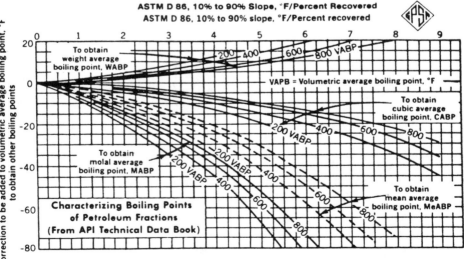

Figure 2-2. Correction to volumetric average boiling points. Courtesy of the Gas Processors Suppliers Association. Published in the GPSA Engineering Data Book, Tenth Edition, 1987.

Step 2. Calculate the 10% to 90% "slope" of the ASTM distillation curve from the following expression:

$$\text{slope} = (t_{90} - t_{10})/80 \qquad (2\text{-}83)$$

Step 3. Enter the value of the slope in the graph and travel vertically to the appropriate set for the type of boiling point desired.

Step 4. Read from the ordinate a correction factor for the VABP and apply the relationship:

$$\text{Desired Boiling Point} = \text{VABP} + \text{Correction Factor} \qquad (2\text{-}84)$$

The use of the graph can best be illustrated by the following examples:

Example 2-7. The following ASTM distillation data for a 55°API gravity petroleum fraction is given:

Cut	Distillation % Over	Temperature, °F
1	IBP*	159
2	10	178
3	20	193
4	30	209
5	40	227
6	50	253
7	60	282
8	70	318
9	80	364
10	90	410
Residue	EP**	475

* Initial boiling point.
** End point.

Calculate:

1. WABP
2. MABP
3. CABP
4. MeABP

Solution.

Step 1. Calculate VABP from Equation 2-82:

$$\text{VABP} = (178 + 209 + 253 + 318 + 410)/5 = 273°F$$

Step 2. Calculate the distillation curve slope from Equation 2-83:

slope = (410 − 178)/80 = 2.9

Step 3. Enter the slope value of 2.9 in Figure 2-2 and move down to the appropriate set of boiling point curves. Read the corresponding correction factors from the ordinate to give:

- Correction factor for WABP = 6°F
- Correction factor for CABP = − 7°F
- Correction factor for MeABP = − 18°F
- Correction factor for MABP = − 33°F

Step 4. Calculate the desired boiling point by applying Equation 2-84

- WABP = 273 + 6 = 279°F
- CABP = 273 − 7 = 266°F
- MeABP = 273 − 18 = 255°F
- MABP = 273 − 33 = 240°F

Molecular Weight

Figure 2-3 shows a convenient graphical correlation for determining the molecular weight of petroleum fractions from their mean average boiling points (MeABP) and API gravities. The following example illustrates the practical application of the graphical method.

Example 2-8. Calculate the molecular weight of the petroleum fraction with an API gravity and MeABP as given in Example 2-7.

Solution. From Example 2-7:

API = 55°
MeABP = 255°F

Enter the above values in Figure 2-3 to give MW = 118.

Critical Temperature

The critical temperature of a petroleum fraction can be determined by using the graphical correlation shown in Figure 2-4. The required correlation parameters are the API gravity and the molal average boiling point (MABP) of the undefined fraction.

M W, Boiling Points, Gravities, Petroleum Fractions

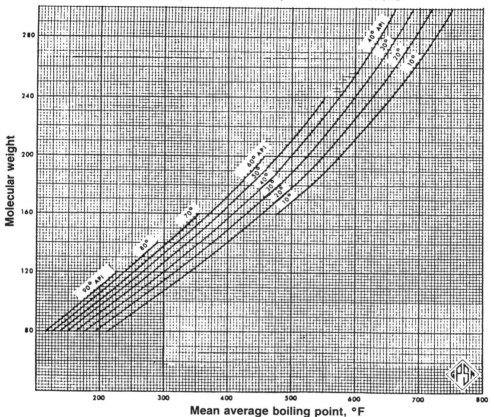

Figure 2-3. Relationship between molecular weight, API gravity, and mean average boiling points. Courtesy of the Gas Processors Suppliers Association. Published in the GPSA Engineering Data Book, Tenth Edition, 1987.

Example 2-9. Calculate the critical temperature of the petroleum fraction with physical properties as given in Example 2-7.

Solution. From Example 2-7:

MABP = 240°F
API = 55°

Enter the above values in Figure 2-4 to give:

$T_c = 600°F$

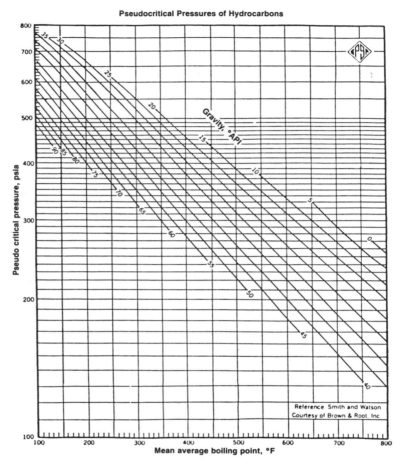

Figure 2-4. Relationship between critical pressure, API gravity, and mean average boiling points. Courtesy of the Gas Processors Suppliers Association. Published in the GPSA Engineering Data Book, Tenth Edition, 1987.

Critical Pressure

Figure 2-5 is a graphical correlation of the critical pressure of the undefined petroleum fractions as a function of the mean average boiling point (MeABP) and the API gravity. The following example shows the practical use of the graphical correlation.

Example 2-10. Calculate the critical pressure of the petroleum fraction from Example 2-7.

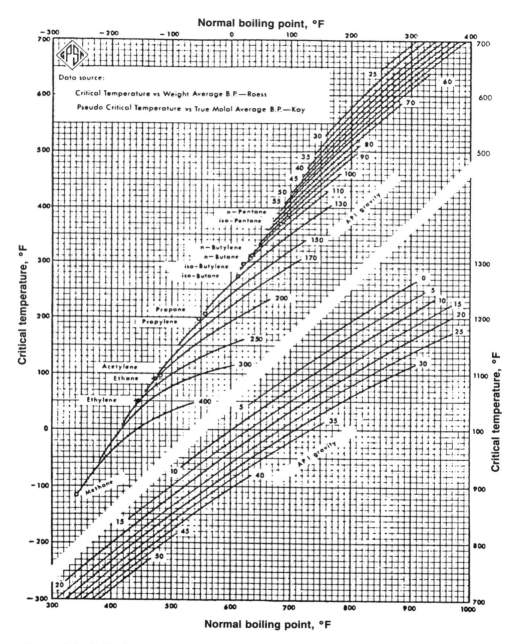

Figure 2-5. Critical temperature as a function of API gravity and boiling points. Courtesy of the Gas Processors Suppliers Association. Published in the GPSA Engineering Data Book, Tenth Edition, 1987.

Solution. From Example 2-7:

API = 55°
MeABP = 240°F

Determine the critical pressure of the fraction from Figure 2-5 to give:

P_c = 460 psia

PROBLEMS

1. If a petroleum fraction has a measured molecular weight of 190 and a specific gravity of 0.8762, characterize this fraction by calculating the boiling point, critical temperature, critical pressure, and critical volume of the fraction. Use the Riazi and Daubert correlation.

2. Calculate the acentric factor and critical compressibility factor of the component in the above problem.

3. Using the Lin and Chao generalized correlation, calculate the physical and critical properties of a component with a MW, ω, and T_b of 200, 0.8251, and 500°R, respectively.

4. A petroleum fraction has the following physical properties:

   API = 50°, T_b = 400°F, MW = 165

   Calculate p_c, T_c, V_c, ω, and Z_c by using the following correlations:

   a. Cavett
   b. Kesler and Lee
   c. Winn-Sim-Daubert
   d. Watansiri-Owens-Starling

5. An undefined petroleum fraction with ten carbon atoms has a measured average boiling point of 791°R and a molecular weight of 134. If the specific gravity of the fraction is 0.78, determine the critical pressure, the critical temperature, and acentric factor of the fraction by using:

   a. Robinson-Peng PNA Method
   b. Bergman PNA Method
   c. Riazi-Daubert Method
   d. Cavett's Correlation
   e. Kesler-Lee Correlation
   f. Willman-Teja

6. A heptanes-plus fraction is characterized by a molecular weight of 200 and specific gravity of 0.810. Calculate P_c, T_c, T_b, and acentric factor of the plus fraction by using:

    a. Riazi-Daubert Method
    b. Rowe's Correlation
    c. Standing's Correlation

7. Using the data given in Problem 6, and the boiling point as calculated by the Riazi and Daubert correlation, determine the critical properties and acentric factor by employing:

    a. Cavett's Correlation
    b. Kesler-Lee Correlation
    c. Compare the results with those obtained in Problem 6.

REFERENCES

1. Ahmed, T., "Composition Modeling of Tyler and Mission Canyon Formation Oils with CO_2 and Lean Gases," final report submitted to Montanans on a New Track for Science (MONTS) (Montana National Science Foundation Grant Program), 1985.

2. Bergman, D. F., Tek, M. R., and Katz, D. L., "Retrograde Condensation in Natural Gas Pipelines," Project PR 2-29 of Pipelines Research Committee, AGA, Jan. 1977.

3. Cavett, R. H., "Physical Data for Distillation Calculations—Vapor-Liquid Equilibrium," Proc. 27th Meeting, API, San Francisco, 1962, pp. 351–366.

4. Edmister, W. C., "Applied Hydrocarbon Thermodynamics, Part 4: Compressibility Factors and Equations of State," *Petroleum Refiner*, April 1958, Vol. 37, pp. 173–179.

5. Edmister, W. C. and Lee, B. I., *Applied Hydrocarbon Thermodynamics*, Vol. 1, Gulf Publishing Company: Houston, 1986.

6. Erbar, J. H., "Prediction of Absorber Oil K-Values and Enthalpies," Research Report 13, GPA, Tulsa, Oklahoma, 1977.

7. Haugen, O. A., Watson, K. M., and Ragatz, R. A., *Chemical Process Principles*, 2nd ed., New York: Wiley, 1959, p. 577.

8. Hopke, S. W. and Lin, C. J., "Application of BWRS Equation to Absorber Oil Systems," proceedings 53rd Annual Convention GPA, Denver, Colorado, March 1974, pp. 63–71.

9. Katz, D. L. and Firoozabadi A., "Predicting Phase Behavior of Condensate/Crude-Oil Systems Using Methane Interaction Coefficients," *JPT*, Nov. 1978, pp. 1649–1655.

10. Kesler, M. G. and Lee, B. I., "Improve Prediction of Enthalpy of Fractions," *Hydrocarbon Processing*, March 1976, pp. 153–158.
11. Lin, H. M. and Chao, K. C., "Correlation of Critical Properties and Acentric Factor of Hydrocarbons and Derivatives," *AIChE Journal*, Vol. 30, No. 6, Nov. 1984, pp. 981–983.
12. Nath, J., "Acentric Factor and the Critical Volumes for Normal Fluids," *Ind. Eng. Chem. Fundam.*, Vol. 21, No. 3, 1985, pp. 325–326.
13. Reid, R., Prausnitz, J. M., and Sherwood, T., *The Properties of Gases and Liquids*, 3rd ed., McGraw-Hill, 1977, p. 21.
14. Riazi, M. R. and Daubert, T. E., "Simplify Property Predictions," *Hydrocarbon Processing*, March 1980, pp. 115–116.
15. Riazi, M. R. and Daubert, T. E., "Characterization Parameters for Petroleum Fractions," *Ind. Eng. Chem. Res.*, Vol. 26, No. 24, 1987, pp. 755–759.
16. Robinson, D. B. and Peng, D. Y., "The Characterization of the Heptanes and Heavier Fractions," Research Report 28, GPA, Tulsa (1978).
17. Rowe, A. M., "Internally Consistent Correlations for Predicting Phase Compositions for use in Reservoir Compositional Simulators," paper SPE 7475 presented at the 53rd Annual Fall Technical Conference and Exhibition.
18. Salerno, S., et al., "Prediction of Vapor Pressures and Saturated Volumes," *Fluid Phase Equilibria*, Vol. 27, June 10, 1986, pp. 15–34.
19. Sim, W. J. and Daubert, T. E., "Prediction of Vapor-Liquid Equilibria of Undefined Mixtures," *Ind. Eng. Chem. Process Des. Dev.*, Vol. 19, No. 3, 1980, pp. 380–393.
20. Standing, M. B., *Volumetric and Phase Behavior of Oil Field Hydrocarbon Systems*, Society of Petroleum Engineers, Dallas, 1977, p. 124.
21. Watansiri, S., Owens, V. H., and Starling, K. E., "Correlations for Estimating Critical Constants, Acentric Factor, and Dipole Moment for Undefined Coal-Fluid Fractions," *Ind. Eng. Chem. Process Des. Dev.*, 1985, Vol. 24, pp. 294–296.
22. Whitson, C. H., "Effect of Physical Properties Estimation on Equation-of-State Predictions," *SPEJ*, Dec. 1984, pp. 685–696.
23. Willman, B. and Teja, A., "Prediction of Dew Points of Semicontinuous Natural Gas and Petroleum Mixtures," *Ind. Eng. Chem. Res.*, 1987, Vol. 226, No. 5, pp. 948–952.
24. Winn, F. W., "Simplified Nomographic Presentation, Characterization of Petroleum Fractions," *Petroleum Refiner*, Vol. 36, No. 2, 1957, p. 157.

3
Properties of Natural Gases

Laws which describe the behavior of gases in terms of pressure p, volume V, and temperature T have been known for many years. These laws are relatively simple for a hypothetical fluid known as a perfect (ideal) gas. This chapter reviews the perfect gas laws and how they can be modified to describe the behavior of real gases which may deviate significantly from these laws under certain conditions of pressure and temperature.

A gas is defined as a homogeneous fluid of low density and viscosity, which has no definite volume but expands to completely fill the vessel in which it is placed. Knowledge of pressure-volume-temperature (PVT) relationships and other physical and chemical properties of gases is essential for solving problems in natural gas reservoir engineering. The physical properties of a natural gas may be obtained directly either by laboratory measurements or by prediction from the known chemical composition of the gas. In the latter case, the calculations are based on the physical properties of individual components of the gas and upon physical laws, often referred to as mixing rules, relating the properties of the components to those of the gas mixture.

BEHAVIOR OF IDEAL GASES

The kinetic theory of gases postulates that the gas is composed of a very large number of particles called molecules. For an ideal gas, the volume of these molecules is insignificant compared with the total volume occupied by the gas. It is also assumed that these molecules have no attractive or repulsive forces between them, and it is assumed that all collisions of molecules are perfectly elastic.

Pure Gases

Based on the above kinetic theory of gases, a mathematical equation called Equation-of-State can be derived to express the relationship existing between pressure, volume, and temperature for a given quantity of gas. This relationship for perfect gases is called the Ideal Gas Law, and is expressed mathematically by the following equation:

$$pV = nRT \tag{3-1}$$

where n is the quantity of gas in moles and is defined by the expression

$$n = \frac{m}{MW} \tag{3-2}$$

Combining Equation 3-2 with 3-1 yields

$$pV = \left(\frac{m}{MW}\right) RT \tag{3-3}$$

in which, for the conventional field units used in the Petroleum industry,

    p = absolute pressure, psia
    V = volume, ft³
    T = absolute temperature, °R
    n = number of moles of gas, lb-mole
    m = weight of gas, lb
    MW = molecular weight, lb/lb-mole
    R = the universal gas constant which, for the above units, has the value 10.730 psia ft³/lb-mole °R

Because the density is defined as the mass per unit volume of the substance, Equation 3-3 can be solved for the density to yield

$$\rho_g = \frac{m}{v} = \frac{p\,MW}{RT} \tag{3-4}$$

where ρ_g = density of the gas, lb/ft³, and lb refers to lbs mass in any subsequent discussions of density in this text.

In dealing with gases at a very low pressure, the ideal gas relationship is a convenient and generally satisfactory tool. For the calculation of the physical properties of natural gases at elevated pressure, the use of the ideal gas equation-of-state may lead to errors as great as 500%, as compared to errors of 2–3% at atmospheric pressure.

Example 3-1. Assuming an ideal gas behavior, calculate the density of propane with a constant temperature of 100°F and 20 psia.

Solution. Applying Equation 3-4 yields

$$\rho_g = \frac{(20)\,(44.097)}{(10.73)(100 + 460)} = 0.1468 \text{ lb/ft}^3$$

IDEAL GAS MIXTURES

Petroleum engineers are usually interested in the behavior of mixtures and rarely deal with pure component gases. Because natural gas is a mixture of hydrocarbon components, the overall physical and chemical properties can be determined from the physical properties of the individual components in the mixture by using appropriate mixing rules.

Conventionally, natural gas compositions are expressed in terms of mole fraction, weight fraction, and volume percent. These are derived as follows:

Mole Fraction: The mole fraction of a particular component, component i, is defined as the number of moles of that component divided by the total number of moles of all the components in the mixture.

$$y_i = \frac{n_i}{n} = \frac{n_i}{\sum_i n_i} \tag{3-5}$$

where y_i = mole fraction of component i in the mixture
   n_i = number of moles of component i
   n = total number of moles in the mixture

Weight Fraction: The weight fraction of any component is defined as the weight of that component divided by the total weight.

$$w_i = \frac{m_i}{m} = \frac{m_i}{\sum_i m_i} \tag{3-6}$$

where w_i = weight fraction of component i
   m_i = weight of component i in the gas phase
   m = total weight of the gas mixture

Volume Fraction: The volume fraction of a specific component in a mixture is defined as the volume of that compound divided by the total volume of the mixture.

$$v_i = \frac{V_i}{V} = \frac{V_i}{\sum_i V_i} \tag{3-7}$$

where v_i = volume fraction of component i in the gas phase
    V_i = volume occupied by component i
    V = total volume of the mixture

It is convenient in many engineering calculations to convert from mole fraction to weight fraction and vice versa. The procedure of converting the composition of the gas phase from mole fraction to weight fraction is summarized in the following steps.

Step 1. Assume that the total number of moles of the gas phase is one, i.e., n = 1.

Step 2. From Equation 3-5, it is apparent that

$$n_i = y_i$$

Step 3. Because the number of moles of a component is equal to the weight of the component divided by the molecular weight of the component, as expressed mathematically by Equation 3-2, the weight of the component can be expressed as

$$m_i = y_i\, MW_i$$

and

$$m = \sum_i m_i = \sum_i y_i\, MW_i$$

Step 4. By recalling the definition of weight fraction

$$w_i = \frac{m_i}{m}$$

then

$$w_i = \frac{y_i\, MW_i}{\sum_i y_i MW_i} \tag{3-8}$$

Similarly, one can convert from weight fraction to mole fraction by applying the following relationship:

$$y_i = \frac{(w_i/MW_i)}{\sum_i (w_i/MW_i)} \tag{3-9}$$

The above procedure is conveniently illustrated through the following examples.

Example 3-2. Determine the composition in weight fraction of the following gas:

Component	Mole Fraction y_i
C_1	0.65
C_2	0.10
C_3	0.10
C_4	0.10
C_5	0.05

Solution.

Component	y_i	MW_i	$m_i = y_i\, MW_i$	$w_i = m_i/m$
C_1	0.65	16.04	10.4260	0.3824
C_2	0.10	30.07	3.0070	0.1103
C_3	0.10	44.10	4.4100	0.1618
C_4	0.10	58.12	5.8120	0.2132
C_5	0.05	72.15	3.6075	0.1323
			$m = 27.2625$	

Example 3-3. Determine the composition in mole fraction of the following gas:

Component	Weight Fraction w_i
C_1	0.40
C_2	0.10
C_3	0.20
C_4	0.20
C_5	0.10

Solution.

Component	w_i	MW_i	$n_i = w_i/MW_i$	$y_i = n_i/n$
C_1	0.40	16.04	0.02494	0.6626
C_2	0.10	30.07	0.00333	0.0885
C_3	0.20	44.10	0.00454	0.1206
C_4	0.20	58.12	0.00344	0.0914
C_5	0.10	72.15	0.00139	0.0369
			$n = 0.03764$	

PROPERTIES OF IDEAL GAS MIXTURES

Usually the petroleum engineer is interested in studying the volumetric behavior and evaluating the basic properties of natural gas mixtures. It is appropriate first to introduce and define the physical properties of ideal gas mixtures.

The basic properties of ideal gases are commonly expressed in terms of the apparent molecular weight, standard volume, density, specific volume, and specific gravity. These properties are defined as follows:

Apparent Molecular Weight

One of the properties that is frequently of interest to engineers is the apparent molecular weight. If y_i represents the mole fraction of the ith component in a gas mixture, the apparent molecular weight is defined mathematically by the following equation

$$MW_a = \sum_{i=1} y_i \cdot MW_i \qquad (3\text{-}10)$$

where MW_a = apparent molecular weight of the gas mixture
MW_i = molecular weight of the ith component in the mixture

Standard Volume

In many natural gas engineering calculations, it is convenient to measure the volume occupied by 1 lb/mole of gas at a reference pressure and temperature. These reference conditions are usually 14.7 psia and 60°F, and commonly referred to as standard conditions. The standard volume is then defined as the volume occupied by 1 lb/mole of an ideal gas at standard conditions. Applying the above conditions to Equation 3-1 and solving for the standard volume yields

$$V_{sc} = \frac{(1)\ R\ T_{sc}}{p_{sc}}$$

Substituting for pressure and temperature, produces

$$V_{sc} = 379.4\ \text{scf/lb-mole} \qquad (3\text{-}11)$$

where V_{sc} = standard volume, scf/lb-mole
scf = standard cubic feet
T_{sc} = standard temperature, °R
p_{sc} = standard pressure, psia

Density

As defined previously by Equation 3-4, the density of an ideal gas mixture is calculated by applying Equation 3-6 and replacing the molecular weight with the apparent molecular weight of the gas mixture to yield:

$$\rho_g = \frac{p\ MW_a}{R\ T} \tag{3-12}$$

where ρ_g = density of the gas mixture, ft³/lb
      MW_a = apparent molecular weight

Specific Volume

The specific volume is defined as the volume occupied by a unit mass of the gas. For an ideal gas, this property can be calculated by applying Equation 3-3:

$$\nu = \frac{V}{m} = \frac{R\ T}{p \cdot MW_a} = \frac{1}{\rho_g} \tag{3-13}$$

where ν = specific volume, ft³/lb
     ρ_g = gas density, lb/ft³

Specific Gravity

The specific gravity is defined as the ratio of the gas density to that of the air. Both densities must be taken at the same temperature and pressure, or:

$$\gamma_g = \frac{\rho_g}{\rho_{air}} \tag{3-14}$$

Assuming the behavior of both the gas mixture and the air is described by the ideal gas equation, the specific gravity can be expressed by the following relationship:

$$\gamma_g = \frac{\dfrac{p\ MW_a}{R\ T}}{\dfrac{p \cdot MW_{air}}{R\ T}} = \frac{MW_a}{MW_{air}} = \frac{MW_a}{28.96} \tag{3-15}$$

where γ_g = gas specific gravity
          ρ_{air} = density of the air
          MW_{air} = apparent molecular weight of the air = 28.96

Example 3-4. A gas has the following composition:

Component	y_i
C_1	0.75
C_2	0.07
C_3	0.05
C_4	0.04
C_5	0.04
C_6	0.03
C_7	0.02

Assuming an ideal gas behavior, calculate the following gas properties at 1,000 psia and 100°F:

a. Apparent molecular weight
b. Specific gravity
c. Gas density
d. Specific volume

Solution.

Component	y_i	MW_i	$y_i \cdot MW_1$
C_1	0.75	16.04	12.030
C_2	0.07	30.07	2.105
C_3	0.05	44.10	2.205
C_4	0.04	58.12	2.325
C_5	0.04	72.15	2.886
C_6	0.03	86.18	2.585
C_7	0.02	100.21	2.004
			MW_a = 26.14

a. Applying Equation 3-10 yields

$MW_a = 26.14$

b. From Equation 3-15

$$\gamma_g = \frac{26.14}{28.96} = 0.903$$

c. Compute the gas density by using Equation 3-12

$$\rho_g = \frac{(1,000)(26.14)}{(10.73)(560)} = 4.35 \text{ lb/ft}$$

d. From Equation 2-13, specific volume $= \dfrac{1}{4.35} = 0.23$ ft³/lb

BEHAVIOR OF REAL GASES

Basically the magnitude of deviations of real gases from the conditions of the ideal gas law increases with increasing pressure and temperature and varies widely with the composition of the gas. Real gases behave differently than ideal gases. The reason for this is that the perfect gas law was derived under the assumption that the volume of molecules is insignificant and no molecular attraction or repulsion exists between them. This is not the case for real gases.

Numerous equations-of-state have been developed in the attempt to correlate the pressure-volume-temperature variables for real gases with experimental data. More recent equations-of-state are presented and discussed in detail in Chapter 6. In order to express a more exact relationship between the variables p, V, and T, a correction factor called the gas compressibility factor, gas deviation factor, or simply the Z-factor, must be introduced into Equation 3-1 to account for the departure of gases from ideality. This is:

$$pV = ZnRT \tag{3-16}$$

where the compressibility factor Z is a dimensionless quantity and is defined as the ratio of the actual volume of n-moles of gas at T and p to the ideal volume of the same number of moles at the same T and p.

$$Z = \frac{V_a}{V_i} \tag{3-17}$$

where V_a = actual gas volume
V_i = ideal gas volume

For a perfect gas, the gas compressibility factor is equal to one. For a real gas, the Z-factor is greater or less than one depending on the pressure, temperature, and the composition of the gas.

The value of Z at any given pressure and temperature can be determined experimentally by measuring the actual volume of some quantity of gas at the specified p and T and solving Equation 3-16 for the compressibility factor Z. A typical curve of the Z-factor for a natural gas is shown in Figure 3-1, where the Z-factor is plotted as a function of pressure for a given con-

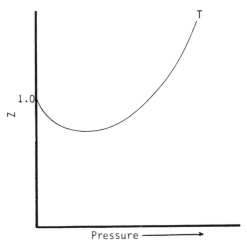

Figure 3-1. A typical Z-p diagram.

stant temperature. For different temperatures, the gas deviation factor curves follow a very definite pattern.

Numerous independent experimental studies of pure gases showed a well-established relationship between the compressibility factors and both pressure and temperature. Selected charts of this relationship for some pure components (methane, ethane, and propane) are given in Figures 3-2 through 3-4. Such experimental determinations of Z-factors for a specific gas as a function of pressure and temperature represent the most reliable method of obtaining the Z, p, and T relationship. However, with the time and expense involved, it is not necessary to conduct such individual experiments because sufficient information is known about the variation of compressibility with pressure and temperature to permit a correlation. This correlation is based on the theory of "Corresponding States." The theory proposes that all gases will exhibit the same behavior, e.g., Z-factor, when viewed in terms of reduced pressure, reduced volume, and reduced temperature. The term "reduced" means that each variable is expressed as a ratio of its critical value. These reduced states can be expressed mathematically by the following relationships:

$$p_r = \frac{p}{p_c} \tag{3-18}$$

$$V_r = \frac{V}{V_c} \tag{3-19}$$

$$T_r = \frac{T}{T_c} \tag{3-20}$$

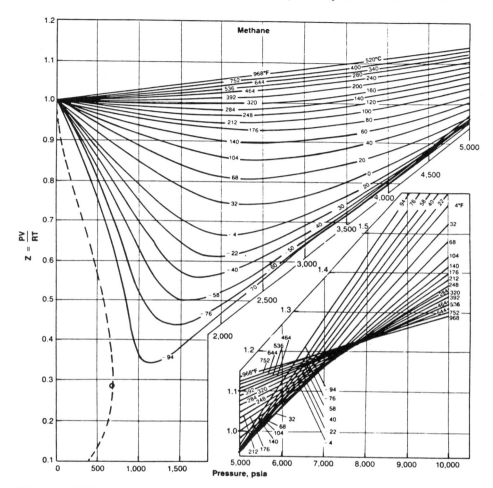

Figure 3-2. Methane compressibility factors chart. Courtesy of the Gas Processors Suppliers Association.

where p_r = reduced pressure
p_c = critical pressure
T_r = reduced temperature
T_c = critical temperature
V_r = reduced volume
V_c = critical volume

Notice that only pure gases possess distinct values of critical properties. Therefore, if the theory of corresponding states can be applied without appreciable error, all gases would have the same value for Z at the same re-

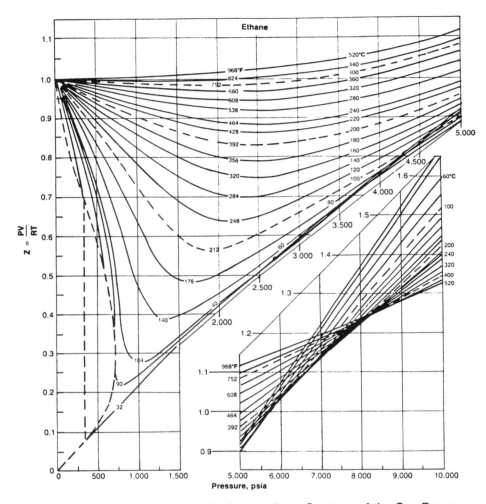

Figure 3-3. Ethane compressibility factors chart. Courtesy of the Gas Processors Suppliers Association.

duced temperature and pressure. This can be observed and appreciated by solving the following example.

Example 3-5. Calculate the compressibility factors of:

a. Methane
b. Ethane
c. Propane

at a reduced pressure and temperature of 2, and 1.6, respectively.

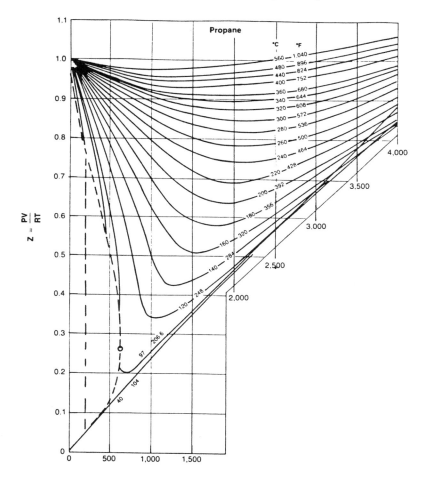

Figure 3-4. Propane compressibility factors chart. Courtesy of the Gas Processors Suppliers Association.

Solution.

a. Z-factor for methane. Because $p_r = p/p_c$ and $T_r = T/T_c$, then

$$p = p_r \cdot p_c = (2)(667.8) = 1{,}335.6 \text{ psia}$$
$$T = T_r \cdot T_c = (1.4)(-116.63 + 460) = 549.392°R$$

From Figure 3-2
   $Z = 0.88$

b. Z-factor for ethane.

$$p = p_r \cdot p_c = (2)(707.8) = 1{,}415.6 \text{ psia}$$
$$T = T_r \cdot T_c = (1.6)(90.09 + 460) = 880.166°R$$

From Figure 3-3

$$Z = 0.882$$

c. Z-factor for propane.

$$p = p_r \cdot p_c = (2)(616.3) = 1{,}232.6 \text{ psia}$$
$$T = T_r \cdot T_c = (1.6)(206.01 + 460) = 1{,}065.616°R$$

From Figure 3-4

$$Z = 0.886$$

The above example shows that at equal values of T_r and p_r the Z-factors for the three substances are very similar, indicating the clear power of the corresponding states principle. The application of the corresponding states principle to mixtures is based on the observation that the compressibility factor is a universal function of reduced pressure and temperature. Thus, the corresponding states principle should be applicable to mixtures if proper values for the critical point properties are used for the mixtures. Kay (1936) introduced the concept of pseudo-critical values which can be used in place of the true critical pressure and temperature of hydrocarbon mixtures. Kay proposed the following mixing rules for calculating the pseudo-critical properties of hydrocarbon mixtures.

$$p_{pc} = \sum_i y_i \, p_{ci} \tag{3-21}$$

$$T_{pc} = \sum_i y_i \, T_{ci} \tag{3-22}$$

where p_{pc} = pseudo-critical pressure, psia
  T_{pc} = pseudo-critical temperature, °R
  p_{ci} = critical temperature of component i, psia
  T_{ci} = critical temperature of component i, °R
  y_i = mole fraction of component i in the gas mixture.

Equations 3-21 and 3-22 are referred to as "Kay's Mixing Rule."
The reduced states for gas mixtures are called the pseudo-reduced pressure and temperature, and they are expressed by the following relationships:

$$p_{pr} = \frac{p}{p_{pc}} \tag{3-23}$$

$$T_{pr} = \frac{T}{T_{pc}} \tag{3-24}$$

where p_{pr} = pseudo-reduced pressure of the gas mixture
    T_{pr} = pseudo-reduced temperature of the gas mixture

Studies of the compressibility factors for natural gases of various compositions have shown that compressibility factors can be generalized with sufficient accuracies for most engineering purposes by introducing the concept of pseudo-reduced pressures and pseudo-reduced temperatures.

Standing and Katz (1942) presented a generalized compressibility factor chart, as shown in Figure 3-5. The chart represents compressibility factors of sweet natural gas as a function of p_{pr} and T_{pr}. This chart is generally reliable for sweet natural gases with minor amounts of non-hydrocarbons. It is one of the most widely accepted correlations in the oil and gas industry.

Equation 3-16 may be written in terms of specific volume or density as follows:

Because

$$pV = Z \left(\frac{m}{MW_a} \right) RT$$

then

$$\nu = \frac{V}{m} = \frac{ZRT}{p\, MW_a} \tag{3-25}$$

$$\rho_g = \frac{1}{\nu} = \frac{p\, MW_a}{ZRT} \tag{3-26}$$

where ν = specific volume, ft³/lb
    ρ_g = density, lb/ft³

Example 3-6. Using the gas composition, pressure, and temperature given in Example 3-4, and assuming a real gas behavior, calculate gas density.

Solution.

Component	y_i	MW_i	$y_i \cdot MW_i$	$T_{ci},°R$	$y_i \cdot T_{ci}$	p_{ci}	$y_i p_{ci}$
C_1	0.75	16.04	12.030	343.5	257.6	673	504.7
C_2	0.07	30.07	2.105	550.1	38.5	708	49.6
C_3	0.05	44.10	2.205	666.2	33.3	617	30.9
n-C_4	0.04	58.12	2.325	765.6	30.6	551	22.0
n-C_5	0.04	72.15	2.886	847.0	33.9	485	19.4
C_6	0.03	86.18	2.585	914.6	27.4	434	13.0
C_7	0.02	100.21	2.004	972.8	19.5	397	7.9
			$MW_a = 26.14$		$T_{pc} = 440.8$		$p_{pc} = 647.5$

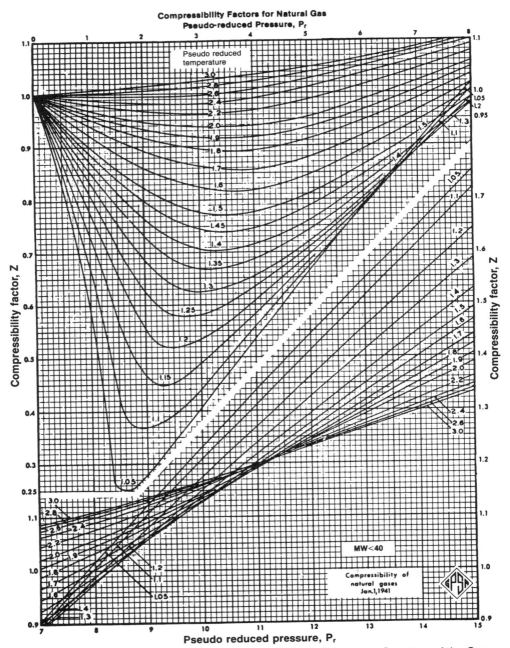

Figure 3-5. Standing and Katz compressibility factors chart. Courtesy of the Gas Processors Suppliers Association. Published in the GPSA Engineering Data Book, Tenth Edition, 1987.

$$p_{pr} = p/p_{pc} = 1,000/647.5 = 1.54$$

$$T_{pr} = T/T_{pc} = (100 + 460)/440.8 = 127$$

From Figure 3-5, $Z = 0.725$
Solve for gas density by applying Equation 3-26.

$$\rho_g = \frac{(1,000)(26.14)}{(0.725)(10.73)(560)} = 6.0 \text{ lb/ft}^3$$

The compressibility factor chart shown in Figure 3-5 is applicable to most gases encountered in petroleum reservoirs and provides satisfactory prediction for all engineering computations. The calculated volumetric behavior of gases containing only minor amounts of non-hydrocarbon can be accurate within 3%.

In cases where the composition of a natural gas is not available, the pseudo-critical properties, i.e., p_{pc} and T_{pc}, can be predicted solely from the specific gravity of the gas. Brown et al. (1948) presented a graphical method for a convenient approximation of the pseudo-critical pressure and pseudo-critical temperature of gases when only the specific gravity of the gas is available. The correlation is presented in Figure 3-6. Standing (1977) expressed this graphical correlation in a mathematical form:

Case 1: Natural Gas Systems

$$T_{pc} = 168 + 325 \, \gamma_g - 12.5 \, \gamma_g^2 \tag{3-27}$$
$$p_{pc} = 677 + 15.0 \, \gamma_g - 37.5 \, \gamma_g^2 \tag{3-28}$$

Case 2: Gas Condensate Systems

$$T_{pc} = 187 + 330 \, \gamma_g - 71.5 \, \gamma_g^2 \tag{3-29}$$
$$p_{pc} = 706 - 51.7 \, \gamma_g - 11.1 \, \gamma_g^2 \tag{3-30}$$

where T_{pc} = pseudo-critical temperature, °R
  p_{pc} = pseudo-critical pressure, psia
  γ_g = specific gravity of the gas mixture

Example 3-7. Using the data given in Example 3-6, recalculate the gas density by estimating the pseudo-critical properties from Equations 3-27 and 3-28.

Solution.

• Calculate the specific gravity of the gas mixture

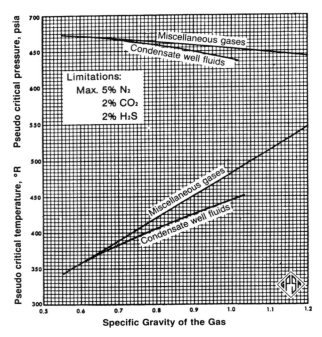

Figure 3-6. Pseudo-critical properties of natural gases. Courtesy of the Gas Processors Suppliers Association. Published in the GPSA Engineering Data Book, Tenth Edition, 1987.

$$\gamma_g = \frac{MW_a}{29.96} = \frac{26.14}{29.96} = 0.903$$

- Solve for T_{pc} and p_{pc} by applying Equations 3-27 and 3-28.

$$T_{pc} = 187 + 330\ (0.903) - 71.5\ (0.903)^2 = 427°R$$
$$p_{pc} = 706 - 51.7\ (0.903) - 11.1\ (0.903)^2 = 650\ psia$$

- Calculate p_{pr} and T_{pr}

$$T_{pr} = \frac{560}{427} = 1.3$$

$$p_{pr} = \frac{1,000}{650} = 1.54$$

- Estimate the compressibility factor from Figure 3-5

$$Z = 0.748$$
$$\rho_g = \frac{(1,000)(26.14)}{(0.748)(10.73)(560)} = 5.82\ lb/ft^3$$

EFFECT OF NON-HYDROCARBON COMPONENTS ON THE Z-FACTOR

Natural gases frequently contain materials other than hydrocarbon components, such as nitrogen, carbon dioxide, and hydrogen sulfide. Hydrocarbon gases are classified as sweet or sour depending on the hydrogen sulfide content. Both sweet and sour gases may contain nitrogen, carbon dioxide, or both. A hydrocarbon gas is termed a sour gas if it contains one grain of H_2S per 100 cubic feet.

The common occurrence of small percentages of nitrogen and carbon dioxide is in part considered in the correlations previously cited. Concentrations of up to 5% of these non-hydrocarbon components will not seriously affect accuracy. Errors in compressibility factor calculations as large as 10% may occur in higher concentrations of non-hydrocarbon components in gas mixtures.

CORRECTION FOR NON-HYDROCARBONS

Wichert-Aziz Correction Method

Natural gases which contain H_2S and/or CO_2 frequently exhibit different compressibility factor behavior than do sweet gases. Wichert and Aziz (1972) developed a simple, easy to use calculation procedure to account for these differences. This method permits the use of the Standing-Katz chart, i.e., Figure 3-5, by using a pseudo-critical temperature adjustment factor, which is a function of the concentration of CO_2 and H_2S in the sour gas. This correction factor is then used to adjust the pseudo-critical temperature and pressure according to the following expressions:

$$Tpc' = T_{pc} - \epsilon \qquad (3\text{-}31)$$

$$p_{pc}' = \frac{p_{pc} \, T_{pc}'}{T_{pc} + B(1 - B) \, \epsilon} \qquad (3\text{-}32)$$

where T_{pc} = pseudo-critical temperature, °R
p_{pc} = pseudo-critical pressure, psia
T_{pc}' = corrected pseudo-critical temperature, °R
p_{pc}' = corrected pseudo-critical pressure, psia
B = mole fraction of H_2S in the gas mixture
ϵ = pseudo-critical temperature adjustment factor and is defined mathematically by the following expression

$$\epsilon = 120 \ (A^{0.9} - A^{1.6}) + 15 \ (B^{0.5} - B^{4.0}) \qquad (3\text{-}33)$$

where the coefficient A is the sum of the mole fraction of H_2S and CO_2 in the gas mixture, or

$$A = y_{H_2S} + y_{CO_2}$$

The computational steps of incorporating the adjustment factor ϵ into the Z-factor calculations are summarized below:

Step 1. Calculate the pseudo-critical properties of the whole gas mixture by applying Equations 3-27 and 3-28 or Equations 3-29 and 3-30.

Step 2. Calculate the adjustment factor from Equation 3-33.

Step 3. Adjust the calculated p_{pc} and T_{pc} (as computed in Step 1) by applying Equations 3-31 and 3-32.

Step 4. Calculate the pseudo-reduced properties from Equations 3-23 and 3-24.

Step 5. Read the compressibility factor from Figure 3-5.

Example 3-8. A sour natural gas has the following composition:

Component	y_i
CO_2	0.10
H_2S	0.20
N_2	0.05
C_1	0.63
C_2	0.02

Determine the density of the gas mixture at 1,000 psia and 110°F

1. Without making any corrections to account for the presence of the non-hydrocarbon components.
2. Using the Wichert-Aziz correction method.

Solution.

Component	y_i	MW_i	$MW_i y_i$	p_{ci}	$y_i p_{ci}$	T_{ci}	$y_i T_{ci}$
CO_2	0.10	44.01	4.401	1,071	107.1	547.57	54.757
H_2S	0.20	34.08	6.816	1,306	261.2	672.37	134.474
N_2	0.05	28.01	1.401	493	24.65	227.29	11.3645
C_1	0.63	16.04	10.105	667.8	420.714	343.06	216.128
C_2	0.02	30.07	0.601	707.8	14.156	549.78	11.00
			$MW_a = 23.324$		827.82		427.72

$$\gamma_g = \frac{23.324}{28.96} = 0.8054$$

$$P_{pc} = 827.82$$

$$T_{pc} = 427.72$$

1. Determination of gas density without corrections:

$$p_{pr} = \frac{1,000}{827.82} = 1.208$$

$$T_{pr} = \frac{570}{427.72} = 1.333$$

- From Figure 3-5, Z = 0.820
- Calculate the gas density by applying Equation 3-26

- $\rho_g = \dfrac{(1,000)(23.324)}{(0.82)(10.73)(570)} = 4.651 \ \text{lb/ft}^3$

2. Determination of gas density with correction:

$$B = y_{H_2S} = 0.2$$

$$A = y_{CO_2} + y_{H_2S} - 0.1 + 0.20 = 0.30$$

- Calculate the correction factor ϵ from Equation 3-33.

$$\epsilon = 29.86$$

- From Equation 3-31, correct the pseudo-critical temperature.

$$T_{pc}' = 427.72 - 29.86 = 397.86$$

- Calculate the corrected p_{pc} by applying Equation 3-32.

- $p_{pc}' = \dfrac{(827.82)\ (397.86)}{427.72 + 0.2\ (1 - 0.2)\ 29.86} = 727.07 \ \text{psia}$

- $p_{pr} = \dfrac{1,000}{727.07} = 1.375$

- $T_{pr} = \dfrac{570}{397.86} = 1.433$

- From Figure 3-5, Z = 0.837

- $\rho_g = \dfrac{(1,000)(23.324)}{(0.837)(10.73)(570)} = 4.56 \ \text{lb/ft}^3$

Carr-Kobayashi-Burrows Correction Method

Carr, Kobayashi, and Burrows (1954) proposed a simplified procedure to adjust the pseudo-critical properties of natural gases when non-hydrocarbon components are present. The method can be used when the composition of the natural gas is not available. The proposed procedure is summarized in the following steps:

Step 1. Knowing the specific gravity of the natural gas, calculate the pseudo-critical temperature and pressure from Figure 3-6, or by applying Equations 3-27 and 3-28.

Step 2. Adjust the estimated pseudo-critical properties by using the following expressions:

$$T_{pc}' = T_{pc} - 80 \, y_{CO_2} + 130 \, y_{H_2S} - 250 \, y_{N_2} \tag{3-34}$$

$$p_{pc}' = p_{pc} + 440 \, y_{CO_2} + 600 \, y_{H_2S} - 170 \, y_{N_2} \tag{3-35}$$

where T_{pc}' = the adjusted pseudo-critical temperature, °R
T_{pc} = the unadjusted pseudo-critical temperature, °R
y_{CO_2} = mole fraction of CO_2
y_{H_2S} = mole fraction of H_2S
y_{N_2} = mole fraction of nitrogen
p_{pc}' = the adjusted pseudo-critical pressure, psia
p_{pc} = the unadjusted pseudo-critical pressure, psia

Step 3. Use the adjusted pseudo-critical temperature and pressure to calculate the pseudo-reduced properties.

Step 4. Calculate the Z-factor from Figure 3-5.

Example 3-9. Recalculate the gas density of Example 3-8 from its specific gravity.

Solution.

- Because the specific gravity of the gas is 0.8054, calculate T_{pc} and p_{pc} by applying Equations 3-27 and 3-28.

  $$T_{pc} = 168 + 325 \, (0.8054) - 12.5 \, (0.8054)^2 = 421.65°R$$

  $$p_{pc} = 677 + 15 \, (0.8054) - 37.5 \, (0.8054)^2 = 650.9 \text{ psia}$$

- Adjust the calculated T_{pc} and p_{pc} by using Equations 3-34 and 3-35.

  $$T_{pc}' = 421.65 - 80(0.10) + 130(0.20) - 250(0.05) = 427.15°R$$

  $$p_{pc}' = 650.9 + 15(0.10) + 600(0.20) - 170(0.05) = 763.9 \text{ psia}$$

- Calculate p_{pr} and T_{pr}

$$p_{pr} = \frac{1,000}{763.9} = 1.31$$

$$T_{pr} = \frac{570}{427.15} = 1.33$$

- From Figure 3-5, estimate the compressibility factor
  $Z = 0.808$

- Solve for the gas density

$$\rho_g = \frac{(1,000)(23.324)}{(0.808)(10.73)(570)} = 4.72 \ \text{lb/ft}^3$$

CORRECTION FOR HIGH-MOLECULAR-WEIGHT GASES

It should be noted that the Standing and Katz compressibility factor chart (Figure 3-5) was prepared from data on binary mixtures of methane with propane, ethane, and butane, and on natural gases, thus covering a wide range in composition of hydrocarbon mixtures containing methane. No mixtures having molecular weights in excess of 40 were included in preparing this plot.

Sutton (1985) evaluated the accuracy of the Standing-Katz compressibility factor chart using laboratory-measured gas compositions and Z-factors, and found that the chart provides satisfactory accuracy for engineering calculations. However, Kay's mixing rules, i.e., Equations 3-21 and 3-22 (or comparable gravity relationships for calculating pseudo-critical pressure and temperature), result in unsatisfactory Z-factors for high molecular weight reservoir gases. The author observed that large deviations occur to gases with high heptanes-plus concentrations. He pointed out that Kay's mixing rules should not be used to determine the pseudo-critical pressure and temperature for reservoir gases with specific gravities greater than about 0.75.

Sutton proposed that this deviation can be minimized by utilizing the mixing rules developed by Stewart et al. (1959), together with empirical adjustment factors related to the presence of the heptane-plus fraction in the gas mixture. The proposed approach is outlined in the following steps:

Step 1. Calculate the parameters J and K from the following relationships:

$$J = \frac{1}{3}\left[\sum_i y_i \left(\frac{T_c}{p_c}\right)_i\right] + \frac{2}{3}\left[\sum_i y_i \left(\frac{T_c}{p_c}\right)_i^{0.5}\right]^2 \tag{3-36}$$

$$K = \sum_i \left[y_i \left(\frac{T_c}{p_c}\right)_i^{0.5} \right] \tag{3-37}$$

where J = Stewart-Burkhardt-Voo correlating parameter, °R/psia

   K = Stewart-Burkhardt-Voo correlating parameter, °R/psia$^{0.5}$

   y_i = mole fraction of component i in the gas mixture

Step 2. Calculate the adjustment parameters F_J, ϵ_J, and ϵ_K from the following expressions:

$$F_J = \frac{1}{3}\left[y \left(\frac{T_c}{p_c}\right) \right]_{C_{7+}} + \frac{2}{3}\left[y \left(\frac{T_c}{p_c}\right)^{0.5} \right]_{C_{7+}}^2 \tag{3-38}$$

$$\epsilon_J = 0.6081\ F_J + 1.1325\ F_J^2 - 14.004\ F_J\ y_{C_{7+}} + 64.434\ F_J\ y_{C_{7+}}^2 \tag{3-39}$$

$$\epsilon_K = \left(\frac{T_c}{p_c^{0.5}}\right)_{C_{7+}} [0.3129\ y_{C_{7+}} - 4.8156\ y_{C_{7+}}^2 + 27.3751\ y_{C_{7+}}^3] \tag{3-40}$$

where $y_{C_{7+}}$ = mole fraction of the heptanes-plus component

   $(T_c)_{C_{7+}}$ = critical temperature of the C_{7+}, °R

   $(p_c)_{C_{7+}}$ = critical pressure of the C_{7+}, psia

Step 3. Adjust the parameters J and K by applying the adjustment factors ϵ_J and ϵ_K according to the relationships:

$$J' = J - \epsilon_J \tag{3-41}$$

$$K' = K - \epsilon_K \tag{3-42}$$

where   J, K = calculated from Equations 3-36 and 3-37

   ϵ_J, ϵ_K = calculated from Equations 3-39 and 3-40

Step 4. Calculate the adjusted pseudo-critical temperature and pressure from the expressions

$$T_{pc} = \frac{(K')^2}{J'} \tag{3-43}$$

$$p_{pc} = \frac{T_{pc}}{J'} \tag{3-44}$$

Step 5. Having calculated the adjusted T_{pc} and p_{pc}, the regular procedure of calculating the compressibility factor from the Standing and Katz chart is followed.

Sutton's proposed mixing rules for calculating the pseudo-critical properties of high-molecular-weight reservoir gases (i.e., $\gamma_g > 1.25$) should significantly improve the accuracy of the calculated Z-factor.

DIRECT CALCULATION OF COMPRESSIBILITY FACTORS

After four decades of existence, the Standing-Katz Z-factor chart is still widely used as a practical source of natural gas compressibility factors. As a result, there was an apparent need for a simple mathematical description of that chart. Several empirical correlations for calculating Z-factors have been developed over the years. Takacs (1976) reviewed the performance of eight of these correlations. The following five empirical correlations are presented here.

- Papay
- Hall-Yarborough
- Dranchuk-Abu-Kassem
- Dranchuk-Purvis-Robinson
- Hankinson-Thomas-Phillips

Papay Method

Papay (1968) proposed the following simplified equation for calculating the compressibility factor:

$$Z = 1 - \frac{p_{pr}}{T_{pr}} \left[0.36748758 - 0.04188423 \left(\frac{p_{pr}}{T_{pr}} \right) \right] \tag{3-45}$$

where p_{pr} = pseudo-reduced pressure
T_{pr} = pseudo-reduced temperature

The above correlation is convenient for application for hand calculations. Takacs pointed out that the proposed correlation produces an average error of -4.8%.

Example 3-10. Using the data given in Example 3-8, re-evaluate the compressibility factor by using the Papay correlation, and calculate the gas density.

Solution.

- From Example 3-8, p_{pr} = 1.375 and T_{pr} = 1.433. Solve Equation 3-45 for the Z-factor.

$$Z = 1 - \frac{1.375}{1.433}\left[0.36748758 - 0.04188423\left(\frac{1.375}{1.433}\right)\right] = 0.687$$

- Calculate the gas density.

$$\rho_g = \frac{(1,000)(23.324)}{(0.687)(10.73)(570)} = 5.56 \text{ lb/ft}^3$$

Hall-Yarborough Method

Hall and Yarborough (1973) presented an equation-of-state that accurately represents the Standing and Katz Z-factor chart. The proposed expression is based on the Starling-Carnahan equation-of-state. The coefficients of the correlation were determined by fitting it to data taken from the Standing and Katz Z-factor chart. Hall and Yarborough proposed the following mathematical form:

$$Z = \left[\frac{0.06125\ p_{pr}\ t}{Y}\right] EXP\left[-1.2\ (1-t)^2\right] \tag{3-46}$$

where P_{pr} = pseudo-reduced pressure
$\quad\quad\ t$ = reciprocal of the pseudo-reduced temperature, i.e., T_{pc}/T
$\quad\quad\ Y$ = the reduced density which can be obtained as the solution of the following equation:

$$F(Y) = -0.06125\ p_{pr}\ t\ EXP\ [-1.2(1-t)^2] + \frac{Y + Y^2 + Y^3 - Y^4}{(1-Y)^3}$$

$$- (14.76t - 9.76t^2 + 4.58t^3)\ Y^2 + (90.7t - 242.2t^2 + 42.4t^3)\ Y^{(2.18 + 2.82t)} = 0 \tag{3-47}$$

This non-linear equation can be conveniently solved for the reduced density Y by using the Newton-Raphson iteration technique. The computational procedure of solving Equation 3-47 is summarized in the following steps.

Step 1. Make an initial guess of the unknown parameter, Y^k, where k is an iteration counter. An appropriate initial guess of Y is given by the following relationship:

$$Y^k = 0.06125\ p_{pr}\ t\ EXP\ [-1.2(1-t)^2]$$

Step 2. Substitute this initial value in Equation 3-47 and evaluate the non-linear function. Unless the correct value of Y has been initially selected, Equation 3-47 will have a non-zero value of $f(Y^k)$.

Step 3. A new, improved estimate of Y, i.e., Y^{k+1}, is calculated from the following expression:

$$Y^{k+1} = Y^k - \frac{f(Y^k)}{f'(Y^k)} \qquad (3\text{-}48)$$

where $f'(Y^k)$ is obtained by evaluating the derivative of Equation 3-47 at Y^k, or

$$f'(Y) = \frac{1 + 4Y + 4Y^2 - 4Y^3 + Y^4}{(1 - Y)^4} - (29.52t - 19.52t^2$$

$$+ 9.16t^3)Y + (2.18 + 2.82t)(90.7t - 242.2t^2$$
$$+ 42.2t^3)Y^{(1.18 + 2.82t)} \qquad (3\text{-}49)$$

Step 4. Steps 1–3 are repeated n times, until the error (i.e., $Y^n - Y^{n-1}$) becomes smaller than a preset tolerance, e.g., 10^{-12}.

Step 5. The correct value of Y is then used to evaluate Equation 3-46 for the compressibility factor Z.

Hall and Yarborough pointed out that the method is not recommended for application if the pseudo-reduced temperature is less than one.

Dranchuk and Abu-Kassem Method

Dranchuk and Abu-Kassem (1975) proposed an eleven-constant equation-of-state for calculating the gas compressibility factors. The authors proposed the following equation:

$$Z = \left[A_1 + \frac{A_2}{T_{pr}} + \frac{A_3}{T_{pr}^3} + \frac{A_4}{T_{pr}^4} + \frac{A_5}{T_{pr}^5}\right] \rho_r$$

$$+ \left[A_6 + \frac{A_7}{T_{pr}} + \frac{A_8}{T_{pr}^2}\right] \rho_r^2 - A_9 \left[\frac{A_7}{T_{pr}} + \frac{A_8}{T_{pr}^2}\right] \rho_r^5$$

$$+ A_{10} (1 + A_{11} \rho_r^2) \frac{\rho_r^2}{T_{pr}^3} \text{EXP}[-A_{11} \rho_r^2] + 1 \qquad (3\text{-}50)$$

where ρ_r = reduced gas density and is defined by the following relationship:

$$\rho_r = \frac{0.27 \, P_{pr}}{Z \, T_{pr}} \qquad (3\text{-}51)$$

The constants A_1 through A_{11} were determined by fitting the equation, using non-linear regression models, to 1,500 data points from the Standing and Katz Z-factor chart. The coefficients have the following values:

$A_1 = 0.3265$ $A_2 = -1.0700$ $A_3 = -0.5339$ $A_4 = 0.01569$
$A_5 = -0.05165$ $A_6 = 0.5475$ $A_7 = -0.7361$ $A_8 = 0.1844$
$A_9 = 0.1056$ $A_{10} = 0.6134$ $A_{11} = 0.7210$

Equation 3-50 can be rearranged by replacing the left hand side of the equation, i.e., Z, with $0.27 p_{pr}/\rho_r T_{pr}$ and solving the resulting equation for ρ_r by using the Newton-Raphson iteration technique. An appropriate initial guess of the solution is given by the following expression:

$$\rho_r = \frac{0.27\ p_{pr}}{T_{pr}} \tag{3-52}$$

Having obtained the correct ρ_r, Equation 3-51 can be solved for the compressibility factor.

The proposed correlation was reported to duplicate compressibility factors from the Standing and Katz chart with an average absolute error of 0.585 percent, and is applicable over the ranges

$0.2 \le p_{pr} < 30$
$1.0 < T_{pr} \le 3.0$

Dranchuk-Purvis-Robinson Method

Dranchuk, Purvis, and Robinson (1974) developed a correlation based on the Benedict-Webb-Rubin type of equation-of-state. The eight coefficients of the proposed equations were optimized by fitting the equation to 1,500 data points from the Standing and Katz Z-factor chart. The equation has the following form:

$$Z = 1 + \left[A_1 + \frac{A_2}{T_{pr}} + \frac{A_3}{T_{pr}^3}\right] \rho_r + \left[A_4 + \frac{A_5}{T_{pr}}\right] \rho_r^2 + \left(\frac{A_5 A_6}{T_{pr}}\right) \rho_r^5$$

$$+ \left[\frac{A_7}{T_{pr}^3} \rho_r^2 (1 + A_8 \rho_r^2)\ \text{EXP}\ (-A_8\ \rho_r^2)\right] \tag{3-53}$$

where ρ_r is defined by Equation 3-51 and the coefficients A_1 through A_8 have the following values:

$A_1 = 0.31506237$ $A_5 = -0.61232032$
$A_2 = -1.0467099$ $A_6 = -0.10488813$
$A_3 = -0.57832729$ $A_7 = 0.68157001$
$A_4 = 0.53530771$ $A_8 = 0.68446549$

The solution procedure of Equation 3-53 is similar to that of Dranchuk and Abu-Kassem.

The method is valid within the following ranges of pseudo-reduced temperature and pressure

$$1.05 \leq T_{pr} < 3.0$$

$$0.2 \leq p_{pr} \leq 3.0$$

Hankinson-Thomas-Phillips Method

Hankinson, Thomas, and Phillips (1969) correlated the compressibility factors for natural gases as a function of the pseudo-reduced temperature and pseudo-reduced pressure by using the Benedict-Webb-Rubin equation-of-state. The proposed equation is expressed in terms of compressibility factor as follows:

$$\frac{1}{Z} - 1 + \left[A_4\, T_{pr} - A_2 - \frac{A_6}{T_{pr}^2} \right] \left(\frac{p_{pr}}{Z^2\, T_{pr}^2} \right) + (A_3\, T_{pr} - A_1) \left(\frac{p_{pr}^2}{Z^3\, T_{pr}^3} \right)$$
$$+ \frac{A_1 A_5 A_7 p_{pr}^5}{Z^6\, T_{pr}^6} \left[1 + \frac{A_8\, p_{pr}^2}{Z^2\, T_{pr}^2} \right] EXP \left[- \frac{A_8\, p_{pr}^2}{Z^2\, T_{pr}^2} \right] = 0 \qquad (3\text{-}54)$$

The accuracy of the data representation is improved considerably by breaking the data into two regions, one region for reduced pressures less than 5.0 and one region for reduced pressures between 5.0 and 15.0. Thus, two sets of coefficients are obtained, one for each pressure range. The two sets of coefficients are presented in Table 3-1.

It is suggested that the proposed correlation is used only at reduced temperatures above 1.1.

Table 3-1
Coefficients for Equation 3-52

Coefficients	p_{pr} from 0.4 to 5.0	p_{pr} from 5 to 15
A_1	0.001290236	0.0014507882
A_2	0.38193005	0.37922269
A_3	0.022199287	0.024181399
A_4	0.12215481	0.11812287
A_5	− 0.015674794	0.037905663
A_6	0.027271364	0.19845016
A_7	0.023834219	0.048911693
A_8	0.43617780	0.0631425417

Equation 3-54 can be solved for the compressibility factor by using the Newton-Raphson iterative technique as outlined previously.

COMPRESSIBILITY OF NATURAL GASES

A knowledge of the variability of fluid compressibility with pressure and temperature is essential in performing many reservoir engineering calculations. For a liquid phase, the compressibility is small and usually assumed to be constant. For a gas phase, the compressibility is neither small nor constant.

By definition, the isothermal gas compressibility is the change in volume per unit volume for a unit change in pressure, or, in equation form:

$$C_g = -\frac{1}{V}\left(\frac{\partial V}{\partial p}\right)_T \qquad (3\text{-}55)$$

where C_g = isothermal gas compressibility, 1/psi
From the real gas equation-of-state

$$V = \frac{nRTZ}{p}$$

Differentiating the above equation with respect to pressure yields

$$\left(\frac{\partial V}{\partial p}\right)_T = nRT\left[\frac{1}{p}\left(\frac{\partial Z}{\partial p}\right) - \frac{Z}{p^2}\right]$$

Substituting into Equation 3-55 produces the following generalized relationship:

$$C_g = \frac{1}{p} - \frac{1}{Z}\left(\frac{\partial Z}{\partial p}\right)_T \qquad (3\text{-}56)$$

For an ideal gas, $Z = 1$ and $(\partial Z/\partial p)_T = 0$, therefore

$$C_g = \frac{1}{p} \qquad (3\text{-}57)$$

Equation 3-57 is useful in determining the expected order of magnitude of the isothermal gas compressibility.

Equation 3-56 can be conveniently expressed in terms of the pseudo-reduced pressure and temperature by simply replacing p with $(p_{pc} \, p_{pr})$, or

$$C_g = \frac{1}{p_{pr} \, p_{pc}} - \frac{1}{Z} \left[\frac{\partial Z}{\partial (p_{pr} \, p_{pc})} \right]_{T_{pr}}$$

Multiplying the above equation by p_{pc} yields

$$C_g \, p_{pc} = C_r = \frac{1}{p_{pr}} - \frac{1}{Z} \left[\frac{\partial Z}{\partial p_{pr}} \right]_{T_{pr}} \tag{3-58}$$

The term C_r is called the isothermal pseudo-reduced compressibility, and is defined by the relationship

$$C_r = C_g \, p_{pc} \tag{3-59}$$

where C_r = isothermal pseudo-reduced compressibility
C_g = isothermal gas compressibility, psi^{-1}
p_{pc} = pseudo-reduced pressure, psi

Values of $[\partial Z / \partial p_{pr}]_{T_{pr}}$ can be calculated from the slope of the T_{pr} isotherm on the Standing and Katz Z-factor chart at the Z-factor of interest.

Example 3-11. Given the following gas composition,

Component	y_i
C_1	0.75
C_2	0.07
C_3	0.05
n-C_4	0.04
n-C_5	0.04
C_6	0.03
C_7	0.02

calculate the isothermal gas compressibility at 1,000 psia and 100°F by assuming:

a. An ideal gas behavior
b. A real gas behavior

Solution.

a. Assuming an ideal gas behavior

- Applying Equation 3-57 yields

$$C_g = \frac{1}{1,000} = 1,000 \times 10^{-6} \text{ psi}^{-1}$$

b. Assuming a real gas behavior

Step 1. Calculate p_{pc} and T_{pc} of the gas mixture

$$p_{pc} = 647.5 \text{ psia}$$
$$T_{pc} = 440.8°R$$

Step 2. Compute p_{pr} and T_{pr}

$$p_{pr} = \frac{1,000}{647.5} = 1.54$$

$$T_{pr} = \frac{560}{440.8} = 1.27$$

Step 3. Estimate the compressibility factor Z from Figure 3-5.

$$Z = 0.725$$

Step 4. Calculate the slope $[\partial Z/\partial p_{pr}]_{T_{pr}=1.27}$ from Figure 3-5

$$\left[\frac{\partial Z}{\partial p_{pr}}\right]_{T_{pr}} = 0.1678$$

Step 5. Solve for C_r by applying Equation 3-58.

$$C_r = \frac{1}{1.54} - \left(\frac{1}{0.725}\right)(-0.1678) = 0.8808$$

Step 6. From Equation 3-59, solve for C_g

$$C_g = \frac{C_r}{p_{pc}} = \frac{0.8808}{647.5} = 1,361.1 \times 10^{-6} \text{ psi}^{-1}$$

Trube (1957) presented graphs from which the isothermal compressibility of natural gases may be obtained. The graphs, as shown in Figures 3-7 and 3-8, give the isothermal pseudo-reduced compressibility as a function of pseudo-reduced pressure and temperature.

Example 3-12. Using Trube's generalized charts, rework Example 3-11.

Solution.

Step 1. From Figure 3-7, find C_r

$$C_r = 0.90$$

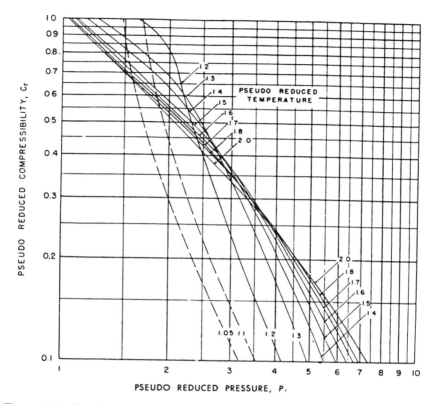

Figure 3-7. Trube's pseudo-reduced compressibility for natural gases. Permission to publish by the Society of Petroleum Engineers of AIME. Copyright SPE-AIME.

Step 2. Solve for C_g by applying Equation 3-58.

$$C_g = \frac{0.90}{647.5} = 1,390 \times 10^{-6} \text{ psi}^{-1}$$

Mattar, Brar, and Aziz (1975) presented an analytical technique for calculating the isothermal gas compressibility. The authors expressed C_r as a function of $[\partial Z/\partial \rho_r]_{T_{pr}}$ rather than $[\partial Z/\partial p_{pr}]_{T_{pr}}$. Equation 3-51 is differentiated with respect to p_{pr} to give

$$\left[\frac{\partial Z}{\partial p_{pr}}\right] = \frac{0.27}{Z\, T_{pr}} \left[\frac{(\partial Z/\partial \rho_r)_{T_{pr}}}{1 + \frac{\rho_r}{Z}(\partial Z/\partial \rho_r)_{T_{pr}}} \right]$$

(3-60)

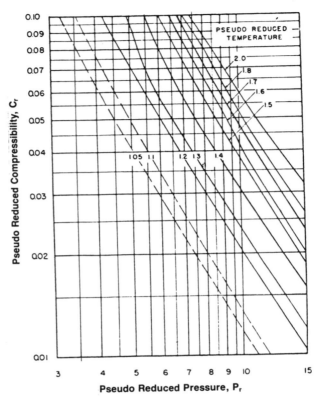

Figure 3-8. Trube's pseudo-reduced compressibility for natural gases. Permission to publish by the Society of Petroleum Engineers of AIME. Copyright SPE-AIME.

Equation 3-60 may be substituted into Equation 3-58 to express the pseudo-reduced compressibility as

$$C_r = \frac{1}{p_{pr}} - \frac{0.27}{Z^2\,T_{pr}} \left[\frac{(\partial Z/\partial \rho_r)_{T_{pr}}}{1 + \frac{\rho_r}{Z}\,(\partial Z/\partial \rho_r)_{T_{pr}}} \right] \tag{3-61}$$

where ρ_r = pseudo-reduced gas density

The partial derivative $[\partial Z/\partial \rho_r]_{T_{pr}}$ appearing in Equation 3-61 is obtained from the Benedict-Webb-Rubin equation-of-state developed by Dranchuk, Purvis, and Robinson (Equation 3-53), to give

$$\left[\frac{\partial Z}{\partial \rho_r}\right]_{T_{pr}} = \left(A_1 + \frac{A_2}{T_{pr}} + \frac{A_3}{T_{pr}^3}\right) + 2\left(A_4 + \frac{A_5}{T_{pr}}\right)\rho_r + 5\ A_5\ A_6\ \frac{\rho_r^4}{T_{pr}}$$

$$+ \frac{2A_7\ \rho_r}{T_{pr}^3}\ (1 + A_8\ \rho_r^2 - A_8^2\ \rho_r^4)\ \mathrm{EXP}\ (-A_8\ \rho_r^2) \tag{3-62}$$

Values of the coefficients A_1–A_8 are given previously in Table 3-1. The computational procedure of estimating C_r through the use of Mattar, Brar, and Aziz is shown by the flow diagram in Figure 3-9.

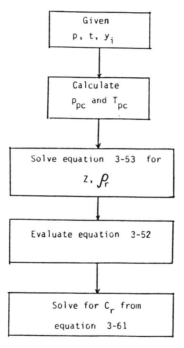

Figure 3-9. Flow diagram for calculating C_r.

GAS FORMATION VOLUME FACTOR

The gas formation volume factor is used to relate the volume of gas, as measured at reservoir conditions, to the volume of the gas as measured at standard conditions, i.e., 60°F and 14.7 psia. This gas property is then defined as the actual volume occupied by a certain amount of gas at a specified pressure and temperature, divided by the volume occupied by the same amount of gas at standard conditions. In an equation form, the relationship is expressed as

$$B_g = \frac{V_{p,T}}{V_{sc}}$$

where B_g = gas formation volume factor, ft^3/scf
$V_{p,T}$ = volume of gas at pressure p and temperature T, ft^3
V_{sc} = volume of gas at standard conditions, scf

Applying the real gas equation-of-state (Equation 3-16) to the above relationship gives

$$B_g = \frac{\dfrac{Z n\ RT}{p}}{\dfrac{Z_{sc}\ n\ R\ T_{sc}}{p_{sc}}} = \frac{p_{sc}}{T_{sc}}\frac{ZT}{p}$$

where Z_{sc} = Z-factor at standard conditions = 1.0
p_{sc}, T_{sc} = standard pressure and temperature, i.e., 14.7 and 60°F

or $B_g = 0.02827\ \dfrac{Z\ T}{p}$, ft^3/scf $\qquad$ (3-63)

In other field units, the gas formation volume factor can be expressed in bbl/scf, to give

$$B_g = 0.005035\ \frac{Z\ T}{p},\ bbl/scf \qquad (3\text{-}64)$$

The reciprocal of the gas formation volume factor is called the Gas Expansion Factor and designated by the symbol E_g, or

$$E_g = 35.37\ \frac{p}{Z\ T},\ scf/ft^3 \qquad (3\text{-}65)$$

In other field units,

$$E_g = 198.6\ \frac{p}{Z\ T},\ scf/bbl \qquad (3\text{-}66)$$

Example 3-13. A gas well is producing at a rate of 15,000 ft^3/day from a gas reservoir at a bottom hole with flowing conditions of 1,000 psia and

100°F. The specific gravity of the gas is 0.903. Calculate the gas flow rate in scf/day.

Solution.

- From Equations 3-27 and 3-28, calculate T_{pc} and p_{pc} of the gas phase.

$$T_{pc} = 427°R$$
$$p_{pc} = 650 \text{ psia}$$

- Compute T_{pr} and p_{pr} and solve Z to give:

$$T_{pr} = 1.3$$
$$p_{pr} = 1.54$$
$$Z = 0.748$$

- Calculate the gas formation volume factor by applying Equation 3-65.

$$B_g = 35.37 \frac{1,000}{(0.748)(560)} = 84.44 \text{ scf/ft}^3$$

- Calculate the gas flow rate in scf/day

gas flow rate = $(15,000)(84.44) = 1,266,592$ scf/day

GAS VISCOSITY

The viscosity of a fluid is a measure of the internal fluid friction (resistance) to flow. If the friction between layers of the fluid is small, i.e., low viscosity, an applied shearing force will result in a large velocity gradient. As the viscosity increases, each fluid layer exerts a larger frictional drag on the adjacent layers and velocity gradient decreases.

The viscosity of a fluid is generally defined as the ratio of the shear force per unit area to the local velocity gradient. Viscosities are expressed in terms of poises, centipoises, or micropoises. One poise equals a viscosity of 1 dyne-sec/cm² and can be converted to other field units by the following relationships.

1 poise = 100 centipoises
      = 1×10^6 micropoises
      = 6.72×10^{-2} lb mass/ft-sec
      = 2.09×10^{-3} lbf-sec/ft²

The gas viscosity is not commonly measured in the laboratory because it can be estimated precisely from empirical correlations. Like all intensive properties, viscosity of a natural gas is completely described by the following function.

$$\mu_g = (p, t, y_i)$$

where μ_g = the viscosity of the gas phase. The above relationship simply states that the viscosity is a function of pressure, temperature, and composition. Many of the widely used gas viscosity correlations may be viewed as modifications of that expression.

METHODS OF CALCULATING THE VISCOSITY OF NATURAL GASES

Carr-Kobayashi-Burrows Method

Carr, Kobayashi, and Burrows (1954) developed graphical correlations for estimating the viscosity of natural gas as a function of temperature, pressure, and gas gravity. The computational procedure of applying the proposed correlations is summarized in the following steps.

Step 1. Calculate the pseudo-critical pressure, pseudo-critical temperature, and apparent molecular weight from the specific gravity or the composition of the natural gas. Corrections to these pseudo-critical properties for the presence of the non-hydrocarbon gases (CO_2, N_2, and H_2S) should be made if they are present in concentration greater than 5 mole percent.

Step 2. Obtain the viscosity of the natural gas at one atmosphere and the temperature of interest from Figure 3-10. This viscosity, as denoted by μ_1, *must be corrected for the presence of non-hydrocarbon components by using the inserts of Figure 3-10.* The non-hydrocarbon fractions tend to increase the viscosity of the gas phase. The effect of non-hydrocarbon components on the viscosity of the natural gas can be expressed mathematically by the following relationship.

$$\mu_1 = (\mu_1)_{\text{uncorrected}} + (\Delta\mu)_{N_2} + (\Delta\mu)_{CO_2} + (\Delta\mu)_{H_2S} \tag{3-67}$$

where
μ_1 = "corrected" gas viscosity at one atmospheric pressure and reservoir temperature, cp
$(\Delta\mu)_{N_2}$ = viscosity corrections due to the presence of N_2
$(\Delta\mu)_{CO_2}$ = viscosity corrections due to the presence of CO_2
$(\Delta\mu)_{H_2S}$ = viscosity corrections due to the presence of H_2S
$(\mu_1)_{\text{uncorrected}}$ = uncorrected gas viscosity, cp

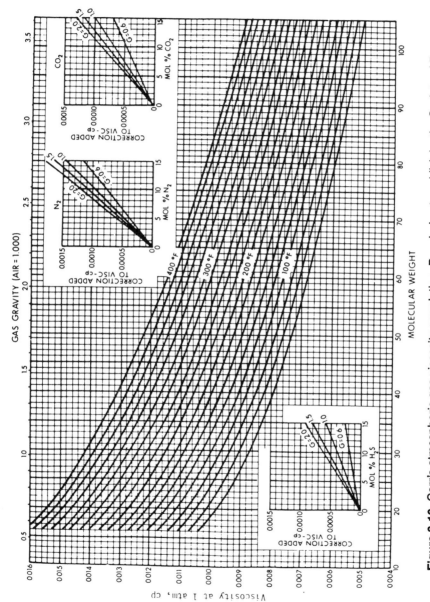

Figure 3-10. Carr's atmospheric gas viscosity correlation. Permission to publish by the Society of Petroleum Engineers of AIME. Copyright SPE-AIME.

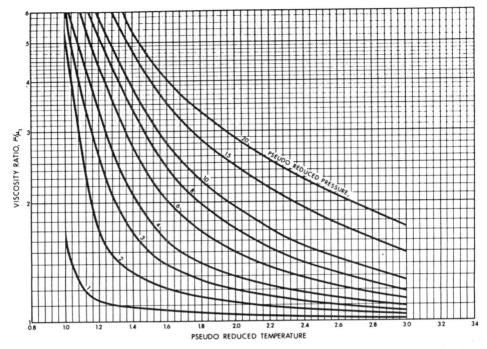

VISCOSITY RATIO, μ_g/μ_1

PSEUDO REDUCED TEMPERATURE

Figure 3-11. Carr's viscosity ratio correlation. Permission to publish by the Society of Petroleum Engineers of AIME. Copyright SPE-AIME.

Step 3. Calculate the pseudo-reduced pressure and temperature.

Step 4. From the pseudo-reduced temperature and pressure obtain the viscosity ratio (μ_g/μ_1) from Figure 3-11. The term μ_g represents the viscosity of the gas at the required conditions.

Step 5. The gas viscosity, μ_g, at the pressure and temperature of interest is calculated by multiplying the viscosity at one atmosphere and system temperature, μ_1, by the viscosity ratio.

The following examples illustrate the use of the proposed graphical correlations.

Example 3-14. Given the following gas composition,

Component	y_i
C_1	0.850
C_2	0.055
C_3	0.035
C_4	0.010

calculate the gas viscosity at 3,000 psia and 150°F.

Solution.

Step 1. From the gas composition, calculate MW_a, μ_g, T_{pc}, and p_{pc}.

Component	y_i	MW_i	$y_i\,MW_i$
C_1	0.850	16.04	13.634
C_2	0.055	30.07	1.654
C_3	0.035	44.09	1.543
C_4	0.010	58.12	0.5812
			MW_a = 17.412

- $\mu_g = \dfrac{17.412}{28.97} = 0.6$

- From Equations 3-27 and 3-28

$$T_{pc} = 358.5$$
$$p_{pc} = 672.5$$

Step 2. Calculate the viscosity of the natural gas at one atmosphere and 150°F from Figure 3-10.

$$\mu_1 = 0.0119 \text{ cp}$$

Step 3. Calculate p_{pr} and T_{pr}

$$p_{pr} = \frac{3,000}{672.5} = 4.46$$

$$T_{pr} = \frac{610}{358.5} = 1.70$$

Step 4. Estimate the viscosity ratio from Figure 3-12 on page 125.

$$\frac{\mu_g}{\mu_1} = 1.7$$

Step 5. Solve for the viscosity of the natural gas μ_g.

$$\mu_g = \frac{\mu_g}{\mu_1}(\mu_1) = (1.7)(0.0119) = 0.0202 \text{ cp}$$

Example 3-15. Given the following gas composition

Component	y_i
N_2	0.05
CO_2	0.05
H_2S	0.02
C_1	0.80
C_2	0.05
C_3	0.03

calculate the gas viscosity at 200°F and 3,500 psia.

Solution.

Component	y_i	MW_i	$y_i\,MW_i$
N_2	0.05	28.02	1.4010
CO_2	0.05	44.01	2.2005
H_2S	0.02	34.08	0.6816
C_1	0.80	16.07	12.8560
C_2	0.05	30.07	1.5035
C_3	0.03	44.09	1.3227
			$MW_a = 19.97$

Step 1. Calculate T_{pc} and p_{pc} by applying Equations 3-27 and 3-28.

$$\gamma_g = \frac{19.97}{28.97} = 0.689$$

$$T_{pc} = 380.43°R$$
$$p_{pc} = 665.11 \text{ psia}$$

Step 2. Correct the calculated pseudo-critical properties to account for the presence of the non-hydrocarbon components using Equations 3-34 and 3-35.

$$T_{pc}' = 380.43 - (80)(0.05) + 130(0.02)$$
$$- 250(0.05) = 366.53°R$$

$$p_{pc}' = 665.11 + 440(0.05) + 600(0.02)$$
$$- 170(0.05) = 690.61 \text{ psia}$$

Step 3. Estimate the gas viscosity of 1 atm and 200°F from Figure 3-10.

$$\mu_1 = 0.0123 \text{ cp}$$

Step 4. Using the inserts in Figure 3-10, estimate $(\Delta\mu)_{N_2}$, $(\Delta\mu)_{CO_2}$, and $(\Delta\mu)_{H_2S}$.

$$(\Delta\mu)_{N_2} = 0.00042 \text{ cp}$$
$$(\Delta\mu)_{CO_2} = 0.00026 \text{ cp}$$
$$(\Delta\mu)_{H_2S} = 0.00005 \text{ cp}$$

Step 5. Calculate the corrected gas viscosity at atmospheric pressure and system temperature by applying Equation 3-67.

$$\mu_1 = 0.0123 + 0.00042 + 0.00025 + 0.00005 = 0.01303 \text{ cp}$$

Step 6. Calculate p_{pr} and T_{pr}

$$p_{pr} = \frac{3,600}{690.61} = 5.21$$

$$T_{pr} = \frac{660}{366.53} = 1.80$$

Step 7. Estimate the viscosity ratio μ_g/μ_1 from Figure 3-12.

$$\frac{\mu_g}{\mu_1} = 1.52$$

Step 8. Calculate the viscosity of the natural gas at 3,500 psia and 200°F.

$$\mu_g = (1.52)(0.01303) = 0.01981 \text{ cp}$$

Standing (1977) proposed a convenient mathematical expression for calculating the viscosity of the natural gas at atmospheric pressure and reservoir temperature, μ_1. Standing also presented equations for describing the effects of N_2, CO_2, and H_2S on μ_1. The proposed relationships are:

$$\mu_1 = (\mu_1)_{\text{uncorrected}} + (\Delta\mu)_{N_2} + (\Delta\mu)_{CO_2} + (\Delta\mu)_{H_2S}$$

$$(\mu_1)_{\text{uncorrected}} = [1.709(10^{-5}) - 2.062(10^{-6}) \ \gamma_g] \ (T - 460)$$
$$+ \ 8.188(10^{-3}) - 6.15(10^{-3}) \ \log (\gamma_g) \tag{3-68}$$

$$(\Delta\mu)_{N_2} = y_{N_2} [8.48(10^{-3}) \log (\gamma_g) + 9.59(10^{-3})] \tag{3-69}$$

$$(\Delta\mu)_{CO_2} = y_{CO_2} [9.08(10^{-3}) \log (\gamma_g) + 6.24(10^{-3})] \tag{3-70}$$

$$(\Delta\mu)_{H_2S} = y_{H_2S} [8.49(10^{-3}) \log (\gamma_g) + 3.73(10^{-3})] \tag{3-71}$$

where μ_1 = viscosity of the gas at atmospheric pressure and reservoir temperature, cp
T = reservoir temperature, °R
γ_g = gas gravity
$y_{N_2}, y_{CO_2}, y_{H_2S}$ = mole fraction of N_2, CO_2, and H_2S, respectively

Dempsey (1965) expressed the viscosity ratio μ_g/μ_1 by the following relationship.

$$\ln \left[T_{pr} \left(\frac{\mu_g}{\mu_1} \right) \right] = a_0 + a_1 \ p_{pr} + a_2 \ p_{pr}^2 + a_3 \ p_{pr}^3$$

$$+ \ T_{pr} \ (a_4 + a_5 \ p_{pr} + a_6 \ p_{pr}^2 + a_7 \ p_{pr}^3)$$

$$+ \ T_{pr}^2 \ (a_8 + a_9 \ p_{pr} + a_{10} \ p_{pr}^2 + a_{11} \ p_{pr}^3)$$

$$+ \ T_{pr}^3 \ (a_{12} + a_{13} \ p_{pr} + a_{14} \ p_{pr}^2 + a_{15} \ p_{pr}^3) \tag{3-72}$$

where T_{pr} = pseudo-reduced temperature of the gas mixture
p_{pr} = pseudo-reduced pressure of the gas mixture
$a_0 - a_{15}$ = coefficients of the equations are given below

$a_0 = -2.46211820$ $a_8 = -7.93385684(10^{-1})$
$a_1 = 2.97054714$ $a_9 = 1.39643306$
$a_2 = -2.86264054(10^{-1})$ $a_{10} = -1.49144925(10^{-1})$
$a_3 = 8.05420522(10^{-3})$ $a_{11} = 4.41015512(10^{-3})$
$a_4 = 2.80860949$ $a_{12} = 8.39387178(10^{-2})$
$a_5 = -3.49803305$ $a_{13} = -1.86408848(10^{-1})$
$a_6 = 3.60373020(10^{-1})$ $a_{14} = 2.03367881(10^{-2})$
$a_7 = -1.044324(10^{-2})$ $a_{15} = -6.09579263(10^{-4})$

Lee-Gonzalez-Eakin Method

Lee, Gonzalez, and Eakin (1966) presented a semi-empirical relationship for calculating the viscosity of natural gases. The authors expressed the gas viscosity in terms of the reservoir temperature, gas density, and the molecular weight of the gas. Their proposed equation is given by

$$\mu_g = 10^{-4} K \; EXP \left[X \left(\frac{\rho_g}{62.4} \right)^Y \right] \tag{3-73}$$

where $$K = \frac{(9.4 + 0.02 \; MW_a) \; T^{1.5}}{209 + 19 \; MW_a + T} \tag{3-74}$$

$$X = 3.5 + \frac{986}{T} + 0.01 \; MW_a \tag{3-75}$$

$$Y = 2.4 - 0.2 \; X \tag{3-76}$$

ρ_g = gas density at reservoir pressure and temperature, lb/ft^3
T = reservoir temperature, °R
MW_a = apparent molecular weight of the gas mixture

The proposed correlation can predict viscosity values with a standard deviation of 2.7% and a maximum deviation of 8.99%. The authors pointed out that the method cannot be used for sour gases.

Example 3-16. Rework Example 3-14 and calculate the gas viscosity by using the Lee-Gonzales-Eakin method.

Solution.

- Using p_{pr} and T_{pr} of 4.46 and 1.7, respectively, calculate the gas compressibility factor from Standing and Katz chart.

  Z = 0.89

- Solve for gas density by applying Equation 3-26.

  $$\rho_g = \frac{(3,000)(17.412)}{(610)(10.73)(0.89)} = 8.97 \text{ lb/ft}^3$$

- Calculate the parameters K, X, and Y by using Equations 3-74, 3-75, and 3-76 respectively.

  $$K = \frac{[9.4 + 0.02\ (17.412)]\ (610)^{1.5}}{209 + 19\ (17.412) + 610} = 127.73$$

  $$X = 3.5 + \frac{986}{610} + 0.01\ (17.412) = 5.291$$

  $$Y = 2.4 - 0.2\ (5.291) = 1.342$$

- Solve for the gas viscosity by applying Equation 3-73.

  $$\mu_g = 10^{-4}\ (127.73)\ \text{EXP}\left[5.291 \left(\frac{8.97}{62.4}\right)^{1.342}\right] = 0.019 \text{ cp}$$

Dean and Stiel Method

Dean and Stiel (1965) proposed the following mathematical expressions for calculating the viscosity of natural gases at atmospheric pressure and reservoir temperature.

$$\mu_1 = 34\ (10^{-5})\ \frac{(T_{pr})^{8/9}}{\xi_m}, \text{ for } T_{pr} \leq 1.5 \tag{3-77}$$

$$\mu_1 = \frac{166.8\ (10^{-5})[0.1338\ T_{pr} - 0.0932]^{5/9}}{\xi_m}, \text{ for } T_{pr} > 1.5 \tag{3-78}$$

where ξ_m is the viscosity parameter of the gas mixture as defined by the following equation.

$$\xi_m = 5.4402\ \frac{(T_{pc})^{1/6}}{(MW_a)^{.5}(p_{pc})^{2/3}} \tag{3-79}$$

Dean and Stiel recommended the following relationship for calculating the viscosity of natural gases at the prevailing reservoir condition.

$$\mu_g = \mu_1 + \frac{10.8(10^{-5})[\text{EXP}(1.439\ \rho_r) - \text{EXP}(-1.111(\rho_r)^{1.888})]}{\xi_m} \quad (3\text{-}80)$$

where μ_g = gas viscosity at reservoir pressure and temperature, cp
$\quad\quad \mu_1$ = gas viscosity at atmospheric pressure and reservoir temperature, cp
$\quad\quad \rho_r$ = reduced gas density as defined by Equation 3-51

The use of the proposed equation can be best illustrated by the following example.

Example 3-17. Rework Example 3-14 by using the Dean and Stiel correlation. The gas composition, pressure, and temperature are given here for convenience.

Component	y_i
C_1	0.850
C_2	0.055
C_3	0.035
C_4	0.010

System pressure = 3,000 psia
System temperature = 150°F

Solution. From Examples 3-14 and 3-16, the following data were obtained.

MW_a = 17.412
p_{pc} = 672.5 psia
T_{pc} = 358.5°R
p_{pr} = 4.46
T_{pr} = 1.70
Z = 0.89

• Calculate the viscosity parameter, ξ_m, by applying Equation 3-79.

$$\xi_m = 5.4402\ \frac{(358.5)^{1/6}}{(17.412)^{.5}(672.5)^{2/3}} = 0.4527$$

• Because $T_{pr} > 1.5$, apply Equation 3-78 to calculate μ_1

$$\mu_1 = \frac{166.8(10^{-5})[0.1338(1.7) - 0.0932]^{5/9}}{0.04527} = 0.0121 \text{ cp}$$

- Calculate the reduced density of the gas mixture from Equation 3-51.

$$\rho_r = \frac{(0.27)(4.46)}{(0.89)(1.70)} = 0.7959$$

- Solve for the gas viscosity by using Equation 3-80 to yield

$$\mu_g = 0.01845 \text{ cp}$$

ENGINEERING APPLICATIONS OF THE NATURAL GAS PVT PROPERTIES

To demonstrate the practical applications of the natural gas physical properties in reservoir engineering, the following sections present and briefly review some of these applications.

Radial Flow of Gases

Henry Darcy, a French engineer, developed a fluid flow equation which has since become one of the standard mathematical tools of the petroleum engineer. The proposed equation, which is known as Darcy's equation, states that the rate of flow of a homogeneous fluid through a homogeneous porous media is proportional to the pressure gradient and the cross-sectional area normal to the direction of the flow; and inversely proportional to the viscosity of the fluid. In a differential form, the equation is expressed by the following relationship:

$$q = \frac{1.1271(2\pi rh)k}{1,000 \, \mu} \frac{dp}{dr} \tag{3-81}$$

where q = fluid flow rate, bbl/day
r = radial radius, ft
h = formation permeability, md
μ = viscosity, cp
p = pressure, psia

Darcy's equation can be applied to determine the gas flow rate in terms of volumetric rate at standard conditions, scf/day, to give

$$Q_g = \frac{1.1271(2\pi rh)k}{1,000 \; \mu_g B_g} \frac{dp}{dr} \tag{3-82}$$

where Q_g = gas flow rate, scf/day
μ_g = gas viscosity, cp
B_g = gas formation volume factor, bbl/scf

The gas formation volume factor, as expressed mathematically by Equation 3-64 is combined with Equation 3-82 to give

$$Q_g = \frac{0.703 \; khr(2)}{T \; Z \; \mu_g} \frac{dp}{dr}$$

Separating variables and integrating the above expression from wellbore conditions to the outer reservoir boundary conditions gives

$$Q_g \int_{r_w}^{r_e} \frac{dr}{r} = \frac{0.703 \; kh}{T} 2 \int_{p_{wf}}^{p_e} \frac{p}{\mu Z} \, dp$$

or

$$Q_g = \frac{0.703 \; kh}{T \; Ln(r_e/r_w)} 2 \int_{p_{wf}}^{p_e} \frac{p}{\mu Z} \, dp \tag{3-83}$$

where r_e = well drainage radius, ft
r_w = wellbore radius, ft
p_{wf} = bottom-hole flowing pressure, psia
p_e = reservoir pressure, psia
T = reservoir temperature, °R

The term $2 \int_{p_{wf}}^{p_e} (p/\mu Z) \, dp$ in Equation 3-83 can be expanded to give

$$2 \int_{p_{wf}}^{p_e} \frac{p}{\mu Z} \, dp = 2 \int_{0}^{p_e} \frac{p}{\mu Z} \, dp - 2 \int_{0}^{p_{wf}} \frac{p}{\mu Z} \, dp \tag{3-84}$$

The expression $2 \int_{0}^{p} (2p/\mu Z) dp$ is commonly called the real gas potential or the real gas pseudo-pressure. It is customarily represented by m(p) or ψ. Thus

$$m(p) = \psi = 2 \int_0^P \frac{p}{\mu Z} \, dp \tag{3-85}$$

Using the concept of the real gas potential in Equation 3-83 gives

$$Q_g = \frac{0.703 \ kh(\psi_e - \psi_{wf})}{T \ Ln(r_e/r_w)} \tag{3-86}$$

where ψ_e = external real gas potential, psi²/cp
ψ_{wf} = bottom-hole real gas potential, psi²/cp

The drop in the real gas potential, i.e., $\psi_e - \psi_{wf}$, can be determined numerically by plotting the term $(2p/\mu Z)$ as a function of pressure and calculating the area under the appropriate part of the resulting curve. Figure 3-12 shows a typical plot of $(2p/\mu Z)$ versus p relationship.

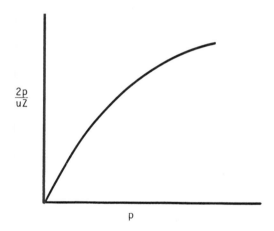

Figure 3-12. A typical plot of the gas pressure function.

The area under the curve, i.e., $\psi_e - \psi_{wf}$, can be numerically determined by employing Weddle's rule, to give

$$\psi_e - \psi_{wf} = 0.3 \ \Delta p \ [Y_1 + 5 \ Y_2 + Y_3 + 6 \ Y_4 + Y_5 + 5 \ Y_6 + Y_7] \tag{3-87}$$

with

$$\Delta p = \frac{p_e - p_{wf}}{6} \tag{3-88}$$

$$Y_i = 2 \ [p_{wf} + (i - 1) \ \Delta p]/(\mu_g \ Z)_i \tag{3-89}$$

where   i = 1,2,...,7

$(\mu_g Z)_i$ = gas viscosity and compressibility factor as evaluated at $[p_{wf} + (i - 1)\, \Delta p]$

Equation 3-83 can be approximated by removing the term $1/(\mu_g Z)$ outside the integral and evaluated at the average pressure, to give

$$Q_g = \frac{0.703\ kh}{T(\mu_g Z)_{avg}\ \text{Ln}(r_e/r_w)}\ 2\ \int_{p_{wf}}^{p_e} p\ dp$$

or

$$Q_g = \frac{0.703\ kh(p_e^2 - p_{wf}^2)}{T(\mu_g Z)_{avg}\ \text{Ln}(r_e/r_w)} \tag{3-90}$$

The expression $(\mu_g Z)_{avg}$ is evaluated at the arithmetic average pressure $(p_e + p_{wf})/2$.

The approximation method represented by Equation 3-90 is commonly known as the pressure-squared approximation method.

Example 3-18. A natural gas with the following composition is flowing from a gas well at a bottom-hole flowing pressure of 2,800 psia.

Component	y_i
C_1	0.850
C_2	0.055
C_3	0.035
C_4	0.010

Given the following additional data,

p_e = 3,200 psia	T = 150°F	r_w = 0.333 ft
k = 50 md	h = 25 ft	r_e = 666 ft

calculate the gas flow rate by using the pressure-squared method.

Solution.

Step 1. Calculate average pressure

$$p_{avg} = (3,200 + 2,800)/2 = 3,000\ \text{psia}$$

Step 2. Calculate the apparent molecular weight of the gas, to give

$$MW_a = 17.412$$

Step 3. Calculate T_{pc} and p_{pc} of the gas, to yield

$$T_{pc} = 358.5$$
$$p_{pc} = 672.5$$

Step 4. Determine the pseudo-reduced temperature and pressure of the gas, to give

$$T_{pr} = 1.7$$
$$p_{pr} = 4.46$$

Step 5. Calculate the gas viscosity and compressibility factor, to yield

$$\mu_g = 0.0202 \text{ cp}$$
$$Z = 0.89$$

Step 6. Calculate the gas flow rate from Equation 3-90.

$$Q_g = 25.331307 \text{ MM scf/day}$$

Gas Reservoir Material Balance

A material balance on a gas reservoir can effectively be used to determine

- The size of the gas reservoir
- The ultimate gas recovery
- The effect of water encroachment in the reservoir

For a volumetric gas reservoir, i.e., a reservoir with no water influx, the material balance on the gas reservoir in terms of moles of gas is expressed as

$$n_p = n_i - n_f \tag{3-91}$$

where n_p = moles of gas produced
n_i = moles of gas initially in the reservoir
n_f = moles of gas remaining in the reservoir

Representing the gas reservoir by an idealized gas container (shown schematically in Figure 3-13), the terms in Equation 3-91 are replaced by their equivalents using the real gas law, to give

$$\frac{p_{sc} \, G_p}{Z_{sc} \, R \, T_{sc}} = \frac{p_i \, V}{Z_i \, R \, T} - \frac{p \, V}{Z \, R \, T}$$

or

$$G_p = \left[\frac{T_{sc}}{p_{sc}}\right]\left[\frac{p_i}{Z_i T}\right] V - \left[\frac{T_{sc}}{p_{sc}}\right]\left[\frac{V}{T}\right]\left[\frac{p}{Z}\right] \tag{3-92}$$

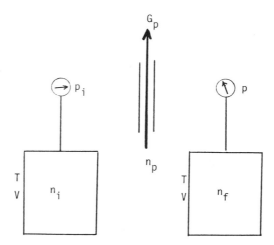

Figure 3-13. A schematic illustration of an idealized volumetric gas reservoir.

where G_p = cumulative gas produced, scf
    T_{sc}, p_{sc} = standard conditions: 520°R and 14.7 psia
    V = gas pore volume, ft^3
    p_i = initial reservoir pressure, psia
    Z_i = initial reservoir pressure, psia
    p = current reservoir pressure, psia
    Z = current gas compressibility factor

Since p_i, T, Z_i, and V are all constants for a volumetric gas reservoir, Equation 3-92 can be expressed as an equation of straight line, or

$$G_p = b - m \frac{p}{Z} \tag{3-93}$$

where   b = 520 p_iV/(14.7 Z_iT)
    m = 520 V/(14.7T)

Equation 3-93 indicates that for a volumetric gas reservoir the graph of the ratio (p/Z) versus cumulative gas production G_p on cartesian coordinate paper is a straight line relationship.

The gas formation volume factor B_g, as expressed in scf/ft^3 by Equation 3-64, can be incorporated into Equation 3-92 to give

$$G_p = B_{gi}V - B_g V \tag{3-94}$$

The term $(B_{gi}V)$ represents the gas-initially-in-place as measured in scf, or

$$G_p = GIIP - B_gV \tag{3-95}$$

where GIIP = gas initially in place, scf.

Equation 3-95 shows that the plot of G_p versus B_g is a straight line with a negative slope of V and with an intercept of GIIP. The areal extent of the reservoir, A, can be calculated from the slope V (gas pore volume) by applying the mathematical definition of V, or

$$V = 43{,}560 \text{ A h } \Phi \ (1 - S_{wc}) \tag{3-96}$$

where V = gas pore volume, ft^3
        A = areal extent of the reservoir, acres
      S_{wc} = connate water saturation
      Φ = reservoir porosity

Example 3-19. A gas reservoir with an areal extent of 1,000 acres has an initial reservoir pressure of 3,000 psia. The reservoir has the following rock and fluid properties:

  h = 10 ft $\Phi = 0.25$ $S_{wc} = 0.15$
  $\gamma_g = 0.601$ $T = 150°F$

Calculate the cumulative gas production when the reservoir pressure drops to 2,500 psia.

Solution.

Step 1. Calculate the past pore volume from Equation 3-96 to give

      V = 92.565 MM ft^3

Step 2. Calculate the gas compressibility factor at 3,000 and 2,500 psia to yield

      $Z_{3,000} = 0.89$
      $Z_{2,500} = 0.86$

Step 3. Solve for G_p by applying Equation 3-92 to give

      $G_p = 2{,}489.67$ MM scf

The material balance as expressed mathematically by Equation 3-92 can be modified to describe the volumetric behavior of gas reservoirs under water drive by incorporating the cumulative water influx W_e into the expression as follows:

$$\frac{p_{sc}\, G_p}{R\, T_{sc}} = \frac{p_i\, V}{Z_i RT} - \frac{p[V - (W_e - W_p)]}{Z\, RT}$$

or

$$G_p = \left[\frac{T_{sc}}{p_{sc}}\right]\left[\frac{p_i}{Z_i T}\right] V - \left[\frac{T_{sc}}{p_{sc}}\right]\left[\frac{p}{ZT}\right][V - W_e + W_p] \tag{3-97}$$

where W_e = cumulative water influx, ft^3
        W_p = cumulative water produced, ft^3

The performance of a water drive gas reservoir can be represented schematically by an idealized gas container as shown in Figure 3-14.

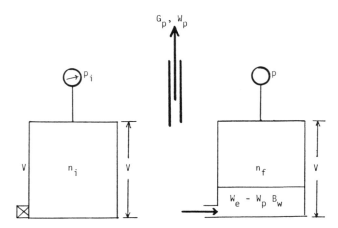

Figure 3-14. A schematic illustration of an idealized water drive gas reservoir.

Equation 3-97 can be expressed in terms of the gas formation volume factor as follows

$$G_p = B_{gi} V - B_g [V - W_e + W_p] \tag{3-98}$$

Since the term $(B_{gi}V)$ represents the gas initially in place, Equation 3-98 can be written as

$$G_p = GIIP - B_g [V - W_e + W_p] \tag{3-99}$$

or

$$G_p = GIIP - B_g \left[\frac{GIIP}{B_{gi}} - W_e + W_p \right] \qquad (3\text{-}100)$$

where GIIP = gas-initially-in-place, scf
  B_{gi} = initial gas formation volume factor, scf/ft³
  B_g = current gas formation volume factor, scf/ft³

PROBLEMS

1. Assuming an ideal gas behavior, calculate the density of n-butane at 220°F and 50 psia.

2. Show that

$$y_i = \frac{(w_i/MW_i)}{\sum_i (w_i/MW_i)}$$

3. Given the following gas

Component	Weight Fraction
C_1	0.65
C_2	0.15
C_3	0.10
n-C_4	0.06
n-C_5	0.04

  calculate

  a. Mole fraction of the gas
  b. Apparent molecular weight
  c. Specific gravity
  d. Specific volume at 300 psia and 120°F by assuming an ideal gas behavior

4. An ideal gas mixture has a density of 1.92 lb/ft³ at 500 psia and 100°F. Calculate the apparent molecular weight of the gas mixture.

5. Using the gas composition as given in Problem 3, and assuming real gas behavior, calculate

  a. Gas density at 2,000 psia and 150°F
  b. Specific volume at 2,000 psia and 150°F
  c. Gas formation volume factor in scf/ft³

6. A natural gas with a specific gravity of 0.75 has a gas formation volume factor of 0.00529 ft^3/scf at the prevailing reservoir pressure and temperature. Calculate the density of the gas.

7. A natural gas has the following composition

Component	y_i
C_1	0.75
C_2	0.10
C_3	0.05
i-C_4	0.04
n-C_4	0.03
i-C_5	0.02
n-C_5	0.01

Reservoir conditions are 3,500 psia and 200°F. Calculate

   a. Isothermal gas compressibility coefficient
   b. Gas viscosity by using
      (1) Carr-Kobayashi-Burrows Method
      (2) Lee-Gonzales-Eakin Method
      (3) Dean-Stiel Method

8. Given the following gas composition

Component	y_i
CO_2	0.06
N_2	0.03
C_1	0.75
C_2	0.07
C_3	0.04
n-C_4	0.03
n-C_5	0.02

If the reservoir pressure and temperature are 2,500 psia and 175°F respectively, calculate

   a. Gas density, by accounting for the presence of non-hydrocarbon components by using
      (1) Wichert-Aziz Method
      (2) Carr-Kobayashi-Burrows Method
   b. Isothermal gas compressibility coefficient
   c. Gas viscosity by using
      (1) Carr-Kobayashi-Burrows Method
      (2) Lee-Gonzales-Eakin Method
      (3) Dean-Stiel Method

9. Forty scf of gas are placed in a 0.5 ft^3 cylinder at a temperature of 80°F. The cylinder is fitted with a pressure gauge. The gauge shows a pressure reading of 1,000 psia. Calculate the gas compressibility factor. What is the gas formation volume factor in ft^3/scf?

10. A gas well is producing at a rate of 1.22 mm scf/day. The specific gravity of the producing gas is 0.74. If the average reservoir pressure and reservoir temperature are 2,000 psia and 125°F, calculate

   a. Gas flow rate in ft^3/day
   b. Gas viscosity under reservoir conditions

11. A high-molecular-weight natural gas has the following composition:

Component	y_i
C_1	0.73
C_2	0.10
C_3	0.05
C_4	0.03
C_5	0.03
C_6	0.02
C_{7+}	0.04

   If the molecular weight and specific gravity of C_{7+} are 135 and 0.81, calculate

   a. Specific gravity of the gas mixture
   b. Density of the gas at 2,000 psia and 120°F

12. A gas well is producing a hydrocarbon gas at a bottom-hole pressure of 3,100 psia. The reservoir pressure and temperature are 3,420 psia and 160°F. Given the following additional well and gas properties data,

   gas gravity = 0.700   k = 29 md r_e = 660 ft
   r_w = 0.25 ft   h = 12 ft

   calculate the gas flow rate by using

   a. The real gas potential approach
   b. The pressure-squared method

13. A volumetric gas reservoir has the following gas and rock properties:

   A = 1,200 acres   h = 26 ft Φ = 0.16
   S_{wc} = 0.24 μ_g = 0.72   T = 130°F
   p_i = 4,100 psia

   Calculate cumulative gas produced when reservoir pressure drops to 3,850 psia.

14. A volumetric gas reservoir has the following pressure-production history:

p psia	G_p MM scf
3,000	100
2,500	500

The reservoir temperature is 170°F. If the specific gravity of the gas is 0.70, calculate

a. Initial gas formation volume factor
b. Initial gas density
c. Gas initially-in-place in scf
d. Gas pore volume
e. Areal extent of the reservoir given

$\Phi = 0.19$
$S_{wc} = 0.24$
$h = 12$ ft

REFERENCES

1. Brown, et al., "Natural Gasoline and the Volatile Hydrocarbons," Tulsa: NGAA, 1948.
2. Carr, N., Kobayashi, R., and Burrows, D., "Viscosity of Hydrocarbon Gases Under Pressure," *Trans. AIME*, 1954, Vol. 201, pp. 270–275.
3. Dean, D. E. and Stiel, L. I., "The Viscosity of Non-polar Gas Mixtures at Moderate and High Pressure," *AIChE Jour.*, 1958, Vol. 4, pp. 430–436.
4. Dempsey, J. R., "Computer Routine Treats Gas Viscosity as a Variable," *Oil and Gas Journal*, Aug. 16, 1965, pp. 141–143.
5. Dranchuk, P. M. and Abu-Kassem, J. H., "Calculation of Z-factors for Natural Gases Using Equations-of-State," *JCPT*, July–September, 1975, pp. 34–36.
6. Dranchuk, P. M., Purvis, R. A., and Robinson, D. B., "Computer Calculations of Natural Gas Compressibility Factors Using the Standing and Katz Correlation," *Inst. of Petroleum Technical Series*, No. IP 74-008, 1974.
7. Hall, K. R. and Yarborough, L., "A New Equation-of-State for Z-factor Calculations," *Oil and Gas Journal*, June 18, 1973, pp. 82–92.
8. Hankinson, R. W., Thomas, L. K., and Phillips, K. A., "Predict Natural Gas Properties," *Hydrocarbon Processing*, April 1969, pp. 106–108.

9. Kay, W. B., "Density of Hydrocarbon Gases and Vapor," *Industrial and Engineering Chemistry*, Vol. 28, 1936, pp. 1014–1019.
10. Lee, A. L., Gonzalez, M. H., and Eakin, B. E., "The Viscosity of Natural Gases," *Journal of Petroleum Technology*, August 1966, pp. 997–1000.
11. Mattar, L. G., Brar, S., and Aziz, K., "Compressibility of Natural Gases," *Journal of Canadian Petroleum Technology*, October–December, 1975, pp. 77–80.
12. Papay, J., "A Termelestechnologiai Parameterek Valtozasa a Gazlelepk Muvelese Soran," *OGIL MUSZ*, Tud, Kuzl., Budapest, 1968, pp. 267–273.
13. Standing, M. B., *Volumetric and Phase Behavior of Oil Field Hydrocarbon Systems*, Dallas: Society of Petroleum Engineers, 1977, pp. 125–126.
14. Standing, M. B., *Volumetric and Phase Behavior of Oil Field Hydrocarbon Systems*, Dallas: Society of Petroleum Engineers of AIME, 1977, pp. 122.
15. Standing, M. B. and Katz, D. L., "Density of Natural Gases," *Tran. AIME*, Vol. 146, 1942, pp. 140–149.
16. Stewart, W. F., Burkhard, S. F., and Voo, D., "Prediction of Pseudo-Critical Parameters for Mixtures," Paper presented at the AIChE Meeting, Kansas City, MO (1959).
17. Sutton, R. P., "Compressibility Factors for High-Molecular-Weight Reservoir Gases," Paper SPE 14265, Presented at the 60th Annual Technical Conference and Exhibition of the Society of Petroleum Engineers, Las Vegas, September 22–25, 1985.
18. Takacs, G., "Comparisons Made for Computer Z-factor Calculation," *Oil and Gas Journal*, Dec. 20, 1976, pp. 64–66.
19. Wichert, E. and Aziz, K., "Calculation of Z's for Sour Gases," *Hydrocarbon Processing*, Vol. 51, No. 5, 1972, pp. 119–122.
20. Trube, A. S., "Compressibility of Natural Gases," *Trans. AIME*, Vol. 210, 1957, pp. 355–357.
21. Yarborough, L. and Hall, K. R., "How to Solve Equation-of-State for Z-factors," *Oil and Gas Journal*, February 18, 1974, pp. 86–88.

4

Phase Behavior of Crude Oils

Petroleum (an equivalent term is "crude oil") is a complex mixture consisting predominantly of hydrocarbons, and containing sulfur, nitrogen, oxygen, and helium as minor constituents. The physical and chemical properties of crude oils vary considerably and are dependent on the concentration of the various types of hydrocarbons and minor constituents present.

An accurate description of physical properties of crude oils is of considerable importance in the fields of both applied and theoretical science and especially in the solution of petroleum reservoir engineering problems. Physical properties of primary interest in petroleum engineering studies include:

- Fluid densities
- Isothermal compressibility
- Solution gas-oil ratios
- Oil formation volume factor
- Fluid viscosities
- Bubble-point pressure
- Surface tension

Data on most of these fluid properties is usually determined by laboratory experiments performed on samples of actual reservoir fluids. In the absence of experimentally measured properties of crude oils, it is necessary for the petroleum engineer to determine the properties from empirically derived correlations.

CRUDE OIL DENSITY AND SPECIFIC GRAVITY

Evaluating the volumetric phase behavior of oil reservoirs requires accurate knowledge of the physical properties of the crude oil at elevated pressure and temperature. Pertinent among the properties of interest are the density and the specific gravity of the crude oil.

136

The crude oil density is defined as the mass of a unit volume of the crude at a specified pressure and temperature. It is usually expressed in pounds per cubic foot. The specific gravity of a crude oil is defined as the ratio of the density of the oil to that of water. Both densities are measured at 60°F and atmospheric pressure.

$$\gamma_o = \frac{\rho_o}{\rho_w} \qquad\qquad\qquad (4\text{-}1)$$

where γ_o = specific gravity of the oil
   ρ_o = density of the crude oil, lb/ft^3
   ρ_w = density of the water, lb/ft^3

Although the density and specific gravity are used extensively in the petroleum industry, the API gravity is the preferred gravity scale. This gravity scale is precisely related to the specific gravity by the following expression:

$$^\circ API = \frac{141.5}{\gamma_o} - 131.5 \qquad\qquad\qquad (4\text{-}2)$$

The API gravities of crude oils usually range from 47°API for the lighter crude oils to 10°API for the heavier asphaltic crude oils.

During the last forty years, numerous methods of calculating the density of crude oils have been proposed. There are two approaches available in the literature to calculate liquid density:

- The equation-of-state approach (discussed in Chapter 6)
- The liquid density-correlation approach.

The second approach of determining liquid density is presented as follows:

METHODS FOR DETERMINING DENSITY OF CRUDE OILS OF KNOWN COMPOSITION

Several reliable methods are available for determining the density of complex hydrocarbon mixtures from their composition. The best known and most widely used methods in the petroleum industry are those of Standing-Katz (1942) and Alani-Kennedy (1960).

The Standing-Katz Method

Standing and Katz (1942) proposed a graphical correlation for determining the density of hydrocarbon liquid mixtures. Standing and Katz developed the correlation from evaluating experimental, compositional, and density data on 15 crude oil samples containing up to 60 mole-percent methane. The proposed method yielded an average error of 1.2% and maximum error of 4% for the data on these crude oils. The original correlation did not have a procedure for handling significant amounts of non-hydrocarbons.

The authors expressed the density of hydrocarbon liquid mixtures as a function of pressure and temperature by the following relationship:

$$\rho_o = \rho_{sc} + \Delta\rho_p - \Delta\rho_T \qquad (4\text{-}3)$$

where ρ_o = crude oil density at p and T, lb/ft^3
ρ_{sc} = crude oil density at standard conditions, i.e., 14.7 psia and 60°F, lb/ft^3
$\Delta\rho_p$ = density correction for compressibility of oils, lb/ft^3
$\Delta\rho_T$ = density correction for thermal expansion of oils, lb/ft^3

Standing and Katz correlated graphically the liquid density at standard conditions with:

• The density of the propane-plus fraction $\rho_{C_{3+}}$
• The weight percent of methane in the entire system $(m_{C_1})_{C_{1+}}$
• The weight percent of ethane in the ethane-plus $(m_{C_2})_{C_{2+}}$

This graphical correlation is shown in Figure 4-1. The following are the specific steps in the Standing and Katz procedure of calculating the liquid density at a specified pressure and temperature.

Step 1. Calculate the total weight and the weight of each component in one lb-mole of the hydrocarbon mixture by applying the following relationships:

$$m_i = x_i \, MW_i \qquad (4\text{-}4)$$

$$m_t = \Sigma \, x_i \, MW_i \qquad (4\text{-}5)$$

where m_i = weight of component i in the mixture, lb/lb-mole
x_i = mole fraction of component i in the mixture
MW_i = molecular weight of component i
m_t = total weight of one lb-mole of the mixture, lb/lb-mole

Step 2. Calculate the weight percent of methane in the entire system and the weight percent of ethane in the ethane-plus from the following expressions:

$$(m_{C_1})_{C_{1+}} = \left[\frac{m_{C_1}}{m_t}\right] 100 \qquad (4\text{-}6)$$

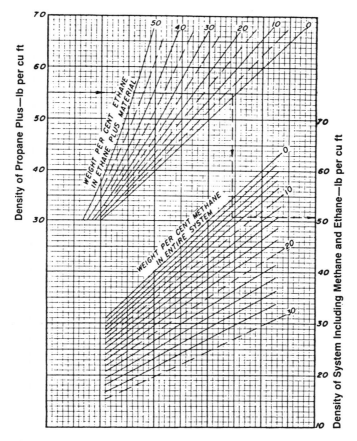

Figure 4-1. Standing's density correlation. From *Volumetric and Phase Behavior of Oil Field Hydrocarbon Systems,* Ninth Edition, Permission to publish by the Society of Petroleum Engineers of AIME. Copyright SPE-AIME, 1981.

and

$$(m_{C_2})_{C_{2+}} = \left[\frac{m_{C_2}}{m_{C_{2+}}}\right] 100 = \left[\frac{m_{C_2}}{m_t - m_{C_1}}\right] 100 \qquad (4\text{-}7)$$

where $(m_{C_1})_{C_{1+}}$ = weight % of methane in the entire system

m_{C_1} = weight of methane in one lb-mole of the mixture

$m_{C_2})_{C_{2+}}$ = weight % of ethane in ethane-plus

m_{C_2} = weight of ethane in one lb-mole of the mixture

Step 3. Calculate the density of the propane-plus fractions at standard conditions by using the following equations:

$$\rho_{C_{3+}} = \frac{m_{C_{3+}}}{V_{C_{3+}}} \tag{4-8}$$

with

$$m_{C_{3+}} = \sum_{i=C_3} m_i \tag{4-9}$$

$$V_{C_{3+}} = \sum_{i=C_3} V_i = \sum_{i=C_3} \frac{m_i}{\rho_{oi}} \tag{4-10}$$

where $\rho_{C_{3+}}$ = density of the propane and heavier components, lb/ft^3

$m_{C_{3+}}$ = weight of the propane and heavier fractions, lb/lb-mole

$V_{C_{3+}}$ = volume of the propane-plus fraction, ft^3/lb-mole

V_i = volume of component i in one lb-mole of the mixture

ρ_{oi} = density of component i at standard conditions, lb/ft^3. Density values for pure components are tabulated in Table 1-1 in Chapter 1, but the density of the plus fraction must be measured.

Step 4. Using Figure 4-1, enter the $\rho_{C_{3+}}$ value into the left ordinate of the chart and move horizontally to the line representing $(m_{C_2})_{C_{2+}}$, then drop vertically to the line representing $(m_{C_1})_{C_{1+}}$. The density of the oil at standard condition is read on the right side of the chart.

Step 5. Correct the density at standard conditions to the actual pressure by reading the additive pressure correction factor, $\Delta\rho_p$, from Figure 4-2.

Step 6. Correct the density at 60°F and pressure to the actual temperature by reading the thermal expansion correction term, $\Delta\rho_T$, from Figure 4-3.

Example 4-1. A crude oil system has the following composition.

Component	x_i
C_1	0.45
C_2	0.05
C_3	0.05
C_4	0.03
C_5	0.01
C_6	0.01
C_{7+}	0.4

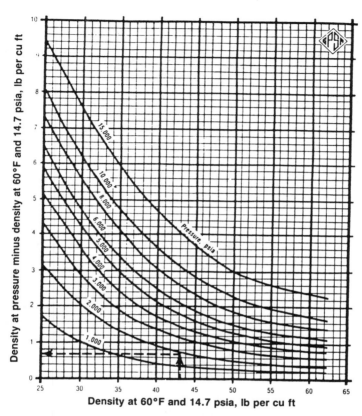

Figure 4-2. Density correction for compressibility of crude oil. Courtesy of the Gas Processors Suppliers Association. Published in the GPSA Engineering Data Book, Tenth Edition, 1987.

If the molecular weight and specific gravity of C_{7+} fractions are 215 and 0.87, respectively, calculate the density of the crude oil at 4,000 psia and 160°F by using the Standing and Katz method.

Solution.

Component	x_i	MW_i	$m_i = x_i MW_i$	ρ_{oi}, lb/ft^3*	$V_i = m_i/\rho_{oi}$
C_1	0.45	16.04	7.218	—	—
C_2	0.05	30.07	1.5035	—	—
C_3	0.05	44.09	2.2045	31.64	0.0697

* From Table 1-1

(*example 4-1 continued on next page*)

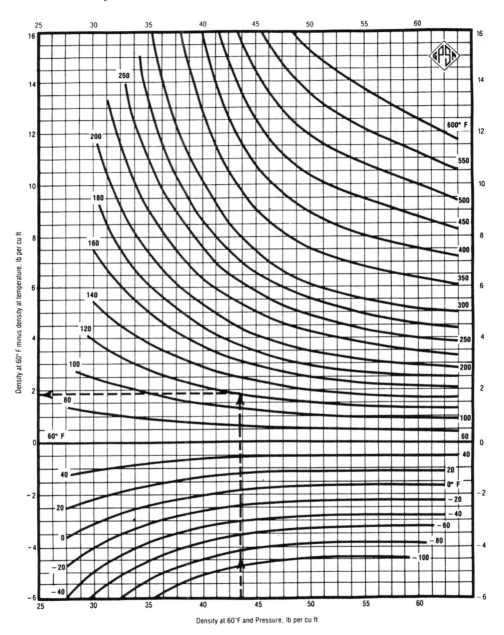

Figure 4-3. Density correction for isothermal expansion of crude oils. Courtesy of the Gas Processors Suppliers Association. Published in the GPSA Engineering Data Book, Tenth Edition, 1987.

Example 4-1. Continued.

Component	x_i	MW_i	$m_i = x_i\,MW_i$	ρ_{oi}, lb/ft^3*	$V_i = m_i/\rho_{oi}$
C_4	0.03	58.12	1.7436	35.71	0.0488
C_5	0.01	72.15	0.7215	39.08	0.0185
C_6	0.01	86.17	0.8617	41.36	0.0208
C_{7+}	0.40	215.0	86.00	54.288**	1.586
			$m_t = 100.253$		$V_{C3+} = 1.7418$

** $\rho_{C_{7+}} = (0.87)(62.4) = 54.288$

Step 1. Calculate the weight % of C_1 in the entire system and the weight % of C_2 in the ethane-plus fraction from Equations 4-6 and 4-7.

$$(m_{C_1})_{C_1+} = \left[\frac{7.218}{100.253}\right] 100 = 7.2\%$$

$$(m_{C_2})_{C_2+} = \left[\frac{1.5035}{100.253 - 7.218}\right] 100 = 1.616\%$$

Step 2. Calculate the density of the propane and heavier by using Equation 4-8.

$$\rho_{C3+} = \frac{100.253 - 7.218 - 1.5035}{1.7418} = 52.55 \text{ lb/ft}^3$$

Step 3. Determine the density of the oil at standard conditions from Figure 4-1.

$$\Delta\rho_{sc} = 47.5 \text{ lb/ft}^3$$

Step 4. Correct for the pressure by using Figure 4-2.

$$\Delta\rho_p = 1.18 \text{ lb/ft}^3$$

Density of the oil at 4,000 psia and 60°F is then calculated by the expression

$$\rho_{p,60} = \rho_{sc} + \Delta\rho_p = 47.5 + 1.18 = 48.68 \text{ lb/ft}^3$$

Step 5. From Figure 4-3, determine the thermal expansion correction factor

$$\Delta\rho_T = 2.45 \text{ lb/ft}^3$$

Step 6. The required density at 4,000 psia and 160°F is

$$\rho_o = 48.68 - 2.45 = 46.23 \text{ lb/ft}^3$$

With the apparent need for a mathematical description of the Standing and Katz charts, Standing (1977) developed the following expressions for determining ρ_{sc}, $\Delta\rho_p$, and $\Delta\rho_T$.

The mathematical expression for ρ_{sc}

$$\rho_{sc} = \rho_{C_{2+}} [1 - 0.012 \, (m_{C_1})_{C_{1+}} - 0.000158 \, (m_{C_1})_{C_{1+}}^2]$$
$$+ 0.0133 \, (m_{C_1})_{C_{1+}} + 0.00058 \, (m_{C_1})_{C_{1+}}^2 \qquad (4\text{-}11)$$

with

$$\rho_{C_{2+}} = \rho_{C_{3+}} [1 - 0.01386 \, (m_{C_2})_{C_{2+}} - 0.000082 \, (m_{C_2})_{C_{2+}}^2]$$
$$+ 0.379 \, (m_{C_2})_{C_{2+}} + 0.0042 \, (m_{C_2})_{C_{2+}}^2 \qquad (4\text{-}12)$$

where $\rho_{C_{2+}}$ is the density of the ethane-plus fraction.

The mathematical expression for $\Delta\rho_p$

$$\Delta\rho_p = [0.167 + (16.181) \, 10^{-0.0425 \, \rho_{sc}}] \left| \frac{p}{1,000} \right|$$
$$- 0.01 \, [0.299 + (263) \, 10^{-0.0603 \, \rho_{sc}}] \left| \frac{p}{1,000} \right|^2 \qquad (4\text{-}13)$$

The mathematical expression for $\Delta\rho_T$

$$\Delta\rho_T = [0.0133 + 152.4 \, (\rho_{sc} + \Delta\rho_p)^{-2.45}] \, (T - 520)$$
$$- [8.1 \, (10^{-6}) - (0.0622) \, 10^{-0.764(\rho_{sc} + \Delta\rho_p)}] \, (T - 520)^2 \qquad (4\text{-}14)$$

where T is the temperature in °R.

To account for the effect of the presence of non-hydrocarbon components (H_2S, CO_2, and N_2) on the Standing and Katz crude oil density, Pedersen et al. (1984) proposed the following correction procedure:

Step 1. Calculate the density of ($H_2S + C_{3+}$) fraction at standard conditions from the following expression

$$\rho(H_2S + C_{3+}) = \frac{\displaystyle\sum_i (x_i \, MW_i)}{\displaystyle\sum \left| \frac{x_i \, MW_i}{\rho_{oi}} \right|} \text{ lb/ft}^3 \qquad (4\text{-}15)$$

where the index i refers to H_2S, C_3, and heavier components.

Step 2. Calculate the weight % of ethane in the ($H_2S + C_{2+}$) from the following relationship:

$$(m_{C_2})_{H_2S + C_{2+}} = \left[\frac{x_{C_2} \, MW_{C_2}}{\sum\limits_{i} (x_i \, MW_i)}\right] 100 \tag{4-16}$$

where MW_{C_2} = molecular weight of ethane
$\quad\quad x_{C_2}$ = mole fraction of ethane
$\quad\quad i$ = refers to H_2S, C_2, and heavier components

Step 3. Calculate the density of $(H_2S + C_{2+})$ fraction at standard conditions from the relationship

$$\rho_{(H_2S + C_{2+})} = \rho_{(H_2S + C_{3+})} - A_0 - A_1 \, a_1 - A_2 \, a_2 \tag{4-17}$$

where $A_0 = 0.1971 \, (m_{C_2})_{H_2S + C_{2+}}$
$\quad\quad A_1 = -0.1612 \, (m_{C_2})_{H_2S + C_{2+}}$
$\quad\quad A_2 = 0.0091 \, (m_{C_2})_{H_2S + C_{2+}}$
$\quad\quad a_1 = 3.3 - 0.0801 \, \rho_{(H_2S + C_{3+})}$
$\quad\quad a_2 = 1 + [0.24038 \, \rho_{(H_2S + C_{3+})} - 6.9]$
$\quad\quad\quad (0.0401 \, \rho_{(H_2S + C_{3+})} - 2.15)$

Step 4. Based on the additive volume concept and a CO_2 density of 51.26 lb/ft^3, the density of $(CO_2 + H_2S + C_{2+})$ fraction is calculated at standard conditions by using the additive volume concept and employing the density of $(H_2S + C_{2+})$ as calculated in the previous step, to give

$$\rho_{CO_2 + H_2S + C_{2+}} = \frac{x_{CO_2} \, MW_{CO_2} + \sum\limits_{i} x_i \, MW_i}{\dfrac{x_{CO_2} \, MW_{CO_2}}{\rho_{CO_2}} + \dfrac{\sum\limits_{i} x_i \, MW_i}{\rho_{H_2S + C_{2+}}}} \tag{4-18}$$

where the index i refers to H_2S, C_2, and heavier components

Step 5. Calculate the weight % of $C_1 + N_2$ in the total mixture from

$$(m_{C_1 + N_2})_{C_{1+}} \left[\frac{x_{C_1} \, MW_{C_1} + x_{N_2} \, MW_{N_2}}{\sum\limits_{i=1} x_i \, MW_i}\right] 100 \tag{4-19}$$

where $\quad\quad\quad\quad x_{C_1}$ = mole fraction of methane in the mixture
$\quad\quad\quad MW_{C_1}$ = molecular weight of methane
$\quad\quad\quad\quad x_{N_2}$ = mole fraction of nitrogen in the mixture
$\quad\quad\quad MW_{N_2}$ = molecular weight of nitrogen
$\quad\quad \Sigma_{i=1} \, x_i \, MW_i$ = total weight of the hydrocarbon system, lb/lb-mole

Step 6. Calculate the density of the crude oil at standard conditions, in pounds per cubic foot, from the following equation:

$$\rho_{sc} = \rho_{CO_2 + H_2S + C_{2+}} - B_0 - B_1\, b_1 \qquad (4\text{-}20)$$

where $B_0 = 5.507112 - 5.95976\, b_2 + 0.46195\, b_3 - 0.37627\, b_4$

$B_1 = 8.86573 - 9.37092\, b_2 + 0.41677\, b_3 + 0.07257\, b_4$

$b_1 = -0.65 + 0.01603\, \rho_{CO_2 + H_2S + C_{2+}}$

$b_2 = 1 - 0.1\, (m_{C_1 + N_2})_{C_{1+}}$

$b_3 = 1 + 0.015\, (m_{C_1 + N_2})^2_{C_{1+}} - 0.3\, (m_{C_1 + N_2})_{C_{1+}}$

$b_4 = 1 - 0.6\, (m_{C_1 + N_2})_{C_{1+}}$
$\qquad + 0.075\, (m_{C_1 + N_2})^2_{C_{1+}} - 0.0025\, (m_{C_1 + N_2})^3_{C_{1+}}$

Step 7. The density of the crude oil at standard conditions can be corrected for an increase in pressure and temperature by applying Equations 4-13 and 4-14 or by using Figures 4-2 and 4-3.

Example 4-2. Calculate the density of the crude oil with the composition given below at 3,000 psia and 120°F.

Component	x_i	Component	x_i
CO_2	0.07	C_3	0.04
N_2	0.03	C_4	0.03
H_2S	0.02	C_5	0.02
C_1	0.41	C_6	0.03
C_2	0.04	C_{7+}	0.31*

* Given: $MW_{C_{7+}} = 200$
$\quad \gamma_{C_{7+}} = 0.8702$

Solution.

Component	x_i	MW_i	$x_i\, MW_i$	ρ_{oi}	$V_i = x_i\, MW_i/\rho_{oi}$
CO_2	0.07	44.01	3.0808	51.26	0.0601
N_2	0.03	28.01	0.8403	—	—
H_2S	0.02	34.08	0.6816	49.30	0.0138
C_1	0.41	16.04	6.5764	—	—
C_2	0.04	30.07	1.2028	—	—
C_3	0.04	44.09	1.7636	31.64	0.0557
C_4	0.03	58.12	1.7436	35.71	0.04883
C_5	0.02	72.15	1.4430	39.08	0.0369
C_6	0.03	86.17	2.5851	41.36	0.0625
C_{7+}	0.31	200	62.00	54.3	1.1418
			$m_t = 81.9171$		1.41963

Step 1. Calculate the density of $(H_2S + C_{3+})$ fraction

$$\rho_{(H_2S + C_{3+})} = \frac{70.2169}{1.41963 - 0.0601} = 51.64 \text{ lb/ft}^3$$

Step 2. Calculate the weight % of C_2 in the $(H_2S + C_{2+})$ fraction from Equation 4-16.

$$(m_{C_2})_{H_2S + C_{2+}} = \left[\frac{1.2028}{71.4197}\right] 100 = 1.68\%$$

Step 3. Calculate the coefficients A_0, A_1, A_2, a_1, and a_2 as illustrated in Step 3 of the procedure.

$A_0 = 0.3311$ $\qquad A_1 = -0.2708$ $\qquad A_2 = 0.0153$

$a_1 = -.83636$ $\qquad a_2 = 0.56315$

Step 4. Calculate the density of $(H_2S + C_{2+})$ fraction by applying Equation 4-17.

$$\rho_{(H_2S + C_{2+})} = 51.64 - .3311 - (-0.2708)(-.83636)$$
$$- (0.0153)(0.56315) = 51.07 \text{ lb/ft}^3$$

Step 5. Calculate the density of $(CO_2 + H_2S + C_{2+})$ fraction by using Equation 4-18, to yield

$$\rho_{CO_2 + H_2S + C_{2+}} = \frac{3.0807 + 71.4197}{\dfrac{3.0807}{51.26} + \dfrac{71.4197}{51.07}} = 51.08 \text{ lb/ft}^3$$

Step 6. Determine the weight % of C_1 and N_2 in the entire system by applying Equation 4-19, to give

$$(m_{C_1 + N_2})_{C_{1+}} = 9.05\%$$

Step 7. Calculate the parameters b_1, b_2, b_3, b_4, B_0, and B_1.

$b_1 = 0.16881$ $\qquad b_2 = 0.095$ $\qquad b_3 = -0.4865$

$b_4 = -0.1404$ $\qquad B_0 = 4.769$ $\qquad B_1 = 7.7625$

Step 8. Calculate the density of the oil mixture of standard conditions by applying Equation 4-20.

$$\rho_{sc} = 51.08 - 4.769 - (7.7625)(0.16881) = 45.0 \text{ lb/ft}^3$$

Step 9. Determine the pressure correction from Figure 4-2.

$$\Delta\rho_p = 1.06 \text{ lb/ft}^3$$

Step 10. Determine the temperature correction from Figure 4-3.

$$\Delta\rho_T = 1.8 \text{ lb/ft}^3$$

Step 11. Calculate the density of the crude at 3,000 psia and 120°F by applying Equation 4-3.

$$\rho_o = 45.0 + 1.06 - 1.8 = 44.26 \text{ lb/ft}^3$$

The Alani-Kennedy Method

Alani and Kennedy (1960) developed an equation to determine the molal liquid volume V_m of pure hydrocarbons over a wide range of temperature and pressure. The equation was then adopted to apply to crude oils with the heavy hydrocarbons expressed as heptanes-plus fraction, i.e., C_{7+}.

The Alani-Kennedy equation is similar in form to the Van der Waals equation, which takes the following form:

$$V_m^3 - \left[\frac{R\,T}{p} + b\right] V_m^2 + \frac{a\,V_m}{p} - \frac{a\,b}{p} = 0 \tag{4-21}$$

where   R = gas constant, 10.73 psia ft³/lb-mole °R
         T = temperature, °R
         p = pressure, psia
         V_m = molal volume, ft³/lb-mole
         a, b = constants for pure substances

Alani and Kennedy considered the constants a and b as functions of temperature and proposed the expressions for calculating the two parameters.

$$a = K\,e^{n/T} \tag{4-22}$$

$$b = m\,T + c \tag{4-23}$$

where K, n, m, and c are constants for each pure component. Values of these constants are tabulated in Table 4-1.

Table 4-1 contains no constants from which the values of the parameters a and b for heptanes-plus can be calculated. Therefore, Alani and Kennedy proposed the following equations for determining a and b of C_{7+}.

$$\ln(a_{C_{7+}}) = 3.8405985\,(10^{-3})(MW)_{C_{7+}} - 9.5638281\,(10^{-4})\left.\left(\frac{MW}{\gamma}\right)\right|_{C_{7+}}$$

$$+ \frac{261.80818}{T} + 7.3104464\,(10^{-6})(MW)_{C_{7+}}^2 + 10.753517 \tag{4-24}$$

Table 4-1
Alani-Kennedy Constants

Component	K	n	$m \times 10^4$	c
C_1 70°–300°F	9,160.6413	61.893223	3.3162472	0.50874303
C_1 301°–460°F	147.47333	3,247.4533	– 14.072637	1.8326695
C_2 100°–249°F	46,709.573	– 404.48844	5.1520981	0.52239654
C_2 250°–460°F	17,495.343	34.163551	2.8201736	0.62309877
C_3	20,247.757	190.24420	2.1586448	0.90832519
$i\text{-}C_4$	32,204.420	131.63171	3.3862284	1.1013834
$n\text{-}C_4$	33,016.212	146.15445	2.902157	1.1168144
$i\text{-}C_5$	37,046.234	299.62630	2.1954785	1.4364289
$n\text{-}C_5$	37,046.234	299.62630	2.1954785	1.4364289
$n\text{-}C_6$	52,093.006	254.56097	3.6961858	1.5929406
H_2S^*	13,200.00	0	17.900	0.3945
N_2^*	4,300.00	2.293	4.490	0.3853
CO_2^*	8,166.00	126.00	1.8180	0.3872

* Values for non-hydrocarbon components as proposed by Lohrenz et al. (1964).

$$b_{C_{7+}} = 0.03499274 \, (MW)_{C_{7+}} - 7.2725403 \, (\gamma)_{C_{7+}} + 2.232395 \, (10^{-4})T$$

$$- 0.016322572 \left.\left|\frac{MW}{\gamma}\right|\right|_{C_{7+}} + 6.2256545 \qquad (4\text{-}25)$$

where $MW_{C_{7+}}$ = molecular weight of C_{7+}
$\gamma_{C_{7+}}$ = specific gravity of C_{7+}
$a_{C_{7+}}, \, b_{C_{7+}}$ = constants of the heptanes-plus fraction.

For hydrocarbon mixtures, the values of a and b are calculated using the following mixing rules

$$a_m = \sum_{i=1}^{C_{7+}} a_i \, x_i \qquad (4\text{-}26)$$

$$b_m = \sum_{i=1}^{C_{7+}} b_i \, x_i \qquad (4\text{-}27)$$

where a_i and b_i refer to pure hydrocarbons at existing temperature and x_i is their mole fraction in the mixture. The values a_m and b_m are then used in Equation 4-21 to solve for the molal volume, V_m. The density of the mixture at pressure and temperature of interest is determined from the following relationship

$$\rho_o = \frac{MW_a}{V_m} \tag{4-28}$$

where ρ_o = density of the crude oil, lb/ft^3
         MW_a = apparent molecular weight
         V_m = molal volume, ft^3/lb-mole

The Alani and Kennedy method for calculating the density of liquids is summarized in the following steps:

Step 1. Calculate the constants a and b for each pure component from Equations 4-22 and 4-23, respectively.
Step 2. Determine $a_{C_{7+}}$ and $b_{C_{7+}}$ by applying Equations 4-24 and 4-25.
Step 3. Solve for a_m and b_m from Equations 4-26 and 4-27.
Step 4. Calculate the molal volume V_m by solving Equation 4-21 for the smallest real root.
Step 5. Compute the apparent molecular weight, MW_a.
Step 6. Determine the density of the crude oil by using Equation 4-28. The average absolute deviation is 1.6% with a maximum error of 4.9%.

Example 4-3. A crude oil system has the composition:

Component	x_i
CO_2	0.0008
N_2	0.0164
C_1	0.2840
C_2	0.0716
C_3	0.1048
i-C_4	0.042
n-C_4	0.042
i-C_5	0.0191
n-C_5	0.0191
C_6	0.0405
C_{7+}	0.3597

Given the following additional data

$MW_{C_{7+}}$ = 252
$\gamma_{C_{7+}}$ = 0.8424
pressure = 1708.7 psia
temperature = 591°R

Calculate the density of the crude oil.

Solution.

Step 1. Calculate the parameters $a_{C_{7+}}$ and $b_{C_{7+}}$ from Equations 4-24 and 4-25, to give

$$a_{C_{7+}} = 229269.9$$

$$b_{C_{7+}} = 4.165811$$

Step 2. Calculate the mixture parameters a_m and b_m from Equations 4-26 and 4-27.

$$a_m = 99111.71$$

$$b_m = 2.119383$$

Step 3. Solve Equation 4-21 for the molal volume.

$$V_m = 2.528417$$

Step 4. Determine the apparent molecular weight of this mixture.

$$MW_a = 113.5102$$

Step 5. Compute the density of the oil system by using Equation 4-28.

$$\rho_o = \frac{113.5102}{2.528417} = 44.896 \text{ lb/ft}^3$$

METHODS FOR DETERMINING DENSITY OF LIQUIDS OF UNKNOWN COMPOSITION

Several empirical correlations for calculating the density of liquids of unknown compositional analysis have been proposed. The correlations employ limited PVT data such as gas gravity, oil gravity, and gas solubility as correlating parameters to estimate liquid density at the prevailing reservoir pressure and temperature.

Katz's Method

The density, in general, can be defined as the mass of a unit volume of material at a specified temperature and pressure. Accordingly, the density of a saturated crude oil at standard conditions can be defined mathematically by the following relationship:

$$\rho_{sc} = \frac{\text{weight of stock tank oil} + \text{weight of solution gas}}{\text{volume of stock tank oil} + \text{increase in stock tank volume due to solution gas}}$$

or,

$$\rho_{sc} = \frac{m_o + m_g}{(V_o)_{sc} + (\Delta V_o)_{sc}}$$

where ρ_{sc} = density of the oil at standard conditions, lb/ft³
         $(V_o)_{sc}$ = volume of oil at standard conditions, ft³/STB
         m_o = total weight of one stock-tank barrel of oil, lb/STB
         m_g = weight of the solution gas, lb/STB
         $(\Delta V_o)_{sc}$ = increase in stock-tank oil volume due to solution gas, ft³/STB

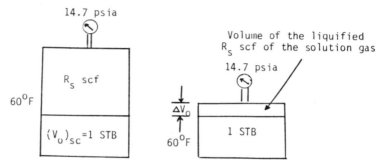

Figure 4-4. Schematic illustration of the Katz's density model at standard conditions.

The procedure of calculating the density at standard conditions is illustrated schematically in Figure 4-4. Katz (1942) expressed the density of crude oil at standard conditions with the following relationship:

$$\rho_{sc} = \frac{m_o + m_g}{(V_o)_{sc} + \dfrac{mg}{\rho_{ga}}} \tag{4-29}$$

where ρ_{ga}, as introduced by Katz, is the apparent density of the liquefied dissolved gas at 60°F and 14.7 psia. Katz correlated the apparent gas density, in lb/ft³, with the specific gravity, the solution gas, and the API gravity of the stock-tank oil as shown graphically in Figure 4-5. The proposed method does not require the composition of the crude oil. The only required properties are the gas gravity, the API gravity, and the gas solubility.
To arrive at the final expression for calculating ρ_{sc}, let:

R_s = gas solubility, scf/STB
γ_g = gas gravity
γ_o = oil gravity

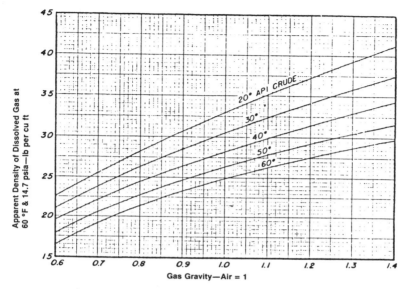

Figure 4-5. Katz's apparent density of dissolved gas. Courtesy of the American Petroleum Institute, Katz, D., *Drilling and Production Practice*, 1942, p. 137.

The weights of the solution gas and the stock-tank oil can be determined in terms of the above defined variables by the following relationships:

$$m_g = \frac{R_s}{379.4}(28.96)(\gamma_g), \text{ lb of solution gas/STB}$$

$$m_o = (5.615)(62.4)(\gamma_o), \text{ lb of oil/STB}$$

Substituting the above terms into Equation 4-29 yields

$$\rho_{sc} = \frac{(5.615)(62.4)(\gamma_o) + \left(\dfrac{R_s}{379.4}\right)(28.96)(\gamma_g)}{5.615 + \left(\dfrac{R_s}{379.4}\right)(28.96)(\gamma_g/\rho_{ga})}$$

or,

$$\rho_{sc} = \frac{350.376\,\gamma_o + \left(\dfrac{R_s\,\gamma_g}{13.1}\right)}{5.615 + \left(\dfrac{R_s\,\gamma_g}{13.1\,\rho_{ga}}\right)} \tag{4-30}$$

The pressure correction adjustment, $\Delta\rho_p$, and the thermal expansion adjustment, $\Delta\rho_T$, for the calculated ρ_{sc} can be made by using Figures 4-2 and 4-3, respectively.

Standing (1981) showed that the apparent liquid density of the dissolved gas as represented by Katz's chart can be closely approximated by the following relationship:

$$\rho_{ga} = (38.52)\ 10^{(-0.00326\ API)} + [94.75 - 33.93\ Log\ (API)]Log(\gamma_g) \quad (4\text{-}31)$$

Example 4-4. A crude oil at its bubble point pressure of 4,000 psia and a temperature of 180°F has an API gravity of 50°, if the gas solubility at the bubble-point pressure and system temperature is 650 scf/STB and the specific gravity of solution gas is 0.7. Calculate the oil density at the specified pressure and temperature by using Katz's method.

Solution.

Step 1. From Figure 4-4, determine the apparent density of dissolved gas.

$$\rho_{ga} = 20.5\ lb/ft^3$$

Step 2. Calculate the stock-tank liquid gravity from Equation 4-2.

$$\gamma_o = \frac{141.5}{API + 131.5} = \frac{141.5}{50 + 131.5} = 0.7796$$

Step 3. Apply Equation 4-30 to calculate the liquid density at standard conditions.

$$\rho_{sc} = \frac{(350.376)(0.7796) + \dfrac{(650)(0.7)}{13.1}}{5.615 + \dfrac{(650)(0.7)}{(13.1)(20.5)}} = 42.12\ lb/ft^3$$

Step 4. Determine the pressure correction factor from Figure 4-2.

$$\Delta\rho_p = 1.55\ lb/ft^3$$

Step 5. Adjust the oil density, as calculated at standard conditions, to reservoir pressure.

$$\rho_{p,60°F} = 42.12 + 1.55 = 43.67\ lb/ft^3$$

Step 6. Determine the isothermal adjustment factor from Figure 4-1.

$$\Delta\rho_T = 3.25\ lb/ft^3$$

Step 7. Calculate the oil density at 4,000 psia and 180°F.

$$\rho_o = 43.67 - 3.25 = 40.42\ lb/ft^3$$

Standing's Method

Standing (1981) proposed an empirical correlation for estimating the oil formation volume factor as a function of the gas solubility R_s, the specific gravity of stock-tank oil γ_o, the specific gravity of solution gas γ_g, and the system temperature T. By coupling the mathematical definition of the oil formation volume factor (as discussed in a later section) with Standing's correlation, the density of a crude oil at a specified pressure and temperature can be calculated from the following expression:

$$\rho_o = \frac{62.4\ \gamma_o + 0.0136R_s\ \gamma_g}{0.972 + 0.000147\left[R_s \left(\frac{\gamma_g}{\gamma_o}\right)^{.5} + 1.25(T - 460)\right]^{1.175}} \tag{4-32}$$

where T = system temperature, °R
γ_o = specific gravity of stock-tank oil

Example 4-5. Rework Example 4-4 and solve for the density by using Standing's Correlation.

Solution. From Equation 4-32:

$$\rho_o = \frac{62.4(.7796) + 0.0136(650)(0.7)}{0.972 + 0.000147\left[650 \left(\frac{.7}{.7796}\right)^{.5} + 1.25(180)\right]^{1.175}} = 39.92\ \text{lb/ft}^3$$

Ahmed's Correlation

Ahmed (1985) developed a correlation for estimating the crude oil density at standard conditions. The correlation is based on calculating the apparent molecular weight of the oil from the readily available PVT on the hydrocarbon system. Ahmed expressed the apparent molecular weight of the crude by the following relationship:

$$MW_a = \frac{0.0763\ R_s\ \gamma_g\ MW_{st} + 350.376\ \gamma_o\ MW_{st}}{0.0026537\ R_s\ MW_{st} + 350.376\ \gamma_o} \tag{4-33}$$

where MW_a = apparent molecular weight of the oil
MW_{st} = molecular weight of the stock-tank oil and can be taken as the molecular weight of the heptanes-plus fraction.
γ_o = specific gravity of the stock-tank oil or the C_{7+} fraction

The density of the oil at standard conditions can then be determined from the expression

$$\rho_{sc} = \frac{0.0763 \, R_s \, \gamma_g + 350.376 \, \gamma_o}{0.0026537 \, R_s + \gamma_o \left(5.615 + \dfrac{199.71432}{MW_{st}}\right)} \qquad (4\text{-}34)$$

If the molecular weight of the stock-tank oil is not available, the density of the oil at standard conditions can be estimated from the following equation:

$$\rho_{sc} = \frac{0.0763 \, R_s \, \gamma_g + 350.4 \, \gamma_o}{0.0027 \, R_s + 2.4893 \, \gamma_o + 3.491} \qquad (4\text{-}35)$$

Example 4-6. Using the data given in Example 4-4, calculate the density of the crude oil by using Equation 4-35.

Solution.

Step 1. Calculate the density of the crude oil at standard conditions from Equation 4-35.

$$\rho_{sc} = \frac{0.0763(650)(0.7) + 350.4(0.7796)}{0.0027(650) + 2.4893(0.7796) + 3.491} = 42.8 \; \text{lb/ft}^3$$

Step 2. Determine $\Delta\rho_p$ from Figure 4-2.

$\Delta\rho_p = 1.5 \; \text{lb/ft}^3$

Step 3. Determine $\Delta\rho_T$ from Figure 4-3.

$\Delta\rho_T = 3.6 \; \text{lb/ft}^3$

Step 4. Calculate ρ_o at 4,000 psia and 180°F.

$\rho_o = 42.8 + 1.5 - 3.6 = 40.7 \; \text{lb/ft}^3$

ISOTHERMAL COMPRESSIBILITY COEFFICIENT
OF UNDERSATURATED CRUDE OILS

Isothermal compressibility coefficients are required in solving many reservoir engineering problems, including transient fluid flow problems, and also they are required in the determination of the physical properties of the undersaturated crude oil.

By definition, the isothermal compressibility of a substance is defined mathematically by the following expression:

$$C = -\frac{1}{V}\left(\frac{\partial V}{\partial p}\right)_T$$

For a crude oil system, the isothermal compressibility coefficient is given by the following equation:

$$C_o = -\frac{1}{V}\left(\frac{\partial V}{\partial p}\right)_T \tag{4-36}$$

where C_o = isothermal compressibility of the crude oil, psi^{-1}
    $(\partial V/\partial p)_T$ = slope of the isothermal pressure-volume curve.

According to Equation 4-36, the isothermal compressibility coefficient is defined as the rate of change in volume with pressure increase per unit volume of liquid, all variables other than pressure being constant.

Generally, isothermal compressibility coefficients of an undersaturated crude oil are determined from a laboratory PVT study. A sample of the crude oil is placed in a PVT cell at the reservoir temperature and at a pressure greater than the bubble-point pressure of the crude oil. At these initial conditions, the reservoir fluid exists as a single-phase liquid. The volume of the oil is allowed to expand as its pressure declines. This volume is recorded and plotted as a function of pressures. A typical pressure-volume relationship is shown in Figure 4-6. If the experimental pressure-volume diagram for the oil is available, C_o can be calculated by graphically determining the volume V and the slope $(\partial V/\partial p)_T$ and then applying these values in Equation 4-36.

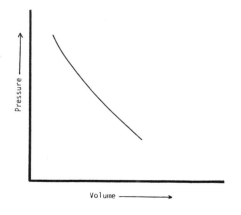

Figure 4-6. A typical p-V diagram for a crude oil system.

Methods of Calculating C_o

1. Trube's Correlation. Trube (1957) presented a correlation for calculating the isothermal pseudo-reduced compressibility C_r of undersaturated crude oils. Trube correlated this property graphically with the pseudo-reduced pressure and temperature, p_{pr} and T_{pr}, as shown in Figure 4-7. The author defined C_r mathematically by the following expression:

$$C_r = C_o \cdot p_{pc} \tag{4-37}$$

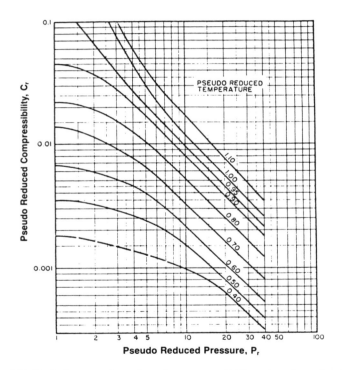

Figure 4-7. Trube's pseudo-reduced compressibility of undersaturated crude oils. Permission to publish by the Society of Petroleum Engineers of AIME. Copyright SPE-AIME.

Additionally, Trube presented graphical correlations, as shown in Figures 4-8 and 4-9, to estimate the pseudo-critical properties of crude oils. The calculation procedure of the proposed method is summarized in the following steps:

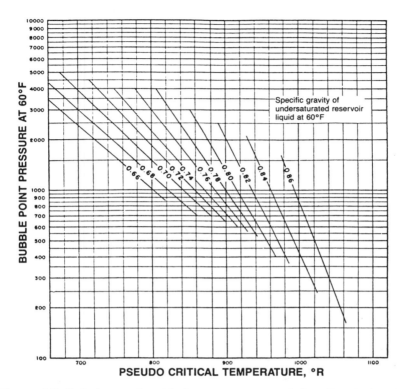

Figure 4-8. Trube's pseudo-critical temperature correlation. Permission to publish by the Society of Petroleum Engineers of AIME. Copyright SPE-AIME.

Step 1. From the bottom-hole pressure measurements and pressure-gradient data, calculate the specific gravity of the reservoir oil from the following expression:

$$(\gamma_o)_T = \frac{dp/dh}{0.433} \tag{4-38}$$

where $(\gamma_o)_T$ = specific gravity at reservoir pressure and temperature T

dp/dh = pressure gradient as obtained from a pressure buildup test, psi/ft

Step 2. Adjust the calculated specific gravity (Equation 4-38) to its value at 60°F by using the following equation.

$$(\gamma_o)_{60} = (\gamma_o)_T + 0.00046 \, (T - 520) \tag{4-39}$$

where $(\gamma_o)_{60}$ = adjusted specific gravity to 60°F
       T = reservoir temperature, °R

Step 3. Determine the bubble-point pressure p_b of the crude oil at reservoir temperature. The following convenient correlation, as proposed by Standing (1981), can be used to estimate p_b if the experimental value is not known.

$$p_b = 18.2 \left[\left(\frac{R_s}{\gamma_g} \right)^{0.83} \left(\frac{10^{0.00091(T-460)}}{10^{0.0125°API}} \right) - 1.4 \right] \quad (4\text{-}40)$$

where R_s = gas solubility, scf/STB
          γ_g = specific gravity of solution gas
          T = temperature, °R
          °API = stock-tank oil API gravity

Step 4. Correct the calculated bubble-point pressure p_b at reservoir temperature to its value at 60°F by using the following equation as proposed by Standing (1962):

$$(p_b)_{60} = \frac{1.134 \, p_b}{10^{0.00091(T-460)}} \quad (4\text{-}41)$$

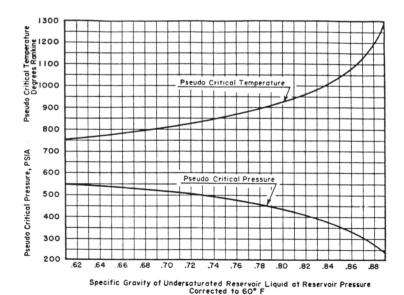

Figure 4-9. Trube's pseudo-critical properties correlation. Permission to publish by the Society of Petroleum Engineers of AIME. Copyright SPE-AIME.

where $(p_b)_{60}$ = bubble-point pressure at 60°F, psi

p_b = bubble-point pressure at reservoir temperature, psia

T = reservoir temperature, °R

Step 5. Enter in Figure 4-8 the values of $(p_b)_{60}$ and $(\gamma_o)_{60}$ and determine the pseudo-critical temperature, T_{pc}, of the crude.

Step 6. Enter the value of T_{pc} in Figure 4-9 and determine the pseudo-critical pressure p_{pc} of the crude.

Step 7. Calculate the pseudo-reduced pressure p_{pr} and temperature T_{pr} from the following relationships:

$$T_{pr} = \frac{T}{T_{pc}} \tag{4-42}$$

$$p_{pr} = \frac{p}{p_{pc}} \tag{4-43}$$

Step 8. Determine C_r by entering Figure 4-7 with the values of T_{pr} and p_{pr}.

Step 9. Calculate C_o from Equation 4-37.

$$C_o = \frac{C_r}{p_{pc}}$$

Trube did not specify the data used to develop the correlation nor did he allude to their accuracy, although the examples presented in his paper showed an average absolute error of 7.9% between calculated and measured values.

Trube's correlation can be best illustrated through the following example:

Example 4-7. Given the following data,
Oil gravity = 45°
Gas solubility = 600 scf/STB
Solution gas gravity = 0.8
Reservoir temperature = 212°F
Reservoir pressure = 2,000 psia
Pressure gradient of reservoir liquid at 2,000 psia and 212°F = 0.32 psi/ft
Find C_o at 2,000, 3,000, and 4,000 psia.

Solution.

Step 1. Determine $(\gamma_o)_T$ from Equation 4-38.

$$(\gamma_o)_T = \frac{0.32}{0.433} = 0.739$$

Step 2. Correct the calculated oil specific gravity to its value at 60°F by applying Equation 4-39.

$$(\gamma_o)_{60} = 0.739 + 0.00046\,(152) = 0.8089$$

Step 3. Calculate the bubble-point pressure from Standing's correlation (Equation 4-40).

$$p_b = 18.2\left[\left(\frac{600}{0.8}\right)^{0.83}\frac{10^{0.00091(T-212)}}{10^{0.0125(45)}} - 1.4\right] = 1{,}866\ \text{psia}$$

Step 4. Adjust p_b to its value at 60°F by applying Equation 4-41.

$$(p_b)_{60} = \frac{1.134\,(1{,}866)}{10^{0.00091(212)}} = 1{,}357.1\ \text{psia}$$

Step 5. Estimate the pseudo-critical temperature T_{pc} of the crude oil from Figure 4-8, to yield

$$T_{pc} = 840°\text{R}$$

Step 6. Estimate the pseudo-critical pressure p_{pc} of the crude oil from Figure 4-9, to give

$$p_{pc} = 500\ \text{psia}$$

Step 7. Calculate the pseudo-reduced temperature from Equation 4-42.

$$T_{pr} = \frac{672}{840} = 0.8$$

Step 8. Calculate the pseudo-reduced pressure p_{pr}, the pseudo-reduced isothermal compressibility coefficient C_r, and the isothermal compressibility coefficient of the oil from Equation 4-43, Figure 4-7, and Equation 4-37, respectively.

p	p_{pr}	C_r	C_o, 10^{-6} psia^{-1}
2,000	4	0.0125	25.0
3,000	6	0.0089	17.8
4,000	8	0.0065	13.00

2. Vasquez-Beggs' Correlation. From a total of 4,036 experimental data points used in a linear regression model, Vasquez and Beggs (1980) correlated the isothermal oil compressibility coefficients with R_s, T, °API, γ_g, and p. They proposed the following expression.

$$C_o = \frac{-1{,}433 + 5R_s + 17.2(T-460) - 1{,}180\,\gamma_{gs} + 12.61°\text{API}}{10^5\ p} \qquad (4\text{-}43)$$

Realizing that the value of the specific gravity of the gas depends on the conditions under which it is separated from the oil, Vasquez and Beggs proposed that the value of the gas specific gravity as obtained from a separator pressure of 100 psig be used in the above equation. This reference pressure was chosen because it represents the average field separator conditions. The authors proposed the following relationship for adjustment of the gas gravity γ_g to the reference separator pressure.

$$\gamma_{gs} = \gamma_g \left[1 + 5.912 \ (10^{-5})(\gamma_o)(T_{sep} - 460) \ \text{Log}\left(\frac{p_{sep}}{114.7}\right) \right] \tag{4-44}$$

where γ_{gs} = gas gravity at the reference separator pressure
γ_g = gas gravity at the actual separator conditions of p_{sep} and T_{sep}
p_{sep} = actual separator pressure, psia
T_{sep} = actual separator temperature, °R

Example 4-8. Rework Example 4-7 by using the Vasquez and Beggs' correlation and assuming the average field separators pressure of 114.7 psia.

Solution.

Step 1. Solve for the gas specific gravity at the reference pressure by applying Equation 4-44.

$$\gamma_{gs} = 0.80 \ [1 + 0] = 0.8$$

Step 2. Calculate C_o from Equation 4-43.

$$C_o = \frac{-1,433 + 5(600) + 17.2(212) - 1,180(0.8) + 12.61(45)}{10^5 \ p}$$

$$C_o = 10^{-5} \ \frac{4,836.85}{p}$$

p	C_o, 10^{-6} psia^{-1}
2,000	24.184
3,000	16.123
4,000	12.092

3. Ahmed's Correlation. Based on 245 experimental data points for the isothermal compressibility coefficients, Ahmed (1985) developed a mathematical expression for estimating C_o by using a non-linear regression model. The proposed correlation uses the gas solubility R_s and the pressure as the only correlating parameters. It should be noted the other correlating parameters such as γ_o, γ_g, and T are implemented in the equation through the gas solubility R_s. The correlation has the following simplified form.

$$C_o = \frac{1}{a_1 + a_2\ R_s}\ \text{EXP}\ (a_3\ p) \tag{4-45}$$

where $a_1 = 24{,}841.0822$
$\quad\quad a_2 = 14.07428745$
$\quad\quad a_3 = -0.00018473$

The proposed relationship produced an average absolute error of 3.9% when tested against the experimental data used in developing the equation.

The isothermal compressibility coefficient can also be determined from the following expression:

$$C_o = \frac{a_1 + a_2\left[R_s\left(\frac{\gamma_g}{\gamma_o}\right)^{.5} + 1.25(T-460)\right]^{1.175}}{a_4\ \gamma_o + a_5\ R_s\ \gamma_g}\ \text{EXP}\ (a_3\ p) \tag{4-46}$$

where $a_1 = 1.026638$
$\quad\quad a_2 = 0.0001553$
$\quad\quad a_3 = -0.0001847272$
$\quad\quad a_4 = 62{,}400$
$\quad\quad a_5 = 13.6$

The above correlation was developed by correlating the isothermal compressibility coefficients with the oil density at the bubble-point pressure, as given by Equation 4-31.

Example 4-9. The data given in Example 4-7 are listed here again for convenience.

$R_s = 600$ scf/STB
$\gamma_g = 0.8$
$\gamma_o = 0.802$
$T = 212°F$

Using Equations 4-45 and 4-46, calculate C_o at 2,000, 3,000, and 4,000 psia.

Solution.

• From Equation 4-45, solve for C_o

$$C_o = \frac{\text{EXP}\ (-0.00018473\ p)}{24{,}841.0822 + 14.07428745\ (600)}$$

$$= 30.0429(10^{-6})\ \text{EXP}\ (-0.00018473\ p)$$

p	C_o, 10^{-6} psi^{-1}
2,000	20.763
3,000	17.26
4,000	14.35

- Applying Equation 4-46 yields

  $C_o = 0.00002589 \ EXP \ (- 0.0001847272p)$

- Substituting for pressure yields

p	C_o, 10^{-6} psi^{-1}
2,000	17.9
3,000	14.88
4,000	12.37

DENSITY OF UNDERSATURATED CRUDE OILS

Figure 4-10 shows a typical liquid density-pressure diagram. As the pressure increases, the gas dissolves in the crude oil and the density decreases. The oil density will continue to decrease until the bubble-point pressure is reached.

A further increase in pressure will cause an increase in density due to compression of the crude oil.

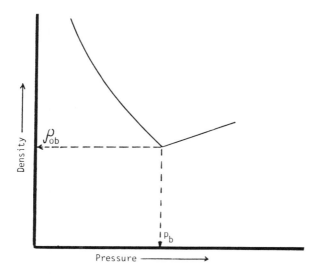

Figure 4-10. A typical liquid density-pressure diagram.

To account for the compression of the oil above the bubble-point pressure, the density of the crude is first calculated at the bubble-point pressure and reservoir temperature. The calculated density is then adjusted by using the isothermal compressibility coefficient as described below.

From the mathematical definition of the density, the volume of m pounds of crude oil with a density of ρ_o is calculated as

$$V = \frac{m}{\rho_o}$$

Differentiating this expression with respect to the pressure yields:

$$\left(\frac{\partial V}{\partial p}\right)_T = \frac{-m}{\rho_o}\frac{\partial \rho_o}{\partial p}$$

Substituting the above relationship into Equation 4-36 gives:

$$C_o = \frac{-\rho_o}{m}\left[\frac{-m}{\rho_o^2}\frac{\partial \rho_o}{\partial p}\right]$$

or

$$C_o = \frac{1}{\rho_o}\frac{\partial \rho_o}{\partial p}$$

Integrating the above expression yields

$$\int_{p_b}^{p} C_o \, dp = \int_{ob}^{\rho_o} \frac{d\,\rho_o}{\rho_o} \tag{4-47}$$

Evaluating C_o at the average pressure, $\bar{p} = p + p_b/2$, gives

$$C_o (p - p_b) = Ln \left|\frac{\rho_o}{\rho_{ob}}\right|$$

or

$$\rho_o = \rho_{ob} \, EXP \, [C_o (p - p_b)] \tag{4-48}$$

where ρ_o = density of the oil at pressure p, lb/ft^3
ρ_{ob} = density of the oil at the bubble-point pressure, lb/ft^3
C_o = isothermal compressibility coefficient at average pressure, psi^{-1}

p = reservoir pressure, psia
p_b = bubble-point pressure, psia
$\bar{p}$ = average pressure, psia

Vasquez-Beggs' isothermal compressibility expression (Equation 4-43), or Ahmed's C_o expressions (Equations 4-45 and 4-46), can be incorporated in Equation 4-47 to give:

For Vasquez-Beggs' C_o equation:

$$\rho_o = \rho_{ob} \ \text{EXP} \left[A \ \text{Ln} \left(\frac{p}{p_b} \right) \right] \qquad (4\text{-}49)$$

where $A = 10^{-5} [- 1{,}433 + 5 \ R_s + 17.2(T - 460) - 1{,}180 \ \gamma_{gs} + 12.61°API]$

For Ahmed's C_o expressions: Using Equation 4-45 gives

$$\rho_o = \rho_{ob} \ \text{EXP} [B \ (\text{EXP} \ (ap) - \text{EXP} \ (ap_b))] \qquad (4\text{-}50)$$

where $B = - (4.588893 + 0.0025999 \ R_s)^{-1}$
$a = - 0.00018473$

Using Equation 4-46 gives

$$\rho_o = \rho_{ob} \ \text{EXP} [D \ (\text{EXP} \ (ep) - \text{EXP} \ (ep_b))] \qquad (4\text{-}51)$$

where $D = \dfrac{a_1 + a_2 \left[R_s \left(\dfrac{\gamma_g}{\gamma_o} \right)^{.5} + 1.25 \ (T - 460) \right]^{1.175}}{a_4 \ \gamma_o + a_5 \ R_s \ \gamma_g}$

$e = - 0.0001847272$
$a_1 = 1.026638$
$a_2 = 0.0001553$
$a_3 = - 0.0001847272$
$a_4 = - 11.526938$
$a_5 = - 0.00251229$

Example 4-10. Data given in Examples 4-7 through 4-9 are summarized below:

$p_b = 1{,}866$ psia   API $= 45°$ $\gamma_o = 0.802$ $\gamma_g = 0.8$
$T = 212°F$ $p_{sep} = 114.7$ psia $R_s = 600$ scf/STB

Calculate the oil density 2,000, 3,000, and 4,000 psia by using

a. Equation 4-48
b. Equation 4-49
c. Equation 4-50
d. Equation 4-51

Solution. Using the available data, the oil density at the bubble-point pressure is calculated by applying Equation 4-32, to give

$$\rho_{ob} = 40.788 \text{ lb/ft}^3$$

a. Solution by applying Equation 4-48. From Example 4-7

$T_{pr} = 0.80$
$p_{pc} = 500$ psia
$p_b = 1,866$

p	$\bar{p}$	p_{pr}	C_r Figure 4-7	C_o 10^{-6}, psi^{-1}	ρ_o, lb/ft^3 from Equation 4-47
2,000	1,933*	3.866	0.013	26.00	41.06
3,000	2,433	4.866	0.0115	23.00	41.98
4,000	2,933	5.866	0.0092	18.4	42.52

* At average pressure, i.e., $\bar{p} = (p + p_b)/2$.

b. Solution by using Equation 4-49

- Calculate the coefficient A of Equation 4-49, to give

    A = 0.048369

- Solve for the density by applying Equation 4-49

p psia	ρ_o, lb/ft^3 from Equation 4-50
2,000	41.06
3,000	41.87
4,000	42.46

c. Solution by applying Equation 4-50

- Calculate the coefficient B of Equation 4-50, to give

    B = 0.162632

- Solve for the density by applying Equation 4-50

p psia	ρ_o, lb/ft^3 (Equation 4-50)
2,000	40.90
3,000	41.59
4,000	42.35

d. Solution by using Equation 4-51

- Calculate the coefficient D of Equation 4-51, to give

    D = 0.1401739

- Solve for the density from Equation 4-51.

p psia	ρ_o, lb/ft^3 (Equation 4-51)
2,000	40.98
3,000	41.66
4,000	42.23

GAS SOLUBILITY

The gas solubility R_s is defined as the number of standard cubic feet of gas which will dissolve in one stock-tank barrel of crude oil at certain pressure and temperature. The solubility of a natural gas in a crude oil is a strong function of the pressure, the temperature, the API gravity, and the gas gravity. The effect of these complex variables on gas solubility is shown in Figures 4-10 through 4-13.

For a particular gas and crude oil to exist at a constant temperature, the solubility increases with pressure until the saturation pressure is reached. At the saturation pressure (bubble-point pressure) all the available gases are dissolved in the oil and the gas solubility reaches its maximum value. Rather than measuring the amount of gas that will dissolve in a given stock-tank crude oil as the pressure is increased, it is customary to determine the amount of gas that will come out of a sample of reservoir crude oil as pressure decreases.

A typical gas solubility curve, as a function of pressure for an undersaturated crude oil, is shown in Figure 4-14. As the pressure is reduced from the initial reservoir pressure p_i, to the bubble-point pressure p_b, no gas evolves from the oil and consequently the gas solubility remains constant at its maximum value of R_{sb}. Below the bubble-point pressure, the solution gas is liberated and the value of R_s decreases with pressure.

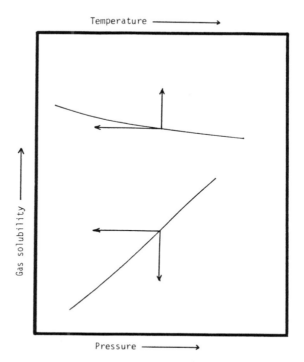

Figure 4-11. R_s vs. p and R_s vs. T relationships.

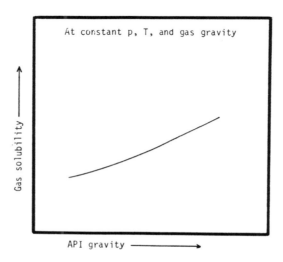

Figure 4-12. API gravity-R_s relationship.

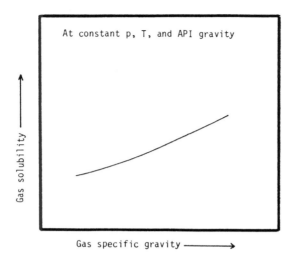

Figure 4-13. Gas gravity-gas solubility relationship.

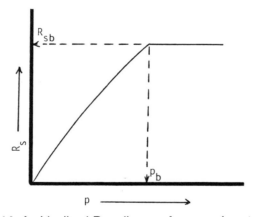

Figure 4-14. An idealized R_s-p diagram for an undersaturated oil.

In determining the PVT relationships (including the gas solubility-pressure relationship) in the laboratory, it is necessary to record the volume of oil and volume of liberated gas as the pressure is reduced below saturation pressure. The manner in which the solution gas is liberated from the oil will significantly affect all the PVT relationships. There are two types of separation (liberation, vaporization) process, namely:

- Flash liberation
- Differential liberation

Flash Liberation

In the flash liberation process, the gas which is liberated from the oil during a pressure decline remains in contact with the oil from which it was liberated. The process, as shown schematically in Figure 4-15, involves the following steps:

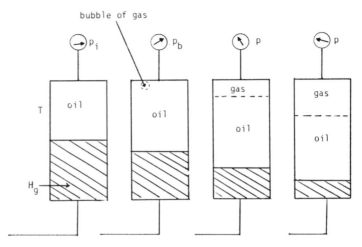

Figure 4-15. A schematic diagram of the flash liberation test.

Step 1. The reservoir fluid sample is charged to a PVT cell which is maintained at reservoir temperature throughout the experiments.

Step 2. The cell pressure is elevated at a pressure higher than the saturation pressure. This can be achieved by injecting mercury (or forcing a piston) into the cell.

Step 3. The cell pressure is lowered in small increments by withdrawing mercury from the cell. The total volume of the hydrocarbon system is recorded at each pressure.

Step 4. A plot of the cell pressure-total hydrocarbon volume is constructed as shown in Figure 4-16.

Step 5. When the cell pressure reaches the bubble-point pressure of the hydrocarbon system, a sign of formation of a gas phase is noted. This stage is marked by a sharp change in the pressure-volume slope (Figure 4-16).

Step 6. As the pressure level is reduced below the bubble-point pressure, the liberated gas is allowed to remain in contact and reach an equilibrium state with the oil phase. This thermodynamic equilibrium is assured by agitating the cell.

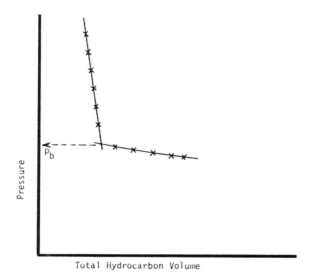

Figure 4-16. Flash liberation p-V diagram.

Step 7. The equilibrium pressure level and the corresponding hydrocarbon total volume is recorded.

Step 8. Steps 6 and 7 are repeated until the capacity of the cell is reached.

The experimental data obtained from the flash liberation test include:

a. The bubble-point pressure
b. The isothermal compressibility coefficient of the liquid phase above the bubble-point pressure
c. Below the bubble point, the two-phase volume is measured as a function of pressure

It should be noted that during the flash liberation process, the overall system composition remains constant because no gas is removed from the PVT cell during pressure reductions. The foregoing process simulates the gas liberation sequence, which is taking place in the reservoir at pressures immediately below the bubble-point pressure. This can be justified by the fact that the liberated gas remains immobile in the pores and in contact with oil until the critical gas saturation is reached at a certain pressure below p_b.

Dodson et al. (1953) pointed out that the flash liberation process best represents the separator type liberation. When entering the separator, the reservoir fluids are in equilibrium due to the agitation occurring in the tubing. In the separator, the two phases are brought to equilibrium and the oil and gas are separated. This behavior follows the flash liberation sequence.

Differential Liberation

In the differential liberation process, the solution gas that is liberated from an oil sample during a decline in pressure is continuously removed from contact with the oil, and before establishing equilibrium with the liquid phase. This type of liberation, as presented schematically in Figure 4-17, is characterized by a varying composition of the total hydrocarbon system.

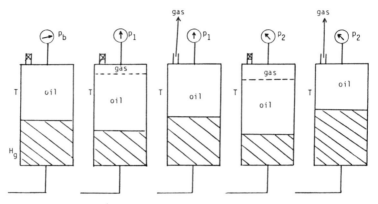

Figure 4-17. A schematic diagram of the differential liberation test.

The experimental procedure of the test is summarized in the following steps:

Step 1. The reservoir fluid sample is placed in a PVT cell at reservoir temperature.

Step 2. The cell is pressurized to saturation by injection of mercury.

Step 3. The volume of the all-liquid sample is recorded.

Step 4. The cell pressure is lowered by removing mercury from the cell.

Step 5. The liberated gas is removed from the cell through the cell flow valve. During this process, the cell pressure is kept constant by reinjecting mercury in the cell at the same rate as the gas discharge rate.

Step 6. The volume of the discharged gas is measured at standard conditions and the volume of the remaining oil is recorded.

Step 7. Steps 5 and 6 are repeated until the cell pressure is lowered to atmospheric pressure.

Step 8. The remaining oil at atmospheric pressure is measured and converted to a volume at 60°F. This final volume is referred to as the residual oil.

The experimental data obtained from the test include:

a. Amount of gas in solution as a function of pressure
b. The shrinkage in the oil volume as a function of pressure
c. Properties of the evolved gas including the composition of the liberated gas, the gas compressibility factor, and the gas specific gravity
d. Density of the remaining oil as a function of pressure

The differential liberation test is considered to better describe the separation process taking place in the reservoir and is also considered to simulate the flowing behavior of hydrocarbon systems at conditions above the critical gas saturation. As the saturation of the liberated gas reaches the critical gas saturation, the liberated gas begins to flow, leaving behind the oil that originally contained it. This is attributed to the fact that gases have, in general, higher mobility than oils. Consequently, this behavior follows the differential liberation sequence.

A comparison of the two different liberation methods for determining the gas solubility as a function of pressure is shown in Figure 4-18. This relationship between the two processes may occur as shown or in reverse, depending upon the composition of the hydrocarbon system.

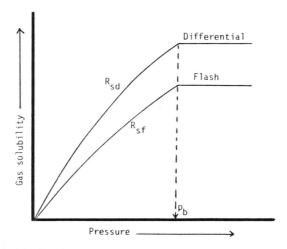

Figure 4-18. Idealized comparison of flash and differential gas solubilities.

Example 4-11. A differential liberation test (discussed in greater detail in Chapter 8) was conducted on a crude oil sample taken from an oil field in Montana. The sample, with a volume of 300cc, was placed in a PVT cell at its bubble-point pressure of 3,000 psia and reservoir temperature of 180°F. The temperature was kept constant and the pressure was reduced to 2,500 psi by removing mercury from the cell. The total volume of the hydrocarbon system was increased to 346.5cc. The gas was bled off at constant pressure (by reinjecting the mercury) and found to occupy a volume of 0.145 scf. The volume of the remaining oil was 290.8cc. The previous process was repeated at 2,000 psia and the remaining oil was flashed through a series of laboratory separators with the separation stage representing stock-tank conditions. The collected experimental data are given below:

Pressure psia	Temperature °F	Total Volume cc	Volume of Liberated Gas scf	Volume of Oil cc
2,000	180	392.3	0.290	281.5
14.7	60°F	—	0.436	230.8

Calculate the gas solubility at 3,000, 2,500, and 2,000 psia.

Solution.

- **Calculation of R_s at 3,000:** By recalling the definition of R_s as the number of scf of gas in solution at p and T per stock-tank barrel of oil, the total scf of gas in solution at 3,000 psia and 180°F is

$$(scf)_{3,000,\ 180°} = 0.145 + 0.290 + 0.436 = 0.871 \text{ scf}$$

$$\begin{matrix} \text{Volume of oil at} \\ \text{standard conditions} \end{matrix} = (V_o)_{sc} = \frac{230.8}{(30.48)^3\ 5.615} = 0.001452 \text{ STB}$$

$$(R_s)_{3,000} = \frac{0.871}{0.001452} = 600 \text{ scf/STB}$$

- **Calculation of R_s at 2,500:** At this pressure, the number of scf of gas in solution is equal to the total scf of gas minus the scf of free gas (liberated gas at 2,500 psia), or

$$(scf)_{2,500} = 0.871 - 0.145 = 0.726 \text{ scf}$$

$$(R_s)_{2,500} = \frac{0.726}{0.001452} = 500 \text{ scf/STB}$$

- **Calculation of R_s at 2,000:**

$$(scf)_{2,000} = 0.871 - 0.145 - 0.29 = 0.436 \text{ scf (in solution)}$$

therefore,

$$(R_s)_{2,000} = \frac{0.436}{0.001452} = 300 \text{ scf/STB}$$

Methods of Calculating R_s

1. Beal's Correlation. Beal (1946) presented a graphical correlation, as shown in Figure 4-19, for estimating the gas solubility as a function of the saturation pressure and the API gravity of the stock-tank oil. The proposed correlation was derived from 508 gas solubility observations taken from 164 crude oil samples. Beal's correlation yields an average deviation of 25%.

2. Standing's Correlation. Standing (1947) proposed a graphical correlation for determining the gas solubility as a function of pressure, gas specific gravity, API gravity, and system temperature. The correlation was developed from a total of 105 experimentally determined data points on 22 hydrocarbon mixtures from California crude oils and natural gases. The proposed correlation has an average error of 4.8%.

The correlation is shown graphically in Figure 4-20. Standing (1981) proposed the following mathematical expression for this graphical correlation:

$$R_s = \gamma_g \left[\left(\frac{p}{18.2} + 1.4 \right) 10^{0.0125 \text{ API} - 0.00091(T - 460)} \right]^{1.2048} \tag{4-52}$$

where T = temperature, °R
　　　 p = system pressure, psia

3. Lasater's Correlation. Lasater (1958) developed a graphical correlation for calculating the gas solubility at the bubble-point pressure. The graphical correlation, as shown in Figure 4-21, was based on 158 experimentally measured bubble-point pressures of 137 independent systems. Vasquez and Beggs (1980) stated that Lasater's correlation is more accurate than Standing's correlation for high-gravity crude oil systems. Standing's correlation is preferred for applications for crudes of API gravity less than 15.

4. Vasquez-Beggs' Correlation. Vasquez and Beggs (1980) presented an improved empirical correlation for estimating R_s. The correlation was obtained by regression analysis using 5,008 measured gas solubility data points. Based on oil gravity, the measured data were divided into two groups. This division was made at a value of oil gravity of 30°API. The proposed equation has the following form:

$$R_s = C_1 \gamma_{gs} p^{C_2} \text{ EXP} \left[C_3 \left(\frac{\text{API}}{T} \right) \right] \tag{4-53}$$

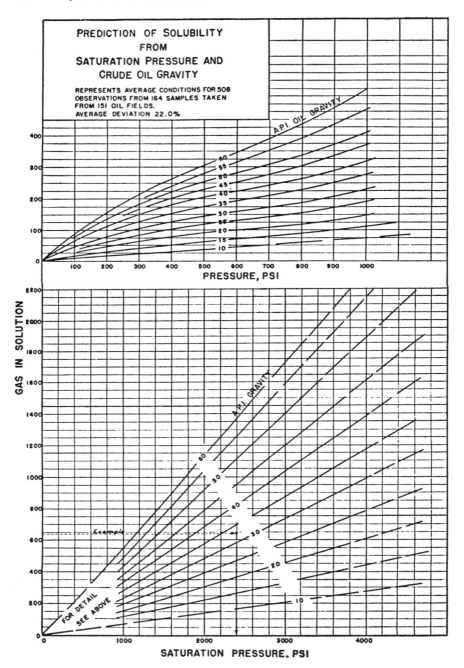

Figure 4-19. Beal's correlation for determining R_s. Permission to publish by the Society of Petroleum Engineers of AIME. Copyright SPE-AIME.

Properties of Natural Hydrocarbon Mixtures of Gas and Liquid Bubble Point Pressure

Example

Required:

Bubble point pressure at 200°F of a liquid having a gas-oil ratio of 350 CFB, a gas gravity of 0.75, and a tank oil gravity of 30°API.

Procedure:

Starting at the left side of the chart, proceed horizontally along the 350 CFB line to a gas gravity of 0.75. From this point drop vertically to the 30°API line. Proceed horizontally from the tank oil gravity scale to the 200°F line. The required pressure is found to be 1,930 PSIA.

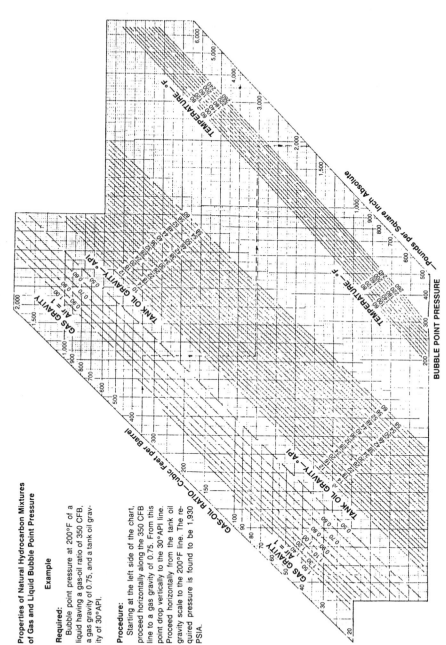

Figure 4-20. Standing's gas solubility correlation. Reprinted by permission of Chevron Corporation.

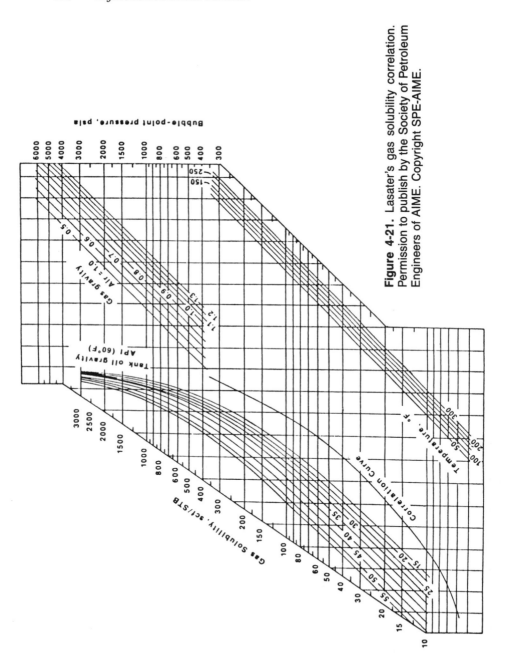

Figure 4-21. Lasater's gas solubility correlation. Permission to publish by the Society of Petroleum Engineers of AIME. Copyright SPE-AIME.

Values for the coefficients are as follows:

Coefficient	API ≤ 30	API > 30
C_1	0.0362	0.0178
C_2	1.0937	1.1870
C_3	25.7240	23.931

The reported average error of the above expression is 0.7%.

The gas specific gravity γ_{gs} at a reference separator pressure of 100 psig was defined previously by Equation 4-44.

5. Glaso's Correlation. Glaso (1980) proposed a correlation for estimating the gas solubility as a function of the API gravity, the pressure, the temperature, and the gas specific gravity. The correlation was developed from studying 45 North Sea crude oil samples. Glaso reported an average error of 1.28% with a standard deviation of 6.98%. The proposed relationship has the following form:

$$R_s = \gamma_g \left[\left(\frac{API^{0.989}}{(T - 460)^{0.172}} \right) (p_b^*) \right]^{1.2255} \tag{4-54}$$

where p_b^* is a correlating number and is defined by the following expression:

$$p_b^* = 10^{[2.8869 - (14.1811 - 3.3093 \, \text{Log} \, (p))^{0.5}]} \tag{4-55}$$

Sutton and Farashad (1984) presented an excellent review of these correlations and documented their associated accuracy. Sutton and Farashad concluded that Glaso's correlation showed the most accuracy and best predicted results for estimating gas solubility. However, the accuracy of Glaso's correlation declines for solution gas-oil ratios in excess of 1,400 scf/STB.

6. Marhoun's Correlation. Marhoun (1988) developed an expression for estimating the saturation pressure of the Middle Eastern crude oil systems. The correlation originates from 160 experimental saturation pressure data. The proposed correlation can be rearranged and solved for the gas solubility to give:

$$R_s = [a \, \gamma_g^b \, \gamma_o^c \, T^d \, P]^e \tag{4-56}$$

where γ_g = gas specific gravity
γ_o = stock-tank oil gravity
T = temperature, °R
a–e = coefficients of the above equation having these values:

$$a = 185.843208$$
$$b = 1.877840$$
$$c = -3.1437$$
$$d = -1.32657$$
$$e = 1.398441$$

Example 4-12. A 38°API crude oil has a bubble-point pressure of 3,811 psia at 180°F. The average gas specific gravity is 0.732. Calculate the gas solubility by using the following correlations:

a. Beal's
b. Standing's
c. Lasater's
d. Vasquez-Beggs'
e. Glaso's
f. Marhoun's

Compare the results with the experimental value of 909 scf/STB

Solution.

a. Beal's Correlation. From Figure 4-19, determine the gas solubility, to give $R_s = 1,250$ scf/STB
b. Standing's Correlation. Solve for R_s by applying Equation 4-50, to yield $R_s = 1,094$ scf/STB
c. Lasater's Correlation. From Figure 4-21, the gas solubility is 900 scf/STB
d. Vasquez-Beggs' Correlation

   • Assuming separator pressure is 100 psig, from Equation 4-46, $\gamma_{gs} = 0.732$
   • Solve for gas solubility by applying Equation 4-53, to give $R_s = 961$ scf/STB

e. Glaso's Correlation

   • Determine the correlating parameter p_b^* from Equation 4-55, to give:

     $$p_b^* = 22.927$$

   • Solve for gas solubility by applying Equation 4-56 to give:

     $$R_s = 935 \text{ scf/STB}$$

f. Marhoun's Correlation

   • Solve for R_s by applying Equation 4-56, to give

     $$R_s = 921 \text{ scf/STB}$$

OIL FORMATION VOLUME FACTOR

The oil formation volume factor, B_o, is defined as the ratio of the volume of oil (plus the gas in solution) at the prevailing reservoir temperature and pressure to the volume of oil at standard conditions. Evidently, B_o is always greater than or equal to unity. The oil formation volume factor can be expressed mathematically as

$$B_o = \frac{(V_o)_{p,T}}{(V_o)_{sc}}$$
(4-57)

where B_o = oil formation volume factor, bbl/STB
  $(V_o)_{p,T}$ = volume of oil under reservoir pressure p and temperature T, bbl
  $(V_o)_{sc}$ = volume of oil as measured under standard conditions, STB

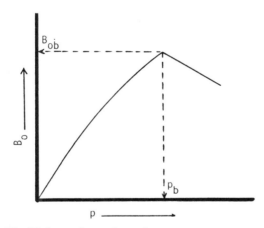

Figure 4-22. Oil formation volume factor versus pressure diagram.

A typical oil formation factor curve, as a function of pressure for an undersaturated crude oil ($p_i > p_b$), is shown in Figure 4-22. As the pressure is reduced below the initial reservoir pressure, p_i, the oil volume increases due to the oil expansion. This behavior results in an increase in the oil formation volume factor and will continue until the bubble-point pressure is reached. At p_b, the oil reaches its maximum expansion and consequently attains a maximum value of B_{ob} for the oil formation volume factor. As the pressure is reduced below p_b, volume of the oil and B_o are decreased as the solution gas is liberated. When the pressure is reduced to atmospheric pressure and the temperature to 60°F, the value of B_o is equal to one.

As in the case of the gas solubility determination, the numerical value of the oil formation volume factor at different pressures will depend upon the method of gas liberation, i.e., flash or differential liberation. Figure 4-23 shows an idealized flash and differential oil formation volume factors.

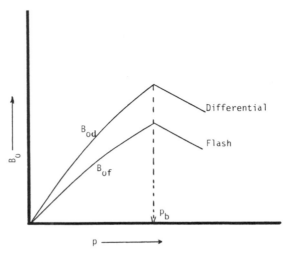

Figure 4-23. Idealized comparison of flash and differential formation volume factors.

Example 4-13. Using the experimental differential liberation data given in Example 4-11, calculate B_o.

Solution. The data given in Example 4-11 are summarized below in the first and second columns. The solution is given in the third column simply by applying Equation 4-56.

Pressure psia	$(V_o)_{p,T}$ cc	B_o bbl/STB
3,000	300	1.30
2,500	290.8	1.26
2,000	281.5	1.22
14.7	230.8*	1.00

* Volume of remaining oil at standard conditions $(V_o)_{sc}$.

Methods of Calculating B_o at Saturation Pressure

1. Standing's Correlation. Standing (1947) presented a graphical correlation for estimating the oil formation volume factor with the gas solubility, gas gravity, oil gravity, and reservoir temperature as the correlating parameters. This graphical correlation, as shown in Figure 4-24, originated from examining a total of 105 experimental data points on 22 different California hydrocarbon systems. An average error of 1.2% was reported for the correlation.

Standing (1981) showed that the B_o chart, i.e., Figure 4-24, can be expressed more conveniently in a mathematical form by the following equation:

$$B_o = 0.9759 + 0.000120 \left[R_s \left(\frac{\gamma_g}{\gamma_o} \right)^{0.5} + 1.25 \, (T - 460) \right]^{1.2} \qquad (4\text{-}58)$$

where T = temperature, °R
γ_o = specific gravity of the stock-tank oil
γ_g = specific gravity of the solution gas

2. Vasquez and Beggs' Correlation. Vasquez and Beggs (1980) developed a relationship for determining B_o as a function of R_s, γ_o, γ_g, and T. The proposed correlation was based on 6,000 measurements of B_o at various pressures. Using the regression analysis technique, Vasquez and Beggs found the following equation to be the best form to reproduce the measured data:

$$B_o = 1.0 + C_1 \, R_s + (T - 520) \left| \frac{API}{\gamma_{gs}} \right| [C_2 + C_3 \, R_s] \qquad (4\text{-}59)$$

where R_s = gas solubility, scf/STB
T = temperature, °R
γ_{gs} = gas specific gravity as defined by Equation 4-44

Values for the coefficients C_1, C_2, and C_3 are given below:

Coefficient	API ≤ 30	API > 30
C_1	4.677×10^{-4}	4.670×10^{-4}
C_2	1.751×10^{-5}	1.100×10^{-5}
C_3	-1.811×10^{-8}	1.337×10^{-9}

Vasquez and Beggs reported an average error of 4.7% for the proposed correlation.

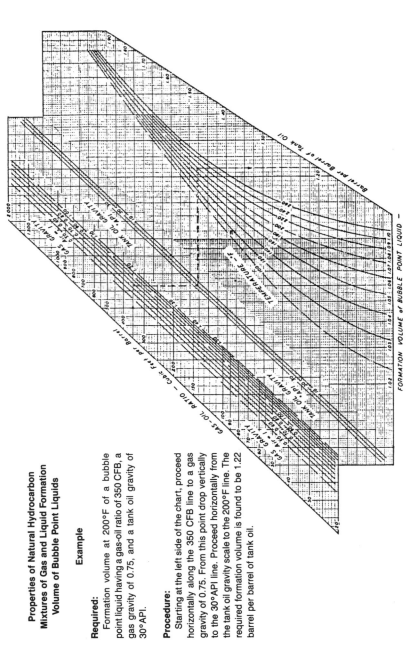

**Properties of Natural Hydrocarbon
Mixtures of Gas and Liquid Formation
Volume of Bubble Point Liquids**

Example

Required:

Formation volume at 200°F of a bubble point liquid having a gas-oil ratio of 350 CFB, a gas gravity of 0.75, and a tank oil gravity of 30°API.

Procedure:

Starting at the left side of the chart, proceed horizontally along the 350 CFB line to a gas gravity of 0.75. From this point drop vertically to the 30°API line. Proceed horizontally from the tank oil gravity scale to the 200°F line. The required formation volume is found to be 1.22 barrel per barrel of tank oil.

Figure 4-24. Standing's oil formation volume factor correlation. Reprinted by permission of Chevron Corporation.

3. Glaso's Correlation. Glaso (1980) proposed the following expressions for calculating the oil formation volume factor:

$$B_o = 1.0 + 10^A \tag{4-60}$$

where $A = -6.58511 + 2.91329 \text{ Log } B_{ob}^* - 0.27683 (\text{Log } B_{ob}^*)^2 \tag{4-61}$

B_{ob}^* is a "correlating number" and defined by the following equation:

$$B_{ob}^* = R_s \left| \frac{\gamma_g}{\gamma_o} \right|^{0.526} + 0.968 \ (T - 460) \tag{4-62}$$

The above correlations were originated from studying PVT data on 45 oil samples. The average error of the correlation was reported at -0.43% with a standard deviation of 2.18%.

Sutton and Farshad (1984) concluded that Glaso's correlation offers the best accuracy when compared with the Standing and Vasquez-Beggs correlations. In general, Glaso's correlation underpredicts formation volume factor. Standing's expression tends to overpredict oil formation volume factors greater than 1.2 bbl/STB. The Vasquez-Beggs correlation typically overpredicts the oil formation volume factor.

4. Marhoun's Correlation. Marhoun (1988) developed a correlation for determining the oil formation volume factor as a function of the gas solubility, stock-tank oil gravity, gas gravity, and temperature. The empirical equation was developed by use of the non-linear multiple regression analysis on 160 experimental data points. The experimental data were obtained from 69 Middle Eastern oil reserves. The author proposed the following expression:

$$
\begin{aligned}
B_o = 0.497069 &+ 0.862963 \times 10^{-3} \ T + 0.182594 \times 10^{-2} \ F \\
&+ 0.318099 \times 10^{-5} \ F^2
\end{aligned} \tag{4-63}
$$

with

$$F = R_s^a \ \gamma_g^b \ \gamma_o^c$$

The coefficients a, b, and c have the following values:

a = 0.742390
b = 0.323294
c = -1.202040

The average absolute error of the correlation was reported at 0.88% with a standard deviation of 1.180%.

5. Arps' Correlation. Realizing that the gas solubility, R_s, is a strong function of pressure, temperature, API gravity, and gas specific gravity, Arps (1962)* proposed the following simplified relationship for approximating the oil formation volume factor of light crude oil systems.

$$B_o = 1.05 + 0.0005 \, R_s \qquad (4\text{-}64)$$

The above linear expression provides a quick approximation of the oil formation volume factor. The proposed relationship can only be used when the necessary PVT data for other equations are not available.

6. Ahmed's Correlation. Using the pressure, temperature, gas specific gravity, oil API gravity, and gas solubility as correlating parameters, Ahmed (1988) proposed the following expression for determining the oil formation volume factor:

$$\begin{aligned}
B_o = F &+ a_1 \, t + a_2 \, t^2 + a_3/t + a_4 \, P + a_5 \, P^2 + a_6/P + a_7 \, R_s \\
&+ a_8 \, R_s^2 + a_9/R_s \qquad (4\text{-}65)
\end{aligned}$$

with

$$F = a_{10} + (R_s^{a11} API^{a12}/\gamma_g^{a13})$$

where t = system temperature, °F
             P = system pressure, psia
             R_s = gas solubility, scf/STB
           API = oil gravity
             γ_g = specific gravity of the gas
    $a_1 - a_{13}$ = coefficients

The values of the coefficients $a_1 - a_{13}$ were determined by using a regression model on the experimental data reported by Marhoun (1988) and Glaso (1980), to give,

$a_1 = -4.5243973 \times 10^{-4}$ $a_2 = 3.9063637 \times 10^{-6}$
$a_3 = -5.5542509$ $a_4 = -5.7603220 \times 10^{-6}$
$a_5 = -3.9528992 \times 10^{-9}$ $a_6 = 16.289473$
$a_7 = 3.8718887 \times 10^{-4}$ $a_8 = 7.0703685 \times 10^{-8}$
$a_9 = -1.4358395$ $a_{10} = -0.12869353$
$a_{11} = 0.023484894$ $a_{12} = 0.015966573$
$a_{13} = 0.021946351$

---

* Frick, T. C., *Petroleum Production Handbook*, Volume II, Dallas: Society of Petroleum Engineers, 1962, Chapter 37, p. 1.

A total of 201 experimental data points were used to generate the optimum values for the coefficients. The proposed correlation produced a mean average deviation of 1.094% and a standard deviation of 2.5% when used to reproduce the experimental oil formation volume factor.

7. Other Methods. The oil formation volume factor can be calculated rigorously from the following data:

- Specific gravity of solution gas, γ_g
- Stock-tank oil gravity, γ_o
- Gas solubility, R_s
- Oil density at the specified pressure and temperature, ρ_o

Following the definition of B_o:

$$B_o = \frac{(V_o)_{p,T}}{(V_o)_{sc}}$$

the oil volume under p and T can be replaced with total weight of the hydrocarbon system divided by the density at the prevailing pressure and temperature.

$$B_o = \frac{\dfrac{m_t}{\rho_o}}{(V_o)_{sc}}$$

where the total weight of the hydrocarbon system is equal to the sum of the stock-tank oil plus the weight of the solution gas, i.e.,

$$m_t = m_o + m_g$$

or

$$B_o = \frac{m_o + m_g}{\rho_o \, (V_o)_{sc}}$$

Given the gas solubility R_s per barrel of the stock-tank oil and the specific gravity of the solution gas, the weight of R_s scf of the gas is calculated as:

$$m_g = \frac{R_s}{379.4} \, (28.96)(\gamma_g)$$

where m_g = weight of solution gas, lb of solution gas/STB

The weight of one barrel of the stock-tank oil is calculated from its specific gravity by the following relationship:

$$m_o = (5.615)(62.4)(\gamma_o)$$

Substituting for m_o and m_g

$$B_o = \frac{(5.615)(62.4)\; \gamma_o + \dfrac{R_s}{379.4}\; (28.96)\; \gamma_g}{5.615\; \rho_o}$$

or

$$B_o = \frac{62.4\; \gamma_o + 0.0136\; R_s\; \gamma_g}{\rho_o} \tag{4-66}$$

The error in calculating B_o by using Equation 4-66 will depend only on the accuracy of the input variables (R_s, γ_g, and γ_o) and the method of calculating ρ_o.

Example 4-14. A crude oil at its bubble-point pressure of 3,410 psia and a temperature of 242°F has an API gravity of 36.6°. The gas solubility and specific gravity of the solution gas are 688 scf/STB and 0.710 respectively. Calculate B_o by using

a. Standings' Correlation
b. Vasquez-Beggs' Correlation (assume an average separator pressure of 100 psig)
c. Glaso's Correlation
d. Marhoun's Correlation
e. Equation 4-65
f. Equation 4-66

Solution.

a. Solution by using Standing's Correlation. Solve for B_o by applying Equation 4-58 to give

$$B_o = 0.9759 + 0.000120 \left[688 \left(\frac{0.710}{0.842}\right)^{0.5} + 1.25\; (242) \right]^{1.2} = 1.42 \text{ bbl/STB}$$

b. Solution by using Vasquez-Beggs' Correlation. Compute B_o by using Equation 4-59

$$B_o = 1 + 4.67(10^{-4})(688) + (702 - 520) \left| \frac{36.6}{0.710} \right| [1.1(10^{-5})$$

$$+ 1.337(10^{-9})688] = 1.433 \text{ bbl/STB}$$

c. **Solution by using Glaso's Correlation**

• Calculate the correlating parameter B_{ob}^* from Equation 4-62

$$B_{ob}^* = 688 \left| \frac{0.71}{0.842} \right|^{0.526} + 0.968 \ (242) = 863.24$$

• Solve for the coefficient A by applying Equation 4-61

$$A = -6.58511 + 2.91329 \ \text{Log}(863.24) - 0.27683 \ [\text{Log}(863.24)]^2$$
$$= -0.41783$$

• Calculate B_o from Equation 4-60

$$B_o = 1.0 + 10^{-0.41783} = 1.39 \text{ bbl/STB}$$

d. **Solution by using Marhoun's Correlation**

• Solve for the correlation parameter F of Equation 4-63, to give
$$F = 140.698$$

• Determine B_o from Equation 4-63, to yield
$$B_o = 1.423 \text{ bbl/STB}$$

e. **Solution by using Equation 4-66**

• Calculate the parameter F of Equation 4-65 to give

$$F = 1.114$$

• Solve Equation 4-65 for B_o to yield

$$B_o = 1.44 \text{ bbl/STB}$$

f. **Solution by using Equation 4-64.** The accuracy of calculating B_o from Equation 4-64 is dependent on the selected method of determining the oil density. To test the sensitivity of Equation 4-64 to the value of the oil density, Standing's correlation and Katz's correlation are used to determine the oil density. The calculated values are then used to compute B_o.

Selected Method	ρ_o lb/ft^3	B_o from Equation 4-64
Katz (Equation 4-31)	42.2	1.402
Standing (Equation 4-32)	41.50	1.426

OIL FORMATION VOLUME FACTOR FOR
UNDERSATURATED OILS

With increasing pressures above the bubble-point pressure, the oil formation volume factor decreases due to the compression of the oil, as illustrated schematically in Figure 4-22. To account for the effects of oil compression on B_o, the oil formation volume factor at the bubble-point pressure is first calculated by using any of the methods previously described. The calculated B_o is then adjusted to account for the effect of increasing the pressure above the bubble-point pressure. This adjustment step is accomplished by using the isothermal compressibility coefficient as described below.

The isothermal compressibility coefficient (as expressed mathematically by Equation 4-36) can be equivalently written in terms of the oil formation volume factor:

$$C_o = \frac{-1}{B_o} \frac{\partial B_o}{\partial p}$$

The above relationship can be rearranged and integrated to produce

$$\int_{p_b}^{p} - C_o \, dp = \int_{B_{ob}}^{B_o} \frac{1}{B_o} \, dB_o \qquad (4\text{-}67)$$

Evaluating C_o at the arithmetic average pressure and concluding the integration procedure to give

$$B_o = B_{ob} \, \text{EXP} \, [- C_o \, (p - p_b)] \qquad (4\text{-}68)$$

where B_o = oil formation volume factor at the pressure of interest, bbl/
         STB
         B_{ob} = oil formation volume factor at the bubble-point pressure, bbl/
         STB
         p = pressure of interest, psia
         p_b = bubble-point pressure, psia

Replacing with the Vasquez-Beggs' C_o expression, i.e., Equation 4-43, and integrating the resulting equation, gives

$$B_o = B_{ob} \, \text{EXP} \left[- A \, \text{Ln} \left(\frac{p}{p_b} \right) \right] \qquad (4\text{-}69)$$

where $A = 10^{-5} [- 1{,}433 + 5 \, R_s + 17.2(T - 460) - 1{,}180 \, \gamma_{gs}$
$+ 12.61 \, °API]$

Replacing C_o in Equation 4-67 with Ahmed's expression for C_o (Equation 4-45), yields

$$B_o = B_{ob} \, EXP \, [D \, [EXP \, (ap) - EXP \, (a \, p_b)]] \qquad (4\text{-}70)$$

where $D = [4.588893 + 0.0025999 \, R_s]^{-1}$
$\quad\;\; a = -0.00018473$

Example 4-15. Calculate B_o at 3,550 psia by using the data given in Example 4-14 and employing Equations 4-69 and 4-70.

Solution.

- From Example 4-14

    $p_b = 3,410$ psia
    $B_{ob} = 1.406$

- Solution by using Equation 4-69
    Calculate the parameter A to give

    $A = 0.05793$

    Solve Equation 4-69 for B_o

    $$B_o = 1.406 \, EXP \left[-0.05793 \, Ln \left| \frac{3,550}{3,410} \right| \right] = 1.403 \text{ bbl/STB}$$

- Solution by using Equation 4-70
    Calculate the parameter D to give

    $D = 0.156798$

    Solve Equation 4-70 for B_o to give

    $B_o = 1.403$ bbl/STB

TOTAL FORMATION VOLUME FACTOR

To describe the pressure-volume relationship of hydrocarbon systems below their bubble-point pressure, it is convenient to express this relationship in terms of the total formation volume factor as a function of pressure. The total formation volume factor defines the total volume of a system regardless of the number of phases present. The total formation volume factor, as denoted by B_t, is defined as the ratio of the total volume of the hydrocarbon system at the prevailing pressure and temperature per unit volume of the stock-tank oil. Because naturally occurring hydrocarbon systems usually exist in either one or two phases, the term "two-phase formation volume factor" has become synonymous with the total formation volume.

Mathematically, B_t is defined by the following relationship:

$$B_t = \frac{(V_o)_{p,T} + (V_g)_{p,T}}{(V_o)_{sc}}$$

where B_t = total formation volume factor, bbl/STB
$(V_o)_{p,T}$ = volume of the oil at p and T, bbl
$(V_g)_{p,T}$ = volume of the liberated gas at p and T, bbl
$(V_o)_{sc}$ = volume of the oil at standard conditions, STB

A typical plot of B_t as a function of pressure for an undersaturated crude oil is shown in Figure 4-25. The oil formation volume factor curve is also included in the illustration. It should be noted that B_o and B_t are identical at pressures above or equal to the bubble-point pressure because only one phase, the oil phase, exists at these pressures. It should also be noted that at pressures below the bubble-point pressure, the difference in the values of the two oil properties represents the volume of the evolved solution gas as measured at system conditions per stock-tank barrel of oil.

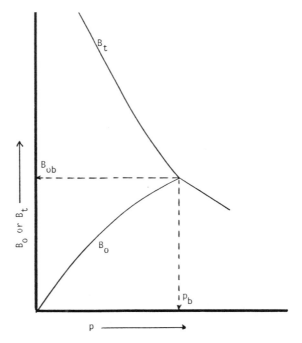

Figure 4-25. B_o and B_t versus p relationships.

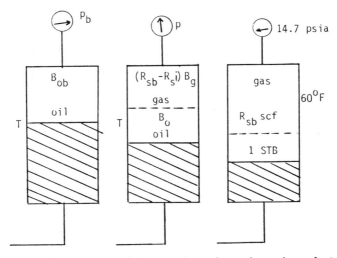

Figure 4-26. The concept of the two-phase formation volume factor.

Consider a crude oil sample placed in a PVT cell at its bubble-point pressure p_b and reservoir temperature, as shown schematically in Figure 4-26. Assume that the volume of the oil sample is sufficient to yield one stock-tank barrel of oil at standard conditions. Let R_{sb} represent the gas solubility at p_b. By lowering the cell pressure to p, a portion of the solution gas is evolved and occupies a certain volume of the PVT cell. Let R_s and B_o represent the corresponding gas solubility and oil formation volume factor at p. Obviously, the term $(R_{sb} - R_s)$ represents the volume of the free gas as measured in scf per stock-tank barrel of the oil. The volume of the free gas at the cell conditions is then

$$(V_g)_{p,T} = (R_{sb} - R_s)B_g$$

where $(V_g)_{p,T}$ = volume of the free gas at p and T, bbl of gas/STB of oil
  B_g = gas formation volume factor, bbl/scf

The volume of the remaining oil at the cell condition is

$$(V)_{p,T} = B_o$$

From the definition of the two-phase formation volume factor

$$B_t = B_o + (R_{sb} - R_s) B_g \tag{4-71}$$

The accuracy of estimating B_t by Equation 4-71 will depend on the accuracy of estimating the PVT parameters B_o, R_{sb}, R_s, and B_g.

Methods of Estimating B_t

1. Standing's Correlation. Standing (1947) proposed a graphical correlation, as shown in Figure 4-27, for estimating the two-phase formation volume factor. To use Figure 4-27, it is necessary to know:

The gas solubility at pressure of interest, R_s
Solution gas gravity, γ_g
API gravity
Reservoir temperature, T
Pressure of interest, p

A total of 387 experimental data points was used in developing the correlation. Standing reported an average error of 5% for the correlation.

2. Glaso's Correlation. The experimental data on 45 crude oil samples were used by Glaso (1980) in developing a generalized correlation for estimating B_t. Using a regression analysis model in developing the correlation, Glaso proposed the following expression for B_t

$$\text{Log}(B_t) = 0.080135 + 0.47257 \, \text{Log}(B_t^*) + 0.17351[\text{Log}(B_t^*)]^2 \qquad (4\text{-}72)$$

where (B_t^*) is a correlating number and is defined by the following equation

$$B_t^* = R_s \frac{(T - 460)^{0.5}}{(\gamma_g)^{0.3}} (\gamma_o)^C \, p^{-1.1089} \qquad (4\text{-}73)$$

where $C = (2.9) \, 10^{-0.00027R_s}$ \hfill (4-74)

Glaso reported a standard deviation of 6.54% for the total formation volume factor correlation.

3. Marhoun's Correlation. Based on 1,556 experimentally determined total formation volume factors, Marhoun (1988) used a non-linear multiple regression model to develop a mathematical expression for B_t. The empirical equation has the following form:

$$B_t = 0.314693 + 0.106253 \times 10^{-4} \, F + 0.18883 \times 10^{-10} \, F^2 \qquad (4\text{-}75)$$

with $F = R_s^a \, \gamma_g^b \, \gamma_o^c \, T^d \, p^e$

Properties of Natural Hydrocarbon
Mixtures of Gas and Liquid Formation
Volume of Gas Plus Liquid Phases

Example

Required:

Formation volume of the gas plus liquid phases of a 1500 CFB mixture, gas gravity of 0.80, tank oil gravity of 40°API, at 200°F and 1000 PSIA

Procedure:

Starting at the left side of the chart, proceed horizontally along the 1500 CFB line to the 0.80 gas gravity line. From this point drop vertically to the 40°API line. Proceed horizontally to 200°F and from that point drop to the 1000 PSIA pressure line. The required formation volume is found to be 5.0 barrels per barrel of tank oil.

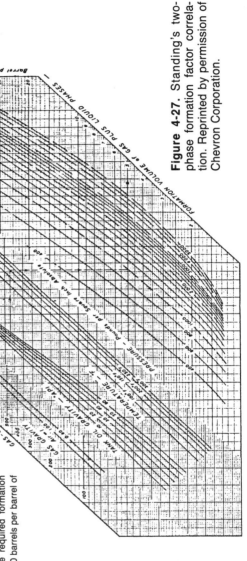

Figure 4-27. Standing's two-phase formation factor correlation. Reprinted by permission of Chevron Corporation.

where   a = 0.644516   b = − 1.079340   c = 0.724874
        d = 2.00621   e = − 0.761910

Marhoun reported an average absolute error of 4.11% with a standard deviation of 4.94% for the correlation.

Example 4-16. Given the following PVT data*

p_b = 2,744 psia	R_{sb} = 603 scf/STB
T = 600°R	API = 36.4°
γ_g = 0.6744	p = 2,000.7 psia
R_s = 443.9 scf/STB	B_o = 1.1752 bbl/STB

Calculate B_t at 2,000.7 psia by using

  a. Equation 4-71
  b. Standing's Correlation
  c. Glaso's Correlation
  d. Marhoun's Correlation

Solution.

a. Solution by using Equation 4-71.

  Step 1. Calculate T_{pc} and p_{pc} of the solution gas from its specific gravity by applying Equations 3-27 and 3-28, to give

$$T_{pc} = 381.49°R$$
$$p_{pc} = 670.06 \text{ psia}$$

  Step 2. Calculate p_{pr} and T_{pr}

$$p_{pr} = \frac{2,000.7}{670.00} = 2.986$$

$$T_{pr} = \frac{600}{381.49} = 1.57$$

  Step 3. Determine the gas compressibility factor from Figure 3-5

$$Z = 0.81$$

  Step 4. Calculate B_g from Equation 3-66

$$B_g = 0.005035 \frac{(0.81)(600)}{2,000.7} = 0.001223 \text{ bbl/scf}$$

---

* The PVT data were reported by Dodson et al. (1953).

Step 5. Solve for B_t by applying Equation 4-71

$$B_t = 1.1752 + 0.0001223 \, (603 - 443.9) = 1.195 \text{ bbl/STB}$$

b. Solution by using Standing's Correlation. Enter Figure 4-27 with the following data:

$R_s = 443.9$
$\gamma_g = 0.6744$
$API = 364$
$T = 140°F$
$p = 2,000.7$ psia

and read B_t to give: $B_t = 1.20$ bbl/STB

c. Solution by using Glaso's Correlation

Step 1. Determine the coefficient C from Equation 4-74

$$C = (2.9) \, 10^{-0.00027(443.9)} = 2.20062$$

Step 2. Calculate the correlating parameter B_t^* from Equation 4-73, to give

$$B_t^* = 0.8927$$

Step 3. Solve for B_t by applying Equation 4-72, to yield

$$B_t = 1.138$$

d. Solution by using Marhoun's Correlation

Step 1. Determine the correlating parameter F of Equation 4-75, to give:

$$F = 78,590.6789$$

Step 2. Solve for B_t from Equation 4-75

$$B_t = 1.2664 \text{ bbl/STB}$$

TOTAL SYSTEM ISOTHERMAL COMPRESSIBILITY COEFFICIENT

All solutions of transient fluid-flow problems contain a parameter called the total system isothermal compressibility, written as C_t. This property of the reservoir fluids and the porous rock is a measure of the change in volume of the fluid content of porous rock with a change in pressure, and it may vary considerably with pressure.

The total system isothermal compressibility is defined mathematically by the following relationship

$$C_t = S_o \, C_o + S_w \, C_w + S_g \, C_g + C_f \tag{4-76}$$

where C_t = total system compressibility, psi^{-1}
        S_o = average oil saturation
        C_o = isothermal compressibility coefficient of the oil phase, psi^{-1}
        S_w = water saturation
        C_w = isothermal compressibility coefficient of the water phase, psi^{-1}
        S_g = gas saturation
        C_g = isothermal compressibility coefficient of the gas phase, psi^{-1}
        C_f = formation compressibility, psi^{-1}

By definition, the isothermal compressibility coefficient of phase i is

$$C_i = \frac{-1}{B_i} \frac{\partial B_i}{\partial p}$$

where B_i is the formation volume factor of phase i.

Accordingly, C_o, C_w, and C_g in Equation 4-76 can be replaced by the above expression, to yield:

$$C_t = S_o \left(\frac{-1}{B_o} \frac{\partial B_o}{\partial p} \right) + S_w \left(\frac{-1}{B_w} \frac{\partial B_w}{\partial p} \right) + S_g \left(\frac{-1}{B_g} \frac{\partial B_g}{\partial p} \right) + C_f \tag{4-77}$$

where B_w = water formation volume factor, bbl/STB
        B_o = oil formation volume factor, bbl/STB
        B_g = gas formation volume factor, bbl/scf

Oil and water formation volume factors contain the effect of solution gas on the change in liquid phase volumes. Thus, terms to reduce the gas phase volume by the quantity of gas going into solution as pressure is increased, must be added, giving

$$C_o = \frac{-1}{B_o} \frac{\partial B_o}{\partial p} + \frac{B_g}{B_o} \frac{\partial R_s}{\partial p} \tag{4-78}$$

$$C_w = \frac{-1}{B_w} \frac{\partial B_w}{\partial p} + \frac{B_g}{B_w} \frac{\partial R_{sw}}{\partial p} \tag{4-79}$$

where R_{sw} is the gas solubility in the water phase, scf/STB.

Therefore, Equation 4-77 can be written in a more generalized form as follows:

$$C_t = S_o \left[\frac{-1}{B_o} \frac{\partial B_o}{\partial p} + \frac{B_g}{B_o} \frac{\partial R_s}{\partial p} \right] + S_w \left[\frac{-1}{B_w} \frac{\partial B_w}{\partial p} + \frac{B_g}{B_w} \frac{\partial R_{sw}}{\partial p} \right]$$

$$+ S_g \left[\frac{-1}{B_g} \frac{\partial B_g}{\partial p} \right] + C_f \tag{4-80}$$

Laboratory PVT data should be used to evaluate all terms of Equation 4-80. If these data are not available, correlations can be used to approximate the terms of Equation 4-80 as follows:

Approximation of $\partial R_s / \partial p$: Ramey (1964) analytically differentiated Standing's correlation, Equation 4-50, with respect to pressure, to produce the following expression:

$$\frac{\partial R_s}{\partial p} = \frac{R_s}{0.83\ p + 21.75} \tag{4-81}$$

Approximation of $\partial B_o / \partial p$: Standing's oil formation volume factor equation, Equation 4-58, can be differentiated with respect to pressure to produce:

$$\frac{\partial B_o}{\partial p} = 0.000144 \left(\frac{\gamma_g}{\gamma_o} \right)^{0.5} \left[R_s \left(\frac{\gamma_g}{\gamma_o} \right)^{0.5} + 1.25\ (T - 460) \right]^{0.12} \left(\frac{\partial R_s}{\partial p} \right) \tag{4-82}$$

Combining Equation 4-81 with Equation 4-82, yields

$$\frac{\partial B_o}{\partial p} = \left[\frac{0.000144\ R_s}{0.83\ p + 21.75} \right] \left(\frac{\gamma_g}{\gamma_o} \right)^{0.5} \left[R_s \left(\frac{\gamma_g}{\gamma_o} \right)^{0.5} + 1.25\ (T - 460) \right]^{0.12} \tag{4-83}$$

Approximation of C_w: Meehan (1980) presented mathematical expressions for determining the isothermal compressibility coefficient of gas-free water and gas-saturated water. Meehan proposed the following relationships:

a. C_w for gas-free water.

$$(C_w)_f = 10^{-6} [A + B\ (T - 460) + C\ (T - 460)^2] \tag{4-84}$$

where $(C_w)_f$ = isothermal compressibility coefficient of the gas-free water, psi^{-1}

$A = 3.8546 - 0.000134\ p$
$B = -0.01052 + 4.77\ (10^{-7})\ p$
$C = 3.9267\ (10^{-5}) - 8.8\ (10^{-10})\ p$
p = pressure, psia
T = temperature, °R

b. C_w for gas-saturated water.

$$(C_w)_g = (C_w)_f [1 + 8.9 \ (10^{-3}) \ R_{sw}] \tag{4-85}$$

where $(C_w)_g$ = isothermal compressibility coefficient of the gas-saturated water, psi^{-1}

R_{sw} = gas solubility in water, scf/STB of water

To account for varying degrees of the salinity of the water on the isothermal compressibility coefficient of the water phase, C_w is adjusted by using a salinity correction factor as follows:

$$C_w = (C_w)_{forg} \ (SC) \tag{4-86}$$

where C_w = isothermal compressibility coefficient of the brine, psi^{-1}

$(C_w)_{forg}$ = isothermal compressibility coefficient of the free-gas or the saturated-gas water

SC = salinity correction factor

Numbere et al. (1977) proposed the following mathematical expression of the salinity correction factor

$$SC = [[- 0.052 + 2.7(10^{-4})(T - 460) - 1.14(10^{-6})(T - 460)^2 + 1.121(10^{-9})(T - 460)^3]]S^{0.7} + 1$$

where S = salinity of the water, weight % of N_aCL

T = temperature, °R

Approximation of R_{sw}. The following correlation* can be used to determine the gas solubility in water:

$$R_{sw} = A + B \ p + C \ p^2 \tag{4-87}$$

where A = $2.12 + 3.45(10^{-3})(T - 460) - 3.59(10^{-5})(T - 460)^2$

B = $0.0107 - 5.26(10^{-5})(T - 460) + 1.48(10^{-7})(T - 460)^2$

C = $- 8.75(10^{-7}) + 3.9(10^{-9})(T - 460) - 1.02(10^{-11})(T - 460)^2$

To correct for the effect of the salinity of the water on the gas solubility, R_{sw} is adjusted by using a salinity correction factor

$$(R_{sw})_b = R_{sw} \cdot SC \tag{4-88}$$

where $(R_{sw})_b$ = gas solubility in brine, scf/STB

R_{sw} = gas solubility in pure water

SC = salinity correction factor and given by the following expression

---

* Hewlett-Packard H.P. 41C Petroleum Fluids PAC manual, 1982.

$$SC = 1 - [0.0753 - 0.000173(T - 460)] \, S \tag{4-89}$$

where S = salinity of the water, weight % of N_aCL

Approximation of the term $\partial R_{sw}/\partial p$. Differentiating Equation 4-87 with respect to pressure yields

$$\frac{\partial R_{sw}}{\partial p} = B + 2 \, C \, P \tag{4-90}$$

where the coefficients B and C are given previously in Equation 4-87. For brine

$$\left(\frac{\partial R_{sw}}{\partial p}\right)_b = \left(\frac{\partial R_{sw}}{\partial p}\right) SC \tag{4-91}$$

where the salinity correction factor is given by Equation 4-89.

Approximation of B_w. The water formation volume factor can be calculated by the following mathematical expression*

$$B_w = A_1 + A_2 \, p + A_3 \, p^2 \tag{4-92}$$

when the coefficients A_1–A_3 are given by

$$A_i = a_1 + a_2(T - 460) + a_3(T - 460)^2$$

with a_1–a_3 given for gas-free and gas-saturated water:

	Gas-free Water		
A_i	a_1	a_2	a_3
A_1	0.9947	$5.8(10^{-6})$	$1.02(10^{-6})$
A_2	$-4.228(10^{-6})$	$1.8376(10^{-8})$	$-6.77(10^{-11})$
A_3	$1.3(10^{-10})$	$-1.3855(10^{-12})$	$4.285(10^{-15})$

	Gas-saturated Water		
A_i	a_1	a_2	a_3
A_1	0.9911	$6.35(10^{-5})$	$8.5(10^{-7})$
A_2	$-1.093(10^{-6})$	$-3.497(10^{-9})$	$4.57(10^{-12})$
A_3	$-5.0(10^{-11})$	$6.429(10^{-13})$	$-1.43(10^{-15})$

* Hewlett-Packard H.P. 41C Petroleum Fluids PAC manual, 1982.

To adjust B_w for the salinity of the water, B_w is corrected by using a salinity correction factor, to give

$$(B_w)_b = B_w \cdot SC \tag{4-93}$$

where $(B_w)_b$ = brine formation volume factor
SC = salinity correction factor; calculated by the following equation as proposed by Numbere et al. (1977):

$$SC = [b_1 p + (b_2 + b_3 p)(T - 520) + (b_4 + b_5 P)(T - 520)^2] S + 1.0$$

where the parameter S is the salinity of water (i.e., weight percent of sodium chloride in the brine.) The coefficients b_1 through b_5 of this expression have the following values:

$b_1 = 5.1(10^{-8})$ $b_2 = 5.47(10^{-6})$ $b_3 = -1.95(10^{-10})$
$b_4 = -3.23(10^{-8})$ $b_5 = 8.5(10^{-13})$

Example 4-17. A depleted reservoir with the following fluids and rock properties is to be water-flooded.

$S_g = 0.10$	$S_o = 0.60$	$S_w = 0.30$
$p = 750$ psia	$T = 155°F$	API $= 30°$
$C_f = 1.5 \times 10^{-6}$ psia	water salinity $= 0$	$\gamma_g = 0.830$

Calculate
a. C_t without gas going back into solution
b. C_t with gas going back into solution

Solution.

a. Calculating C_t without gas going back into solution.

Step 1. Calculate R_s from Equation 4-51, to give

$$R_s = 145.9 \text{ scf/STB}$$

Step 2. Determine B_o from Equation 4-57, to give

$$B_o = 1.109 \text{ bbl/STB}$$

Step 3. Compute $\partial B_o / \partial p$ from Equation 4-83

$$\frac{\partial B_o}{\partial p} = 0.000105$$

Step 4. Determine C_w from Equation 4-84, to yield

$C_w = 3.1(10^{-6})$ psia^{-1}

Step 5. Calculate C_g as discussed in Chapter 3, to give

$p_{pc} = 663.6$ psia $\qquad\qquad T_{pc} = 429.1°R$
$p_{pr} = 1.13 \qquad\qquad\qquad T_{pr} = 1.433$

From Figure 3-7 $C_r = 1.1$

$$C_g = \frac{1.1}{663.6} = 165.7(10^{-6}) \text{ psia}$$

Step 6. Solve for C_t by applying Equation 4-77, to yield

$C_t = 298.7(10^{-6})$ psia

b. Calculation of C_t with gas going back into solution.

Step 1. Calculate $\partial R_s/\partial p$ from Equation 4-81

$$\frac{\partial R_s}{\partial p} = 0.22646 \text{ scf/STB/psia}$$

Step 2. Determine $\partial R_{sw}/\partial p$ from Equation 4-90, to yield

$$\frac{\partial R_{sw}}{\partial p} = 0.0063 \text{ scf/STB of water/psia}$$

Step 3. Compute B_w by applying Equation 4-92, to give

$B_w = 1.0175$ bbl/STB

Step 4. Calculate the gas formation volume factor from Equation 3-66.

$B_g = 0.003533$ bbl/scf

Step 5. Solve for C_t by applying Equation 4-80.

$C_t = 730.86(10^{-6})$ psia^{-1}

CRUDE OIL VISCOSITY

Crude oil viscosity is an important physical property that controls and influences the flow of oil through porous media and pipes. The viscosity, in general, is defined as the internal resistance of the fluid to flow.

The oil viscosity is a strong function of the temperature, the pressure, the oil gravity, the gas gravity, and the gas solubility. Figure 4-28 shows the general effect changing these parameters has on the viscosity of the oil.

Gas Solubility or Temperature ⟶

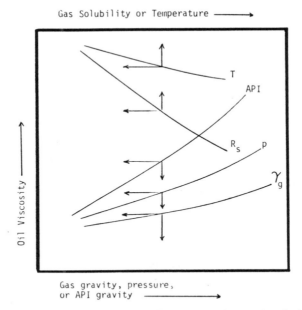

Gas gravity, pressure,
or API gravity ⟶

Figure 4-28. Effect of p, R_s, API, T, and $_g$ on the crude oil viscosity.

Whenever possible, oil viscosity should be determined by laboratory measurements at reservoir temperature and pressure. The viscosity is usually reported in standard PVT analyses. If such laboratory data are not available, engineers may refer to published correlations, which usually vary in complexity and accuracy depending upon the available data on the crude oil.

Sutton and Farashad (1986), and recently Khan et al. (1987), evaluated and presented a review of the most widely used viscosity correlations. According to the pressure, the viscosity of crude oils can be classified into three categories:

- Dead Oil Viscosity
  The dead oil viscosity is defined as the viscosity of crude oil at atmospheric pressure (no gas in solution) and system temperature.
- Saturated Oil Viscosity
  The saturated (bubble-point) oil viscosity is defined as the viscosity of the crude oil at the bubble-point pressure and reservoir temperature.
- Undersaturated Oil Viscosity
  The undersaturated oil viscosity is defined as the viscosity of the crude oil at a pressure above the bubble-point pressure and reservoir temperature.

METHODS OF CALCULATING THE VISCOSITY OF THE DEAD OIL

Beal's Correlation

From a total of 753 values for dead oil viscosity at and above 100°F, Beal (1946) developed a graphical correlation for determining the viscosity of the dead oil as a function of temperature and the API gravity of the crude. This graphical correlation is shown in Figure 4-29. Standing (1981) expressed the proposed graphical correlation in a mathematical relationship as follows:

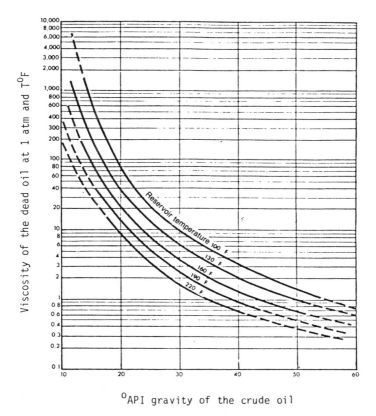

Figure 4-29. Beal's dead oil correlation. Permission to publish by the Society of Petroleum Engineers of AIME. Copyright SPE-AIME.

$$\mu_{od} = \left(0.32 + \frac{1.8(10^7)}{API^{4.53}}\right)\left(\frac{360}{T - 260}\right)^a \tag{4-94}$$

with

$$a = 10^{(0.43 + 8.33/API)}$$

where μ_{od} = viscosity of the dead oil as measured at 14.7 psia and reservoir
temperature, cp
T = temperature, °R

Beal's graphical correlation is composed of five isotherms covering the temperature range from 100°F. Beal's correlation reproduces the original data with an average error of 24.2%.

Beggs-Robinson Correlation

Beggs and Robinson (1975) developed an empirical correlation for determining the viscosity of the dead oil. The correlation originated from analyzing 460 dead oil viscosity measurements. The proposed relationship is expressed mathematically as follows:

$$\mu_{od} = 10^X - 1 \tag{4-95}$$

where $X = Y (T - 460)^{-1.163}$
$Y = 10^Z$
$Z = 3.0324 - 0.02023°API$

An average error of -0.64% with a standard deviation of 13.53% was reported for the correlation when tested against the data used for its development. Sutton and Farshad (1986) reported an error of 114.3% when the correlation was tested against 93 cases from the literature.

Glaso's Correlation

Glaso (1980) proposed a generalized mathematical relationship for computing the dead oil viscosity. The relationship was developed from experimental measurements on 26 crude oil samples. The correlation has the following form:

$$\mu_{od} = [3.141(10^{10})] (T - 460)^{-3.444} [Log(API)]^a \tag{4-96}$$

where a = 10.313 [Log(T − 460)] − 36.447

The above expression can be used within the range of 50–300°F for the system temperature and 20.1–48.1° for the API gravity of the crude.

Sutton and Farashad (1986) concluded that Glaso's correlation showed the best accuracy of the three previous correlations.

METHODS OF CALCULATING THE VISCOSITY OF THE SATURATED CRUDE OIL

Chew-Connally Correlation

Chew and Connally (1959) presented a graphical correlation to adjust the dead oil viscosity according to the gas solubility at saturation pressure. The correlation, as shown in Figure 4-30, was developed from 457 crude oil sam-

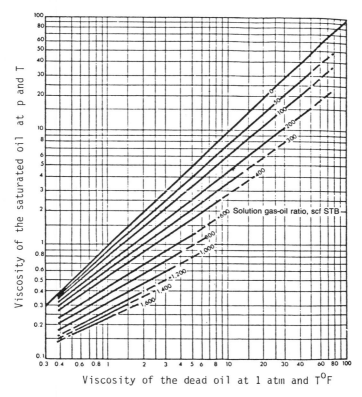

Figure 4-30. Chew and Connally viscosity correlation. Permission to publish by the Society of Petroleum Engineers of AIME. Copyright SPE-AIME.

ples. Standing (1981) expressed the correlation in a mathematical form as follows:

$$\mu_{ob} = (10)^a \, (\mu_{od})^b \tag{4-97}$$

with

$$a = R_s \, [2.2(10^{-7}) \, R_s - 7.4(10^{-4})]$$

$$b = \frac{0.68}{10^c} + \frac{0.25}{10^d} + \frac{0.062}{10^e}$$

$$c = 8.62(10^{-5})R_s$$

$$d = 1.1(10^{-3})R_s$$

$$e = 3.74(10^{-3})R_s$$

where μ_{ob} = viscosity of the oil at the bubble-point pressure, cp
$\quad\quad \mu_{od}$ = viscosity of the dead oil at 14.7 psia and reservoir temperature, cp

Chew and Connally's development data encompassed the following ranges of values for the independent variables:

Pressure, psia: 132–5,645
Temperature, °F: 72–292
Gas solubility, scf/STB: 51–3,544
Dead oil viscosity, cp: 0.377–50

Beggs-Robinson Correlation

From 2,073 saturated oil viscosity measurements, Beggs and Robinson (1975) proposed an empirical correlation for estimating the saturated oil viscosity. The proposed mathematical expression has the following form:

$$\mu_{ob} = a(\mu_{od})^b \tag{4-98}$$

where $a = 10.715(R_s + 100)^{-0.515}$
$\quad\quad\quad b = 5.44(R_s + 150)^{-0.338}$

The reported accuracy of the correlation is $- 1.83\%$ with a standard deviation of 27.25%.

The ranges of the data used to develop Beggs and Robinson's equation are:

Pressure, psia: 132—5,265
Temperature, °F: 70—295
API gravity: 16—58
Gas solubility, scf/STB: 20—2,070

Khan's Correlation

Based on data from Saudi Arabian crude oils, Khan et al. (1987) proposed empirical equations for estimating oil viscosity at or below the bubble-point pressure. A total of 75 bottom-hole samples with 1,841 viscosity data points from 62 fields was used in developing the following equations:

Viscosity at the bubble-point pressure:

$$\mu_{ob} = \frac{0.09 \ (\gamma_g)^{0.5}}{(R_s)^{1/3} \ \theta_r^{4.5}(1 - \gamma_o)^3} \tag{4-99}$$

Viscosity below the bubble-point pressure:

$$\mu_b = \mu_{ob} \ (p/p_b)^{-0.14} \ EXP \ (- 2.5(10^{-4})(p - p_b)) \tag{4-100}$$

where μ_{ob} = oil viscosity at the bubble-point, cp
μ_b = oil viscosity below the bubble-point, cp
T = system temperature, °R
θ_r = T/460
γ_o = specific gravity of the stock-tank oil
γ_g = specific gravity of the solution gas
p_b = bubble-point pressure, psi

The correlation proposed by Khan and his co-workers can be used within the following ranges of values:

Pressure, psia: 100—4,315
Temperature, °F: 75—240
API gravity: 14.3—44.6
Gas solubility, scf/STB: 24—1,901

The reported average relative error of the correlation is − 1.33%.

METHODS OF CALCULATING THE VISCOSITY OF THE UNDERSATURATED CRUDE OIL

At pressures greater than the bubble-point pressure of the crude oil, a further adjustment should be made to the bubble-point oil viscosity, μ_{ob}, to account for the compression and the degree of undersaturation in the reservoir.

Beal's Correlation

Beal (1946) presented a graphical correlation, as shown in Figure 4-31, for estimating the viscosity of undersaturated oil. The correlation was generated from analyzing 52 viscosity observations taken from 26 crude oil sam-

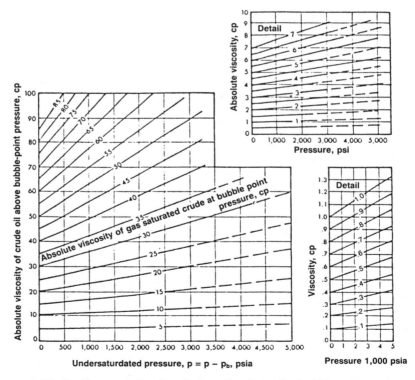

Figure 4-31. Beal's correlation for oil viscosity above the bubble-point pressure. Permission to publish by the Society of Petroleum Engineers of AIME. Copyright SPE-AIME.

ples. Standing (1981) expressed Beal's chart in the following mathematical form:

$$\mu_o = \mu_{ob} + 0.001(p - p_b)(0.024\, \mu_{ob}^{1.6} + 0.038\, \mu_{ob}^{0.56}) \qquad (4\text{-}101)$$

where μ_o = viscosity of the undersaturated oil, cp
The reported average error of Beal's correlation is 2.7%.

Vasquez-Beggs Correlation

From a total of 3,593 data points, Vasquez and Beggs (1976) proposed the following expression for estimating the viscosity of undersaturated crude oil:

$$\mu_o = \mu_{ob} \left(\frac{p}{p_b} \right)^m \qquad (4\text{-}102)$$

where $m = 2.6\, p^{1.187}\, 10^a$
  with $a = -3.9(10^{-5})\, p - 5$

The data used in developing the above correlation have the following ranges:

Pressure, psia: 141–9,515
Gas solubility, scf/STB: 9.3–2,199
Viscosity, cp: 0.117–148
Gas gravity: 0.511–1.351
API gravity: 15.3–59.5

The average error of the viscosity correlation is reported as -7.54%.

Khan's Correlation

From a total of 1,503 experimental data points on Saudi Arabian crude oils, Khan et al. (1987) developed the following equation for determining the viscosity of the undersaturated crude oil:

$$\mu_o = \mu_{ob}\, EXP\, [9.6(10^{-5})(p - p_b)] \qquad (4\text{-}103)$$

The authors reported an average absolute relative error of 2% for the correlation.

In general, all the above correlations provide satisfactory results for predicting the undersaturated oil viscosity.

Example 4-18. Given the following reservoir and PVT data,

p = 4,474.7 psia	p_b = 2,744.7 psia	T = 600°R
R_s = 603 scf/STB	API = 35°	γ_g = 0.774

calculate the viscosity of the crude oil at 4,474.7 psia by using:

a. Beal and Chew-Connally Correlations
b. Beggs-Robinson Correlation
c. Glaso's Correlation
d. Khan's Correlation

Compare the results with the experimental viscosity of 0.6475 cp.

Solution.

a. **Solution by Using Beal and Chew-Connally Correlations.**

Step 1. Solve for the coefficient a of Equation 4-94, to give

$$a = 4.6061$$

Step 2. Calculate the viscosity of the dead oil by applying Equation 4-94, to yield

$$\mu_{od} = 2.5842 \text{ cp}$$

Step 3. Calculate the coefficients a–e of Equation 4-97, to give

$$a = -0.3662$$
$$b = 0.6579$$
$$c = 0.05198$$
$$d = 0.6633$$
$$e = 2.2552$$

Step 4. Solve for the viscosity of the saturated oil by applying Equation 4-97

$$\mu_{ob} = 0.8036 \text{ cp}$$

Step 5. Calculate the viscosity of the undersaturated crude by solving Equation 4-101, to give

$$\mu_o = 0.7664 \text{ cp}$$

b. **Solution by Using the Beggs-Robinson Equation.**

Step 1. Calculate the coefficients Z, Y, and X of Equation 4-95, to yield

Z = 2.310189
Y = 204.2627
X = 0.652

Step 2. Determine the viscosity of the dead oil by applying Equation 4-95

μ_{od} = 3.4874 cp

Step 3. Compute the coefficients a and b of Equation 4-98

a = 0.36628
b = 0.57975

Step 4. Solve for the viscosity of the saturated crude oil by applying Equation 4-98, to yield

μ_{ob} = 0.75567 cp

Step 5. Calculate the coefficients a and m of Equation 4-102, to give

a = − 5.1745
m = 0.3749

Step 6. Determine the viscosity of the undersaturated crude oil from Equation 4-102.

μ_o = 0.9076 cp

c. **Solution by Using Glaso's Correlation.**

Step 1. Calculate the coefficient a of Equation 4-96

a = − 14.314

Step 2. Solve for the viscosity of the dead oil by using Equation 4-96

μ_{od} = 2.348 cp

Step 3. Calculate the coefficients a and b of Equation 4-97, to give

a = − 0.3662
b = 0.6579

Step 4. Solve for the viscosity of the saturated oil by applying Equation 4-97.

μ_{ob} = 0.7545 cp

Step 5. Determine the viscosity of the undersaturated crude oil from Equation 4-101, to give

μ_o = 0.8371 cp

d. Solution by Using Khan's Correlation.

Step 1. Calculate the temperature parameter Θ_r

$$\Theta_r = 1.3043$$

Step 2. Solve for the viscosity of the saturated oil from Equation 4-99, to yield

$$\mu_{ob} = 0.7809 \text{ cp}$$

Step 3. Determine the viscosity of the undersaturated crude oil from Equation 4-103, to give

$$\mu_o = 0.922 \text{ cp}$$

Summary of the Results

Method	μ_{od}	μ_{ob}	μ_o
Beal	2.5842	0.8036	0.9121
Beggs-Robinson	3.4874	0.75567	0.9076
Glaso	2.348	0.7545	0.8371
Khan	—	0.7809	0.922

CALCULATING VISCOSITIES OF CRUDE OILS FROM THEIR COMPOSITION

Like all intensive properties, viscosity is a strong function of the pressure, the temperature, and the composition of the oil. In general, this relationship can be expressed mathematically by the following function:

$$\mu = f\,(p,\ T,\ x_1,\ \ldots,\ x_n)$$

where $\displaystyle\sum_{i=1}^{n} x_i = 1$

In compositional material balance computations, the compositions of the reservoir gases and oils are known. The calculation of the viscosities of these fluids from their compositions is required for a true and complete compositional material balance.

There are two widely used empirical correlations of calculating the viscosity of crude oils from their compositions. These are:

- Lohrenz-Bray-Clark Correlation
- Little-Kennedy Correlation

Lohrenz, Bray-Clark Correlation

Lohrenz, Bray, and Clark (1964) developed an empirical correlation for determining the viscosity of the saturated oil from its composition. The proposed correlation has been enjoying great acceptance and application by engineers in the petroleum industry.

The authors proposed the following generalized form of the equation:

$$\mu_{ob} = \mu_{oL} + (\xi_m)^{-1}\,[[a_0 + a_1\,\rho_r + a_2\,\rho_r^2 + a_3\,\rho_r^3 + a_4\,\rho_r^4]^4 - 0.0001] \tag{4-104}$$

with the mixture viscosity parameter ξ_m defined mathematically by

$$\xi_m = \frac{5.4402\,(T_{pc})^{1/6}}{(MW_a)^{0.5}(p_{pc})^{2/3}} \tag{4-105}$$

where μ_{oL} = "Life" oil viscosity at system temperature and atmosphere pressure, cp
$\quad\quad a_0 = 0.1023$
$\quad\quad a_1 = 0.023364$
$\quad\quad a_2 = 0.058533$
$\quad\quad a_3 = -\,0.040758$
$\quad\quad a_4 = 0.0093724$
$\quad\quad T_{pc}$ = pseudo-critical temperatures of the crude, °R
$\quad\quad p_{pc}$ = pseudo-critical pressure of the crude, psia
$\quad\quad MW_a$ = apparent molecular weight of the mixture
$\quad\quad \rho_r$ = reduced oil density and is given by the following mathematical expression:

$$\rho_r = (MW_a)^{-1}\,\rho_o\left[\sum_{\substack{i=1 \\ i \neq C_{7+}}}^{n}[(x_i\,MW_i\,V_{ci})] + x_{C_{7+}}\,V_{C_{7+}}\right] \tag{4-106}$$

where ρ_o = oil density at the prevailing system condition, lb/ft³
$\quad\quad x_i$ = mole fraction of component i
$\quad\quad MW_i$ = molecular weight of component i
$\quad\quad V_{ci}$ = critical volume of component i, ft³/lb
$\quad\quad x_{C_{7+}}$ = mole fraction of C_{7+}
$\quad\quad V_{C_{7+}}$ = critical volume of C_{7+}, ft³/lb-mole
$\quad\quad n$ = number of components in the mixture

Lohrenz and co-workers proposed the following expression for calculating $V_{C_{7+}}$:

$$V_{C_{7+}} = 21.573 + 0.015122 \ MW_{C_{7+}} - 27.656 \ \gamma_{C_{7+}}$$
$$+ 0.070615 \ MW_{C_{7+}} \ \gamma_{C_{7+}} \qquad (4\text{-}107)$$

where $MW_{C_{7+}}$ = molecular weight of C_{7+}
$\gamma_{C_{7+}}$ = specific gravity of C_{7+}

The authors employed the following relationship to determine the viscosity of the life oil:

$$\mu_{oL} = \frac{\displaystyle\sum_{i=1}^{n} (x_i \ \mu_i \ MW_i^{0.5})}{\displaystyle\sum_{i=1}^{n} (x_i \ MW_i^{0.5})} \qquad (4\text{-}108)$$

where μ_i is the viscosity of component i in the mixture at the atmospheric pressure and system temperature and is calculated from the following equations:

$$\mu_i = \frac{34(10^{-5})(T_{ri})^{0.94}}{\xi_i}, \ \text{for } T_{ri} \leq 1.5 \qquad (4\text{-}109)$$

$$\mu_i = \frac{17.78(10^{-5})[4.58 \ T_{ri} - 1.67]^{0.625}}{\xi_i}, \ \text{for } T_{ri} > 1.5 \qquad (4\text{-}110)$$

where T_{ri} = reduced temperature of component i; T/T_{ci}
ξ_i = viscosity parameter of component i and is given by:

$$\xi_i = \frac{5.4402 \ (T_{ci})^{1/6}}{(MW_i)^{0.5}(p_{ci})^{2/3}} \qquad (4\text{-}111)$$

It should be noted that when applying Equation 4-106, the viscosity of the nth component, i.e., C_{7+}, must be calculated by any of the dead oil viscosity correlations discussed before.

Experience with the Lohrenz et al. equation has shown that the correlation is extremely sensitive to the density of the oil and the critical volume of C_{7+}. When observed viscosity data are available, the critical volume of the plus fraction can be adjusted until a match with the experimental data is achieved.

In cases where the adjustment of critical volume of the plus-fraction is inadequate, the values of the coefficients a_0–a_4 should be modified.

Little-Kennedy Correlation

Little and Kennedy (1968) proposed an empirical equation for predicting the viscosity of the saturated crude oil. The correlation was originated from studying the behavior of 828 distinct crude oil systems representing 3,349 viscosity measurements.

The equation is similar in form to Van der Waals' equation-of-state. The authors expressed the equation in the following form:

$$\mu_{ob}^3 - \left(b_m + \frac{p}{T}\right)\mu_{ob}^2 + \left(\frac{a_m}{T}\right)\mu_{ob} - \left(\frac{a_m\,b_m}{T}\right) = 0 \tag{4-112}$$

where μ_{ob} = viscosity of the saturated crude oil, cp
    p = system pressure, psia
    T = system temperature, °R
    a_m, b_m = mixture coefficient parameters

Little and Kennedy proposed the following relationships for calculating the parameters a_m and b_m:

$$a_m = EXP\ (A) \tag{4-113}$$

$$b_m = EXP\ (B) \tag{4-114}$$

where A and B are defined by the equations

$$\begin{aligned}
A = A_0 &+ A_1/T + A_2(MW)_{C7+} + A_3(MW/\gamma)_{C7+} + A_4(\rho_o/T) \\
&+ A_5(\rho_o/T)^2 + A_6(MW_a) + A_7(MW_a)^3 + A_8(MW_a\,\rho_o) \\
&+ A_9(MW_a\,\rho_o)^3 + A_{10}(\rho_o)^2
\end{aligned} \tag{4-115}$$

$$\begin{aligned}
B = B_0 &+ B_1/T + B_2/T^4 + B_3(\gamma)_{C7+}^3 + B_4(\gamma)_{C7+}^4 \\
&+ B_5(MW/\gamma)_{C7+}^4 + B_6(\rho_o/T)^4 + B_7(MW_a) + B_8(MW_a\,\rho_o) \\
&+ B_9(MW_a\,\rho_o)^4 + B_{10}(\rho_o)^3 + B_{11}(\rho_o)^4
\end{aligned} \tag{4-116}$$

where $(MW)_{C7+}$ = molecular weight of C_{7+}
    γ_{C7+} = specific gravity of C_{7+}
    ρ_o = density of the saturated crude oil, lb/ft^3
    T = temperature, °R
    MW_a = apparent molecular weight of the oil
    A_0–A_{10}, B_0–B_{11} = coefficients of Equations 4-115 and 4-116, and are given in Table 4-2.

Table 4-2
Coefficients for Equations 4-113 and 4-114

$A_0 = 21.918581$	$B_0 = -2.6941621$
$A_1 = -16815.621$	$B_1 = 3757.4919$
$A_2 = 0.023315983$	$B_2 = -0.31409829 \ (10^{12})$
$A_3 = -0.019218951$	$B_3 = -33.744827$
$A_4 = 479.783669$	$B_4 = 31.333913$
$A_5 = -719.808848$	$B_5 = 0.24400196 \ (10^{-10})$
$A_6 = -0.096858449$	$B_6 = 4.632634 \ (10^4)$
$A_7 = 0.54324554 \ (10^{-6})$	$B_7 = -0.037022195$
$A_8 = 0.0021040196$	$B_8 = 0.0011348044$
$A_9 = -0.4332274341 \ (10^{-11})$	$B_9 = -0.0547665355 \ (10^{-15})$
$A_{10} = -0.0081362043$	$B_{10} = 0.0893548761 \ (10^{-3})$
	$B_{11} = -2.05018084 \ (10^{-6})$

Equation 4-110 can be solved for the viscosity of the saturated crude oil and the pressure and temperature of interest by extracting the minimum real root of the equation. An iterative technique, e.g., Newton-Raphson iterative method, can be employed to solve the proposed equation.

The correlation is reported to produce an average error of 9.9% when tested against the experimental data used in developing the equation.

BUBBLE-POINT PRESSURE

The bubble-point pressure p_b of a hydrocarbon system is defined as the highest pressure at which a bubble of gas is first liberated from the oil. This important property can be measured experimentally for a crude oil system by conducting a constant-composition expansion test (i.e., flash liberation test).

In the absence of the experimentally measured bubble-point pressure, it is necessary for the engineer to make an estimate of this crude oil property from the readily available measured producing parameters. Several graphical and mathematical correlations for determining p_b have been proposed during the last four decades. These correlations are essentially based on the assumption that the bubble-point pressure is a strong function of gas solubility, gas gravity, oil gravity, and temperature, or

$$p_b = f \ (R_s, \ \gamma_g, \ API, \ T)$$

Several ways of combining the above parameters in a graphical form or a mathematical expression are presented below:

Standing's Correlation

Based on 105 experimentally measured bubble-point pressures on 22 hydrocarbon systems from California oil fields, Standing (1947) proposed a graphical correlation for determining the bubble-point pressure of crude oil systems. The correlation is shown in Figure 4-20. The correlating parameters in the proposed correlation are the gas solubility R_s, gas gravity γ_g, oil API gravity, and the system temperature. The reported average error is 4.8%.

In a mathematical form, Standing (1981) expressed the graphical correlation by the following expression:

$$p_b = 18.2 \left[(R_s/\gamma_g)^{0.83} (10)^a - 1.4 \right] \tag{4-117}$$

with $a = 0.00091 (T - 460) - 0.0125 (API)$

where p_b = bubble-point pressure, psia

$\quad\quad T$ = system temperature, °R

Standing's correlation should be used with caution if nonhydrocarbon components are known to be present in the system.

Lasater's Correlation

Lasater (1958) presented a graphical correlation for estimating the bubble-point pressure. The correlation was developed from a total of 158 experimentally measured saturation pressures of crude oil samples. The samples were collected from oil reservoirs in Canada, Western and Mid-Continental United States, and South America. The correlation is presented in Figure 4-21. Lasater employed the same correlating parameters used by Standing in developing his correlation.

The following steps summarize the procedure of using Lasater's graphical correlation:

Step 1. Enter the left side of the graph with the gas solubility of the crude oil system.

Step 2. Proceed horizontally to the API gravity of the oil.

Step 3. Drop vertically to the correlating curve.

Step 4. Move horizontally to the system temperature line.

Step 5. Rise vertically to the gas gravity line.

Step 6. Move horizontally to the right and read the required bubble-point pressure.

It should be noted that the natural gases associated with the crude oils samples used in Lasater's study were essentially free of nonhydrocarbons. Lasater reported an average error of 3.8% between measured and calculated bubble-point pressure.

Vasquez-Beggs' Correlation

Vasquez and Beggs' gas solubility correlation as presented by Equation 4-53 can be solved for the bubble-point pressure p_b to give:

$$p_b = [(C_1 \, R_s / \gamma_{gs})(10)^a]^{C_2} \tag{4-118}$$

with $a = -C_3 \, API/T$

The gas specific gravity γ_{gs} at the reference separator pressure is defined by Equation 4-44. The coefficients C_1, C_2, and C_3 have the following values:

Coefficient	API $\leq$ 30	API > 30
C_1	27.62	56.18
C_2	0.914328	0.84246
C_3	11.172	10.393

Glaso's Correlation

Glaso (1980) used 45 oil samples, mostly from the North Sea hydrocarbon system, to develop an accurate correlation for bubble-point pressure prediction. Glaso proposed the following expression:

$$\text{Log} \, (p_b) = 1.7669 + 1.7447 \, \text{Log} \, (p_b^*) - 0.30218[\text{Log} \, (p_b^*)]^2 \tag{4-119}$$

where p_b^* is a "correlating number" and defined by the following equation:

$$p_b^* = (R_s / \gamma_g)^a (t)^b (API)^c \tag{4-120}$$

where R_s = gas solubility, scf/STB
           t = system temperature, °F
           γ_g = average specific gravity of the total surface gases
    a, b, c = coefficients of the above equation having the following values: a = 0.816, b = 0.172, c = -0.989

For volatile oils, Glaso recommends that the temperature exponent b of Equation 4-120 be slightly changed, to the value of 0.130. To account for the presence of nonhydrocarbon components in the crude oil system, Glaso devised a correction procedure to adjust p_b as calculated from Equation 4-119. The procedure is summarized in the following steps:

Step 1. Calculate the nitrogen correction factor from the following expression:

$$(\text{Corr})_{N_2} = 1 + [(a_1\text{API} + a_2)t + a_3(\text{API}) - a_4]y_{N_2}$$
$$+ [a_5(\text{API})^{a_6}t + a_6(\text{API})^{a_7} - a_8](y_{N_2})^2 \qquad (4\text{-}121)$$

where $(\text{Corr})_{N_2}$ = nitrogen correction factor
   t = system temperature, °F
   y_{N_2} = mole fraction of nitrogen in total surface gases
   $a_1 - a_8$ = coefficients of the correlation having the following values

   $a_1 = -2.65 \times 10^{-4}$ $a_2 = 5.5 \times 10^{-3}$
   $a_3 = 0.0391$ $a_4 = 0.8295$
   $a_5 = 1.954 \times 10^{-11}$ $a_6 = -4.699$
   $a_7 = 0.027$ $a_8 = 2.366$

Step 2. Compute the carbon dioxide correction factor as follows:

$$(\text{Corr})_{CO_2} = 1.0 - 693.8 \ (y_{CO_2})(t)^{-1.553}$$

where $(\text{Corr})_{CO_2}$ = CO_2 correction factor
   y_{CO_2} = mole fraction of CO_2 in total surface gases

Step 3. Determine the hydrogen sulfide correction factor from the following expression:

$$(\text{Corr})_{H_2S} = 1.0 - (0.9035 + 0.0015 \ \text{API})y_{H_2S}$$
$$+ 0.019 \ (45 - \text{API}) \ (y_{H_2S})^2 \qquad (4\text{-}122)$$

where $(\text{Corr})_{H_2S}$ = H_2S correction factor
   y_{H_2S} = mold fraction of H_2S in total surface gases

Step 4. Calculate the corrected bubble-point pressure of the crude oil from the expression:

$$(p_b)_c = p_b \cdot (\text{Corr})_{N_2} \cdot (\text{Corr})_{CO_2} \cdot (\text{Corr})_{H_2S} \qquad (4\text{-}123)$$

where $(p_b)_c$ = corrected bubble-point pressure, psia
   p_b = uncorrected bubble-point pressure as calculated from Equation 4-119

Glaso reported an average error of 1.28% with a standard deviation of 6.98% for the correlation.

Marhoun's Correlation

Marhoun (1988) used 160 experimentally determined bubble-point pressures from the PVT analysis of 69 Middle Eastern hydrocarbon mixtures to develop a correlation for estimating p_b. Marhoun proposed the following expression:

$$p_b = a\ R_s^b\ \gamma_g^c\ \gamma_o^d\ T^e \tag{4-124}$$

where T = temperature, °R
γ_o = stock-tank oil gravity
a–e = coefficients of the correlation having the following values

a = 5.38088 × 10^{-3}	b = 0.715082
c = − 1.87784	d = 3.1437
e = 1.32657	

The reported average absolute relative error for the correlation is 3.66% when compared with the experimental data used to develop the correlation.

Sutton and Farshad (1986) compared the performance of the first four correlations against the bubble-point pressure data of Gulf of Mexico crude oil systems, and concluded that Glaso's correlation yields the best results of p_b predictions.

Example 4-19. A crude oil system exists at a pressure higher than the bubble-point pressure. The gas-oil ratio and system temperature are 580 scf/STB and 131°F. The API gravity of the oil is 39.7°. The average specific gravity of the separator gas is 1.075. Calculate the bubble-point pressure by using the correlations of:

a. Standing
b. Lasater
c. Vasquez-Beggs
d. Glaso
e. Marhoun

Compare the results with the experimental p_b of 1,709 psia.

Solution.

a. Standing's Correlation.

- Calculate the coefficient a of Equation 4-117 to give

$$a = - 0.377$$

- Solve for p_b by applying Equation 4-117, to yield

  $p_b = 1,389$ psia

b. Lasater's Correlation.

- From Figure 4-21, $p_b = 1,550$ psia

c. Vasquez-Beggs' Correlation.

- Calculate the correlating parameter a of Equation 4-18, to give

  $a = -0.6981$

- $p_b = 1,537$ psia

d. Glaso's Correlation.

- Determine the correlating number p_b^* from Equation 4-120, to give

  $p_b^* = 10.287$

- Calculate p_b from Equation 4-119, to give

  $p_b = 1,673$ psia

e. Marhoun's Correlation.

- Apply Equation 4-125, to give

  $p_b = 1,160$ psia

SURFACE TENSION

The surface tension is defined as the force exerted on the boundary layer between a liquid phase and a vapor phase per unit length. This force is caused by differences between the molecular forces in the vapor phase and those in the liquid phase, and also by the imbalance of these forces at the interface. The surface can be measured in the laboratory and is usually expressed in dynes per centimeter.

The surface tension is an important property in reservoir engineering calculations and designing enhanced oil recovery projects.

Calculation of the Surface Tension

Sugden (1924) suggested a relationship between the surface tension of a pure liquid in equilibrium with its own vapor and densities of both phases. The relationship is expressed mathematically as follows:

$$\sigma = \left[\frac{P_{ch}\ (\rho_L - \rho_V)}{MW}\right]^4 \tag{4-125}$$

where σ is the surface tension and P_{ch} is a temperature independent parameter and is called the *parachor*.

The parachor is a dimensionless constant characteristic of a pure compound and is calculated by imposing experimental surface tension and density data on Equation 4-125 and solving for P_{ch}. Parachor, values for selected pure compounds, as given by Weinaug and Katz (1943), are given in Table 4-3.

Table 4-3
Parachors for Pure Substances

Component	Parachor	Component	Parachor
CO_2	78.0	n-C_4	189.9
N_2	41.0	i-C_5	225.0
C_1	77.0	n-C_5	231.5
C_2	108.0	n-C_6	271.0
C_3	150.3	n-C_7	312.5
i-C_4	181.5	n-C_8	351.5

Figure 4-32 correlates the parachor with molecular weights for paraffins and mixtures (plus-fractions). The parachor-molecular weight of hydrocarbon mixtures can be closely described by the following equation:

$$(P_{ch})_{c+} = a_1 + a_2(MW)_{c+} + a_3(MW)^2_{c+} + a_4/(MW)_{c+} \tag{4-126}$$

where $(P_{ch})_{c+}$ = parachor of the plus-fraction, or the hydrocarbon mixture
$(MW)_{c+}$ = molecular weight of the plus-fraction
a_1-a_4 = coefficients of the equation and have the following values:

$a_1 = -4.6148734$ $\qquad\qquad$ $a_2 = 2.558855$
$a_3 = 3.4004065(10^{-4})$ $\qquad$ $a_4 = 3.767396(10^3)$

Fanchi (1985) used a regression analysis model to develop a relationship between parachor and the critical properties of pure substances. Fanchi used the critical volume V_c, the critical temperature T_c, and the critical pressure p_c as correlating parameters for parachor-critical properties relationships. The following expression is proposed:

$$P_{ch} = a_0 + a_1(V_cMW) + a_2\ T_c + a_3\ H + a_4\ H^2 + a_5\ H^3 + \frac{a_6}{H} \tag{4-127}$$

where H denotes Herzog parameter and is defined by:

$$H = (V_c MW)^{5/6} (T_c)^{0.25}$$
V_c = critical volume, ft^3/lb
T_c = critical temperature, °R
MW = molecular weight

Parachors for Paraffins and Mixtures

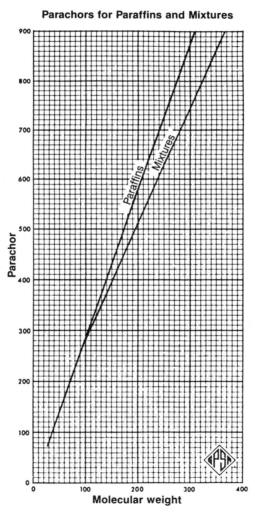

Figure 4-32. Weinaug and Katz parachor vs. molecular weight relationship. Courtesy of the Gas Processors Suppliers Association. Published in the GPSA Engineering Data Book, Tenth Edition, 1987.

The regression coefficients a_0–a_6 have the following values:

$a_0 = -72.765582$ $a_1 = -572.7095626$ $a_2 = -0.5988159389$
$a_3 = 188.1552808$ $a_4 = -0.8211355803$ $a_5 = 0.0128436293$
$a_6 = 653.6146935$

For a complex hydrocarbon mixture, Katz et al. (1943) employed the Sugden correlation for mixtures by introducing the compositions of the two phases into Equation 4-115. The modified expression has the following form:

$$\sigma^{1/4} = \sum_{i=1}^{n} [(P_{ch})_i \, (Ax_i - By_i)] \tag{4-128}$$

with

$$A = \frac{\rho_o}{62.4(MW_a)_L}$$

$$B = \frac{\rho_g}{62.4(MW_a)_g}$$

where
ρ_o = density of the oil phase, lb/ft^3
$(MW_a)_L$ = apparent molecular weight of the oil phase
ρ_g = density of the gas phase, lb/ft^3
$(MW_a)_g$ = apparent molecular weight of the gas phase
x_i = mole fraction of component i in the oil phase
y_i = mole fraction of component i in the gas phase
n = total number of components in the system

Example 4-19. The composition of a crude oil and the associated equilibrium gas is given below. The reservoir pressure and temperature are 4,000 psia and 160°F, respectively. Calculate the surface tension of the mixture.

Component	x_i	y_i
C_1	0.45	0.77
C_2	0.05	0.08
C_3	0.05	0.06
$n\text{-}C_4$	0.03	0.04
$n\text{-}C_5$	0.01	0.02
C_6	0.01	0.02
C_{7+}	0.40	0.01

$$MW_{C_{7+}} = 215$$

Solution.

Step 1. Calculate the density and the apparent molecular weight of the oil phase.

$\rho_o = 46.23$ lb/ft^3 (from Standing's correlation)
$(MW_a)_L = 100.253$

Step 2. Calculate the density and the apparent molecular weight of the gas phase.

$\rho_g = 18.21$ lb/ft^3
$(MW_a)_g = 24.99$

Step 3. Calculate the coefficients A and B of Equation 4-128.

$$A = \frac{46.23}{62.4(100.253)} = 0.00739$$

$$B = \frac{18.21}{62.4(24.99)} = 0.01168$$

Step 4. Calculate the parachor of C_{7+} from Equation 4-126 to give

$(P_{ch})_{C_{7+}} = 551.7$

Step 5. Calculate σ from Equation 4-129.

$\sigma^{1/4} = 0.90359$
$\sigma = 0.6666$ dynes/cm

APPLICATION OF THE CRUDE OIL PVT PROPERTIES IN RESERVOIR ENGINEERING

The importance of determining the PVT properties of the hydrocarbon system under consideration is illustrated by their use in determining the oil flow rate and estimating the initial hydrocarbons in place in the reservoir.

RADIAL FLOW OF CRUDE OILS

When Darcy's Law in differential form (as expressed by Equation 3-81 from Chapter 3) is applied to a crude oil flowing in a radial system, it gives

$$q_o = \frac{1.1271(2\pi rh)kh}{1,000 \; \mu_o} \frac{dp}{dr}$$

where q_o is the oil flow rate in bbl/day.

To express the oil flow rate in STB/day, the oil formation volume factor B_o is incorporated in the equation, to give

$$Q_o = \frac{1.1271(2\pi rh)kh}{1,000\ \mu_o\ B_o} \frac{dp}{dr} \tag{4-129}$$

Separating variables and integrating Equation 4-129 by imposing the inner and outer boundaries, gives

$$Q_o \int_{r_w}^{r_e} \frac{dr}{r} = \frac{1.1271(2\pi h)k}{1,000\ \mu_o\ B_o} \int_{p_{wf}}^{p_e} dp$$

or

$$Q_o = \frac{0.00708\ kh\ (p_e - p_{wf})}{\mu_o\ B_o\ Ln(r_e/r_w)} \tag{4-130}$$

where Q_o = oil flow rate, STB/day
h = pay zone thickness, ft
k = permeability, md
p_e = reservoir pressure, psia
μ_o = oil viscosity, cp
p_{wf} = bottom-hole flowing pressure, psia
B_o = oil formation volume factor, bbl/STB
r_w = wellbore radius, ft
r_e = drainage radius, ft

The application of Equation 4-130 requires the following conditions:

• Steady-state flow
• Laminar flow
• Incompressible fluid
• Homogeneous fluid
• Homogeneous reservoir

The physical properties of the crude oil (included in Equation 4-130) can be determined by the correlation previously presented if the experimental data are not available.

Example 4-20. A well has a drainage radius of 660 ft and a wellbore radius of 0.50 ft. The reservoir pressure and the bottom-hole flowing pressure are 4,000 and 3,800 psia. Given the following additional data

k = 110 md   h = 27 ft
μ_o = 2.1 cp B_o = 1.15 bbl/STB

calculate the oil flow rate, assuming a steady-state flow.

Solution. Calculate the oil flow rate by applying Equation 4-130

$$Q_o = \frac{(0.00708)(110)(27)(4{,}000 - 3{,}800)}{(2.1)(1.15)\ Ln(660/0.5)} = 242 \text{ bbl/day}$$

THE MATERIAL BALANCE EQUATION FOR OIL RESERVOIRS

The material balance equation is one of the reservoir engineer's most useful tools. In a volumetric basis, it states that since the pore volume of a reservoir is constant, the algebraic sum of the volume changes of the oil, the free gas, and the water volumes in the reservoir must be zero. Neglecting the change in the reservoir porosity with the change of the reservoir pressure and assuming the production of certain amounts of reservoir fluids, the material balance equation can be written in a generalized form as follows:

Pore volume occupied by the oil initially in place + pore volume occupied by the gas in the gas cap zone = Pore volume occupied by the remaining oil + pore volume occupied by gas in the gas cap zone + pore volume occupied by the evolved solution gas + pore volume occupied by the new water influx   (4-131)

Capsulizing the reservoir pore volume as an idealized container, as shown schematically in Figure 4-33, the different volumetric terms composing Equation 4-121 can be determined mathematically by employing the oil and gas PVT properties as follows:

Pore Volume Occupied by the Oil Initially in Place

Volume occupied by initial oil in place = N B_{oi} (4-132)

where   N = oil initially in place, STB
        B_{oi} = oil formation volume factor at initial reservoir pressure p_i, bbl/
           STB

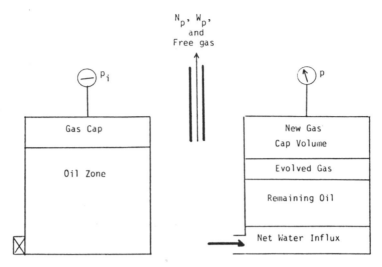

Figure 4-33. A schematic illustration of an idealized oil reservoir.

Pore Volume Occupied by the Gas in the Gas Cap

$$\text{Volume of gas cap} = m \, N \, B_{oi} \qquad (4\text{-}133)$$

where m is a dimensionless parameter and defined as the ratio of gas cap volume to the oil zone volume.

Pore Volume Occupied by the Remaining Oil

$$\text{Volume of the remaining oil} = (N - N_p) \, B_o \qquad (4\text{-}134)$$

where N_p = cumulative oil production, STB
B_o = oil formation volume factor at reservoir pressure p, bbl/STB

Pore Volume Occupied by the Gas Cap at Reservoir Pressure p

As the reservoir pressure drops to a new level p, the gas in the gas cap expands and occupies a larger volume. Assuming no gas is produced from the gas cap during the pressure declines, the new volume of the gas cap can be determined as:

Volume of the gas cap at $p = \left[\dfrac{m\,N}{B_{gi}}\right] B_g$ $\hspace{2cm}$ (4-135)

where B_{gi} = gas formation volume factor at initial reservoir pressure, bbl/scf

$\hspace{1.3cm}$ B_g = current gas formation volume factor, bbl/scf

Pore Volume Occupied by the Evolved Solution Gas

This volumetric term can be determined by applying the following material balance on the solution gas:

$$\left[\begin{array}{c} \text{volume of the evolved} \\ \text{solution gas} \end{array}\right] = \left[\begin{array}{c} \text{volume of gas initially} \\ \text{in solution} \end{array}\right]$$

$$- \left[\begin{array}{c} \text{volume of gas} \\ \text{produced} \end{array}\right]$$

$$- \left[\begin{array}{c} \text{volume of gas} \\ \text{remaining in solution} \end{array}\right]$$

or

$$\left[\begin{array}{c} \text{volume of the evolved} \\ \text{solution gas} \end{array}\right] = [N\,R_{si} - N_p\,R_p - (N - N_p)\,R_s]B_g \hspace{1cm} (4\text{-}136)$$

where N_p = cumulative oil produced, STB

$\hspace{1.3cm}$ R_p = net cumulative produced gas-oil ratio, scf/STB

$\hspace{1.3cm}$ R_s = current gas solubility factor, scf/STB

$\hspace{1.3cm}$ B_g = current gas formation volume factor, bbl/scf

$\hspace{1.3cm}$ R_{si} = gas solubility at initial reservoir pressure, scf/STB

Pore Volume Occupied by the Net Water Influx

$$\text{new water influx} = W_e - W_p\,B_w \hspace{2cm} (4\text{-}137)$$

where W_e = cumulative water influx, bbl

$\hspace{1.3cm}$ W_p = cumulative water produced, STB

$\hspace{1.3cm}$ B_w = water formation volume factor, bbl/STB

Combining Equations 4-131 through 4-137 with Equation 4-131 and re-arranging, gives

$$N = \frac{N_p [B_o + B_g (R_p - R_s)] - (W_e - W_p B_w)}{(B_o - B_{oi}) + B_g (R_{si} - R_s) + [m B_{oi} [B_g - B_{gi}]/B_{gi}]} \qquad (4\text{-}138)$$

The above relationship is called the material balance equation. A more convenient form of the material balance equation can be arrived at by introducing the two-phase formation volume factor B_t in the equation. Since

$$B_t = B_o + (R_{si} - R_s) B_g$$

incorporating the above relationship into Equation 4-138, gives

$$N - \frac{N_p [B_t + (R_p - R_{si}) B_g] - (W_e - W_p B_w)}{(B_t - B_{ti}) + [m B_{ti} (B_g - B_{gi})/B_{gi}]} \qquad (4\text{-}139)$$

Rearranging the above equation gives

$$\frac{N(B_t - B_{ti})}{A} + \frac{NmB_{ti}(B_g - B_{gi})/B_{gi}}{A} + \frac{W_e - W_p B_w}{A} = 1 \qquad (4\text{-}140)$$

with $A = N_p [B_t + (R_p - R_{si}) B_g]$ $\qquad (4\text{-}141)$

Examination of Equation 4-140 reveals that there are three major mechanisms by which oil may be recovered from a naturally occurring reservoir. These three driving forces are:

a. Depletion Drive. Depletion drive is the oil recovery mechanism wherein the production of the oil from its reservoir rock is achieved by the expansion of the original oil volume with all its original dissolved gas. This driving mechanism is represented mathematically by the first term of Equation 4-133, or

$$DDI = N (B_t - B_{ti})/A \qquad (4\text{-}142)$$

where DDI is called the depletion drive index.

b. Segregation Drive. Segregation Drive (gas cap drive) is the mechanism wherein the displacement of oil from the formation is accomplished by the expansion of the original free gas cap. This driving force is described by the second term of Equation 4-140, or

$$SDI = [NmB_{ti} \, (B_g - B_{gi})/B_{gi}]/A \qquad (4\text{-}143)$$

where SDI is termed the segregation drive index.

c. Water Drive. Water drive is the mechanism wherein the displacement of the oil is accomplished by the net encroachment of water into the oil zone. This mechanism is represented by the third term of Equation 4-140, or

$$WDI = (W_e - W_p \, B_w)/A \qquad (4\text{-}144)$$

Using the above driving indices in Equation 4-140, gives

$$DDI + SDI + WDI = 1.0 \qquad (4\text{-}145)$$

Since the sum of the driving indices is equal to one, it follows that if the magnitude of one of the index terms is reduced, then one or both of the remaining terms must be correspondingly increased.

The previous development of the material balance expression as described by Equations 4-132 and 4-138 assumes that the reservoir pore volume does not change as reservoir pressure declines. In some reservoirs, the compressibility of the reservoir rock is of such magnitude that it should be considered in the material balance equation. The effects of connate water expansion and rock compressibility are important in undersaturated depletion drive reservoirs and in water drive reservoirs. However, the effect of pore volume reduction due to the rock expansion as well as the expansion of the connate water can be generally neglected for solution gas drive reservoirs.

The combined effects of connate water and formation expansion can be expressed mathematically by the relationship:

$$E_{f,w} = N \, B_{oi} \, (1 + m) \left[\frac{C_w \, S_{wc} + C_f}{1 - S_{wc}} \right] (p_i - p) \qquad (4\text{-}146)$$

where $E_{f,w}$ = total reduction in the hydrocarbon pore volume due to connate water and formation expansions
  N = initial oil in place, STB
  B_{oi} = initial oil formation volume factor
  C_w = isothermal compressibility coefficient of the connate water
  S_{wc} = connate water saturation
  C_f = isothermal compressibility coefficient of the formation
  p_i = initial reservoir pressure
  p = current reservoir pressure

Equation 4-146 is incorporated with Equation 4-132 and rearranged to produce the following generalized material balance equation:

$$N = \frac{N_p [B_o + B_g (R_p - R_s)] - (W_e - W_p B_w)}{E_{f,w} + (B_o - B_{oi}) + B_g (R_{si} - R_s) + [m \, B_{oi} (B_g - B_{gi})/B_{gi}]} \quad (4\text{-}147)$$

Hall (1953) presented a graphical correlation for determining the formation isothermal compressibility coefficient C_f from the reservoir porosity. The graphical correlation can be closely approximated by the following expression:

$$C_f = a_1 + a_2 \, \theta + a_3 \, \theta^2 + a_4/\theta \quad (4\text{-}148)$$

where C_f = formation isothermal compressibility coefficient, psi^{-1}
$\quad\quad\quad \theta$ = porosity
$\quad\quad\quad a_1\text{-}a_4$ = coefficients and given by
$\quad\quad\quad\quad\quad a_1 = 5.2793512 \times 10^{-6}$ $\quad a_2 = -1.7433696 \times 10^{-5}$
$\quad\quad\quad\quad\quad a_3 = 3.3117681 \times 10^{-5}$ $\quad a_4 = 1.0155993 \times 10^{-7}$

For a volumetric undersaturated oil reservoir, m, W_e, and W_p all equal zero, and since the reservoir pressure is above the bubble-point pressure, gas solubilities (R_{si} and R_s) and R_p are all equal. Imposing these conditions on Equation 4-147 gives:

$$N = \frac{N_p B_o}{E_{f,w} + (B_o - B_{oi})} \quad (4\text{-}149)$$

Notice that m is zero, therefore the total expansion term $E_{f,w}$ is given by

$$E_{f,w} = N \, B_{oi} \left[\frac{C_w \, S_{wc} + C_f}{1 - S_{wc}}\right] (p_i - p) \quad (4\text{-}150)$$

Equation 4-149 can be solved for the oil recovery factor r, to give

$$\frac{N_p}{N} = r = [E_{f,w} + (B_o - B_{oi})]/B_o \quad (4\text{-}151)$$

where   r = recovery factor, dimensionless
$\quad\quad\quad N_p$ = cumulative oil produced, STB

From the mathematical expression of the isothermal compressibility, coefficient of the oil in terms of formation volume factor C_o can be expressed as:

$$C_o = \frac{1}{B_o}\left[\frac{B_o - B_{oi}}{p_i - p}\right] \tag{4-152}$$

combining Equation 4-152 with 4-151 gives

$$r = [E_{f,w}/B_o] + C_o\,(p_i - p) \tag{4-153}$$

PROBLEMS

1. An undersaturated* reservoir has an initial pressure of 3,000 psia and temperature of 234°F. The bubble-point pressure of the crude oil is 2,761 psia. The composition of the crude oil is given below:

Component	x_i
C_1	0.3808
C_2	0.0933
C_3	0.0885
$i\text{-}C_4$	0.0300
$n\text{-}C_4$	0.0300
$i\text{-}C_5$	0.0189
$n\text{-}C_5$	0.0189
C_6	0.0356
C_{7+}	0.3043

If molecular weight and specific gravity of the C_{7+} are 200 and 0.8366, calculate

A. Oil density at the bubble-point pressure by using

1. Standing-Katz correlation
2. Alani-Kennedy correlation

Compare the results with the experimental value of 38.01 lb/ft³.

B. If the API gravity of the stock-tank oil is 42.1° and the specific gravity of the solution gas is 0.996, calculate the initial gas solubility by using Equation 4-56.

C. Calculate the oil density at 3,000 psia.

2. A crude oil system* exists at its bubble-point pressure of 1,708.7 psia and a temperature of 131°F. Given the following data:

* Coats, K. H. and Smart, G. T., "Application of Regression-Based EOS PVT Program to Laboratory Data," *SPE Reservoir Engineering*, May 1986, pp. 277–299.

API = 39.81°
specific gravity of separator gas = 1.075

A. Calculate R_{sb} by using:

1. Beal's correlation
2. Standing's correlation
3. Lasater's correlation
4. Vasquez-Beggs' correlation (assuming an average separator pressure of 114.7 psia)
5. Glaso's correlation

Compare the results with the experimental value of 557 scf/STB.

B. Calculate B_{ob} by applying:

1. Standing's correlation
2. Vasquez-Beggs' correlation
3. Glaso's correlation
4. Equation 4-65
5. Equation 4-66

Compare the results with the experimental value of 1.324 bbl/STB.

C. Calculate the two-phase formation volume factor at 1,600 psia and 1,200 psia.

3. A crude oil system* exists at an initial reservoir pressure of 3,000 psia and temperature of 250°F. The experimental bubble-point pressure of the crude is 2,562 psia. The hydrocarbon mixture has the following composition:

Component	x_i	Component	x_i
CO_2	0.0235	n-C_4	0.02535
N_2	0.0011	i-C_5	0.02615
C_1	0.3521	n-C_5	0.02615
C_2	0.0672	C_6	0.0410
C_3	0.0624	C_{7+}	0.3497
i-C_4	0.02535		

The reported molecular weight and specific gravity of C_{7+} are 213 and 0.8405, respectively. The API gravity of the stock-tank oil is 36.75°. If the specific gravity of the solution gas is 0.82, calculate the viscosity of the crude oil by using:

---

* Coats, K. H. and Smart, G. T., "Application of Regression-Based EOS PVT Program to Laboratory Data," *SPE Reservoir Engineering*, May 1986, pp. 277–299.

1. Beal's correlation
2. Beggs-Robinson correlation
3. Glaso's correlation
4. Khan's correlation
5. Lohrenz-Bray-Clark correlation
6. Little-Kennedy correlation

4. A two-phase hydrocarbon system exists in equilibrium at 2,500 psia and 120°F. The composition of both phases is given below:

Component	x_i	y_i
CO_2	0.0217	0.010
N_2	0.0034	0.005
C_1	0.7064	0.350
C_2	0.1076	0.070
C_3	0.0494	0.050
$i-C_4$	0.0151	0.050
$n-C_4$	0.0151	0.050
$i-C_5$	0.00675	0.030
$n-C_5$	0.00675	0.030
C_6	0.0090	0.040
C_{7+}	0.0588	0.315

The molecular weight and specific gravity of C_{7+} are 135 and 0.80. Calculate the surface tension.

5. A 300 cc crude oil sample was placed in a PVT cell at its bubble-point pressure of 3,000 psig and reservoir temperature of 250°F. A differential liberation was then performed with the recorded measurements as given below:

Cell Pressure psig	Total Volume cc	Volume of Liquid cc	Volume of Liberated Gas scf	Specific Gravity of Liberated Gas
2,547	300	300	0	—
2,143	323.2	286.4	0.1627	0.824
1,645	375.2	271.5	0.184	0.823
1,150	483.2	257.8	0.1671	0.845
647	785.4	245.1	0.1569	0.924

The oil was then subjected to separator tests and remaining oil volume and the volume of the liberated gas were recorded at standard conditions as 179.53 cc and 0.3817 scf, respectively. The API gravity of the stock-tank oil was 41°. Calculate and plot the following properties as a function of cell pressures:

a. Oil formation volume factor
b. Gas solubility factor
c. Total formation volume factor
d. Oil density
e. Gas compressibility factor
f. Oil viscosity
g. Apparent molecular weight of the crude oil (at each pressure)
h. Gas formation volume factor in ft^3/scf
i. Gas viscosity

6. A PVT laboratory test is conducted on a crude oil sample with the following results:

p psig	B_o bbl/STB	R_s scf/STB	γ_o gm/cc
3,100	1.852	1,230	—
2,746	1.866	1,230	0.6090
2,400	1.771	1,059	0.6240
2,200	1.725	972	0.6314
1,600	1.599	737	0.6543
1,000	1.488	529	0.6767
394	1.371	321	0.7028
0*	1.0	0	0.7687

* Temperature at this pressure is 60°F.

The bubble point pressure and system temperature are 2,746 psig and 234°F. Calculate

a. Oil density at 3,100 psig
b. Total formation volume factor at each pressure
c. Gas compressibility factors at cell pressures of 2,400 to 1,000 psig.

7. A reservoir with the PVT properties as given in Problem 6 is to be water flooded at 1,000 psi. Given the following rock and connate water properties at 1,000 psia:

$S_g = 0.11$
$S_w = 0.24$
$C_f = 1.2(10^{-6})$ psia
water salinity = 0

Calculate

a. The total isothermal compressibility coefficient without gas going back in solution
b. The total isothermal compressibility coefficient with gas going back in solution.

8. If T denotes the system temperature, what is the difference between $\partial B_o/\partial T$ and $\partial B_{ob}/\partial T$? Which is greater?

9. Why is differential separation more efficient than flash separation?

10. An oil well has a drainage radius of 660 ft and a wellbore radius of 0.333 ft. The well is producing a crude oil with 40°API gravity. The current reservoir pressure is 3,800 psia and the reservoir temperature is 150°F. The following additional oil and rock data are available:

    k = 75 md
    h = 36 ft
    p_{wf} = 3,450 psia
    solution gas gravity = 0.70

    If the bubble-point pressure is 3,400 psia, calculate

    a. Oil flow rate in STB/day
    b. Oil density at 3,800 psia

11. A volumetric-undersaturated oil reservoir has the following oil and reservoir properties:

    Initial reservoir pressure = 4,200 psia
    Bubble-point pressure = 3,700 psia
    Oil gravity = 42°API
    Solution gas gravity = 0.65
    Reservoir temperature = 200°F

    Calculate the oil recovery factor at 4,000, at 3,900, and at 3,800 psia (neglect $E_{f,w}$ term).

12. A volumetric oil reservoir has an oil bulk volume of 11 MMMbbl with a bulk volume for the gas cap as 1.3 MMMbbl. The initial reservoir pressure is 3,000 psia and the reservoir temperature is 120°F. The following additional data are available:

    Producing oil gravity = 45°API
    Current reservoir pressure = 2,500 psia
    Net cumulative produced gas-oil ratio = 900 scf/STB
    Connate water saturation = 0.20
    Porosity = 0.16
    Specific gravity of the solution gas = 0.72

    Assuming that the average net cumulative producing gas-oil ratio R_p will remain constant at 900 scf/STB, calculate total oil production and the driving indices at 2,500, at 2,000, at 1,500, and at 1,000 psia.

REFERENCES

1. Ahmed, T., "Compositional Modeling of Tyler and Mission Canyon Formation Oils with CO_2 and Lean Gases," final report submitted to Montanans on a New Track for Science (MONTS) (Montana National Science Foundation Grant Program), 1985–1988.
2. Alani, G. H. and Kennedy, H. T., "Volume of Liquid Hydrocarbons at High Temperatures and Pressures," *Trans. AIME* (1960), Vol. 219, pp. 288–292.
3. Baker, O. and Swerdloff, W., "Calculations of Surface Tension-3: Calculations of Surface Tension Parachor Values," *OGJ*, December 5, 1955, Vol. 43, p. 141.
4. Beal, C., "The Viscosity of Air, Water, Natural Gas, Crude Oils and its Associated Gases at Oil Field Temperatures and Pressures," *Trans. AIME* (1946), Vol. 165, pp. 94–112.
5. Beggs, H. D. and Robinson, J. R., "Estimating the Viscosity of Crude Oil Systems," *JPT*, September 1975, pp. 1140–1141.
6. Chew, J. and Connally, Jr., C. A., "A Viscosity Correlation for Gas-Saturated Crude Oils," *Trans. AIME* (1959), Vol. 216, pp. 23–25.
7. Dodson, C. R., Goodwill, D., and Mayer, E. H., "Application of Laboratory PVT Data to Reservoir Engineering Problems," *Trans. AIME* (1953), Vol. 198, pp. 287–298.
8. Fanchi, J. R., "Calculation of Parachors for Composition Simulation," *JPT*, November, 1985, pp. 2049–2050.
9. Glaso, O., "Generalized Pressure-Volume-Temperature Correlations," *JPT*, May 1980, pp. 785–795.
10. Hall, H. N., "Compressibility of Reservoir Rocks," *Trans. AIME* (1953), Vol. 198, p. 309.
11. Katz, D. L., "Prediction of the Shrinkage of Crude Oils," *Drilling and Production Practice*, American Petroleum Institute, 1942, Vol. 137.
12. Katz, D. L., et al., *Handbook of Natural Gas Engineering*, McGraw-Hill Book Company: New York, 1959.
13. Khan, S. A., et al., "Viscosity Correlations for Saudi Arabian Crude Oils," SPE Paper 15720, Presented at the Fifth SPE Middle East Conference held in Manama, Bahrain, March 7–10, 1987.
14. Lasater, J. A., "Bubble-Point Pressure Correlation," *Trans. AIME* (1958), Vol. 213, pp. 379–381.
15. Little, J. E. and Kennedy, H. T., "A Correlation of the Viscosity of Hydrocarbon Systems with Pressure, Temperature, and Composition," *SPEJ*, June 1968, pp. 157–162.
16. Lohrenz, J., Bra, B. G., and Clark, C. R., "Calculating Viscosities of Reservoir Fluids From Their Compositions," *J. Pet. Tech.*, October 1964, pp. 1171–1176.

17. Marhoun, M. A., "PVT Correlation for Middle East Crude Oils," *JPT*, May 1988, pp. 650–665.
18. Meehan, D. N., "A Correlation for Water Compressibility," *Petroleum Engineer*, November 1980, pp. 125–126.
19. Numbere, D., Brigham, W., and Standing, M. B., "Correlations for Physical Properties of Petroleum Reservoir Brines," *Petroleum Research Institute*, Stanford University, November 1977, p. 17.
20. Pedersen, K. S., Thomassen, P., and Fredenslund, A., "Thermodynamics of Petroleum Mixtures Containing Heavy Hydrocarbon," *Ind. Eng. Process Des. Dev.*, 1986, Vol. 23, pp. 566–573.
21. Ramey, H. J., "Rapid Methods for Estimating Reservoir Compressibilities," *JPT*, April, 1964, pp. 447–454.
22. Standing, M. B. and Katz, D. L., "Density of Crude Oils Saturated with Natural Gas," *Trans. AIME* (1942), Vol. 146, pp. 159–165.
23. Standing, M. B., "A Pressure-Volume-Temperature Correlation for Mixtures of California Oils and Gases," *Drilling and Production Practice*, API, 1957, pp. 275–287.
24. Standing, M. B., *Volumetric and Phase Behavior of Oil Field Hydrocarbon Systems*, 9th ed. Dallas: Society of Petroleum Engineers, 1981.
25. Sugden, S., "The Variation of Surface Tension, VI. The Variation of Surface Tension with Temperature and Some Related Functions," *J. Chem. Soc.* (1924), Vol. 125, pp. 32–39.
26. Sutton, R. P. and Farashad, F. F., "Evaluation of Empirically Derived PVT Properties for Gulf of Mexico Crude Oils," Paper SPE 13172, Presented at the 59th Annual Technical Conference, Houston, Texas, 1984.
27. Trube, A. S., "Compressibility of Undersaturated Hydrocarbon Reservoir Fluids," *Trans. AIME* (1957), Vol. 210, pp. 341–344.
28. Vasquez, M. and Beggs, H. D., "Correlations for Fluid Physical Property Prediction," *JPT*, June 1980, pp. 968–970.
29. Weinaug, C. and Katz, D. L., "Surface Tension of Methane-Propane Mixtures," *Ind. Eng. Chem.*, 1943, Vol. 25, pp. 35–43.

5
Vapor-Liquid Phase Equilibria

A phase is defined as that part of a system which is uniform in physical and chemical properties, homogeneous in composition, and separated from other coexisting phases by definite boundary surfaces. The most important phases occurring in petroleum production are the hydrocarbon liquid phase and the gas phase. Water is also commonly present as an additional liquid phase. These can coexist in equilibrium when the variables describing change in the entire system remain constant with time and position. The chief variables that determine the state of equilibrium are system temperature, system pressure, and composition.

The conditions under which these different phases can exist is a matter of considerable practical importance in designing surface separation facilities and developing compositional models. These types of calculations are based on the concept of equilibrium ratios.

EQUILIBRIUM RATIOS

The equilibrium ratio K_i of a given component is defined as the ratio of the mole fraction of the component in the gas phase y_i to the mole fraction of the component in the liquid phase x_i. Mathematically, the relationship is expressed as

$$K_i = \frac{y_i}{x_i} \tag{5-1}$$

At pressures below 100 psia, Raoult's and Dalton's Laws for ideal solutions provide a simplified means of predicting equilibrium ratios. Raoult's Law states that the partial pressure p_i of a component in a multi-component system is the product of its mole fraction in the liquid phase and the vapor pressure of the component p_{vi}, or

$$p_i = x_i \, p_{vi} \tag{5-2}$$

where p_i = partial pressure of component i, psia
p_{vi} = vapor pressure of component i, psia
x_i = mole fraction of component i in the liquid phase

Dalton's Law states that the partial pressure of a component is the product of its mole fraction in the gas phase and the total pressure of the system, or

$$p_i = y_i \, p \tag{5-3}$$

where p = total system pressure, psia

At equilibrium and in accordance with the above stated laws, the partial pressure exerted by a component in the gas phase must be equal to the partial pressure exerted by the same component in the liquid phase. Therefore, combining the equations describing the two laws yields

$$x_i \, p_{vi} = y_i \, p$$

Rearranging the above relationship and introducing the concept of the equilibrium ratio gives

$$\frac{y_i}{x_i} = \frac{p_{vi}}{p} = K_i \tag{5-4}$$

Equation 5-4 shows that for ideal solutions and regardless of the composition of the hydrocarbon mixture, the equilibrium ratio is only a function of the system pressure and the temperature. (As indicated in Chapter 1, the vapor pressure of a component is only a function of temperature.)

It is appropriate at this stage to introduce and define the following nomenclatures:

z_i = mole fraction of component i in the entire hydrocarbon mixture
n = total number of moles of the hydrocarbon mixture, lb-mole
n_L = total number of moles in the liquid phase
n_v = total number of moles in the vapor (gas) phase

By definition

$$n = n_L + n_v \tag{5-5}$$

Equation 5-5 indicates that the total number of moles in the system is equal to the total number of moles in the liquid phase plus the total number of moles in the vapor phase.

A material balance on the ith component results in

$$z_i n = x_i\, n_L + y_i\, n_v \tag{5-6}$$

where $z_i n$ = total number of moles of component i in the system
$x_i\, n_L$ = total number of moles of component i in the liquid phase
$y_i\, n_v$ = total number of moles of component i in the vapor phase

Also by the definition of mole fraction, we may write

$$\sum_i z_i = 1 \tag{5-7}$$

$$\sum_i x_i = 1 \tag{5-8}$$

$$\sum_i y_i = 1 \tag{5-9}$$

It is convenient to perform all the phase-equilibria calculations on the basis of one mole of the hydrocarbon mixture, i.e., n = 1. That assumption reduces Equations 5-5 and 5-6 into

$$n_L + n_v = 1 \tag{5-10}$$

$$x_i n_L + y_i n_v = z_i \tag{5-11}$$

Combining Equations 5-4 and 5-11 to eliminate y_i from Equation 5-11 gives

$$x_i n_L + (x_i K_i)\, n_v = z_i$$

Solving for x_i yields

$$x_i = \frac{z_i}{n_L + n_v K_i} \tag{5-12}$$

Equation 5-11 can also be solved for y_i by combining it with Equation 5-4 to eliminate x_i, to give

$$y_i = \frac{z_i K_i}{n_L + n_v K_i} \tag{5-13}$$

Combining Equation 5-12 with 5-8 and Equation 5-13 with 5-9 results in

$$\sum_i x_i = \sum_i \frac{z_i}{n_L + n_v K_i} = 1 \tag{5-14}$$

$$\sum_i y_i = \sum_i \frac{z_i K_i}{n_L + n_v K_i} = 1 \tag{5-15}$$

Since

$$\sum_i y_i - \sum_i x_i = 0$$

Therefore

$$\sum_i \frac{z_i K_i}{n_L + n_v K_i} - \sum_i \frac{z_i}{n_L + n_v K_i} = 0$$

or

$$\sum_i \frac{z_i(K_i - 1)}{n_L + n_v K_i} = 0$$

Replacing n_L with $(1 - n_v)$ yields

$$f(n_v) = \sum_i \frac{z_i(K_i - 1)}{n_v(K_i - 1) + 1} = 0 \tag{5-16}$$

FLASH CALCULATIONS

Flash calculations are an integral part of all reservoir and process engineering calculations. They are required whenever it is desirable to know the amounts (in moles) of hydrocarbon liquid and gas coexisting in a reservoir or a vessel at a given pressure and temperature. These calculations are also performed to determine the composition of the existing hydrocarbon phases.

Flash calculations are then needed for determining:

- Moles of the gas phase n_v
- Moles of the liquid phase n_L
- Composition of the liquid phase x_i
- Composition of the gas phase y_i

The computational steps in determining n_L, n_v, y_i, and x_i of a hydrocarbon mixture with a known overall composition, i.e., z_i, and with a calculated set of equilibrium ratios, i.e., K_i, are summarized on the next page.

Step 1. Calculation of n_v

Equation 5-16 can be solved for n_v by using the Newton-Raphson iteration techniques. In applying this iterative technique:

- A value of n_v is assumed. A good assumed value is calculated from the following relationship:

$$n_v = A/(A - B)$$

where $A = \sum_i [z_i(K_i - 1)]$

$$B = \sum_i [z_i(K_i - 1)/K_i]$$

- Evaluate the function $f(n_v)$ as given by Equation 5-16 using the assumed value of n_v.
- If the absolute value of the function $f(n_v)$ is smaller than a preset tolerance, e.g., 10^{-15}, then the assumed value of n_v is the desired solution.
- If the absolute value of $f(n_v)$ is greater than the preset tolerance, then a new value of n_v is calculated from the following expression:

$$(n_v)_n = n_v - f(n_v)/f'(n_v)$$

where $(n_v)_n$ = new value of n_v
$f'(n_v)$ = first derivative of $f(n_v)$ with respect to n_v, given by

$$f'(n_v) = -\sum_i \left[z_i(K_i - 1)^2/[n_v(K_i - 1) + 1]^2 \right]$$

- The above procedure is repeated with the new values of n_v until convergence is achieved.

Step 2. Calculation of n_L

Calculate the number of moles of the liquid phase from Equation 5-10 to give

$$n_L = 1 - n_v$$

Step 3. Calculation of x_i

Calculate the composition of the liquid phase by applying Equation 5-12.

Step 4. Calculation of y_i

Determine the composition of the gas phase from Equation 5-13.

Example 5-1. A hydrocarbon mixture with the following overall composition is flashed in a separator at 50 psia and 100°F.

Component	z_i
C_3	0.20
i-C_4	0.10
n-C_4	0.10
i-C_5	0.20
n-C_5	0.20
C_6	0.20

Assuming an ideal solution behavior, perform flash calculations.

Solution.

Step 1. Determine the vapor pressure for the Cox chart (see Chapter 1) and calculate the equilibrium ratios from Equation 5-4.

Component	z_i	p_{vi} at 100°F	K_i
C_3	0.20	190	3.80
i-C_4	0.10	72.2	1.444
n-C_4	0.10	51.6	1.032
i-C_5	0.20	20.44	0.4088
n-C_5	0.20	15.57	0.3114
C_6	0.20	4.956	0.09912

Step 2. Solve Equation 5-16 for n_v by using the Newton-Raphson method, to give

Iteration	n_v	$f(n_v)$
0	0.08196579	3.073 E-02
1	0.1079687	8.894 E-04
2	0.1086363	7.60 E-07
3	0.1086368	1.49 E-08
4	0.1086368	0.0

Step 3. Solve for n_L

$$n_L = 1 - 0.1086368 = 0.8913631$$

Step 4. Solve for x_i and y_i, to yield

Component	x_i Equation 5-12	y_i Equation 5-13
C_3	0.1534	0.5827
i-C_4	0.0954	0.1378
n-C_4	0.0997	0.1028
i-C_5	0.2137	0.0874
n-C_5	0.2162	0.0673
C_6	0.2216	0.0220

EQUILIBRIUM RATIOS FOR REAL SOLUTIONS

The equilibrium ratios, which indicate the partitioning of each component between the liquid phase and gas phase, as calculated by Equation 5-4 in terms of vapor pressure and system pressure, proved to be inadequate. The basic assumptions behind Equation 5-4 are that:

- The vapor phase is an ideal gas as described by Dalton's Law
- The liquid phase is an ideal solution as described by Raoult's Law

The above combination of assumptions is unrealistic and results in inaccurate predictions of equilibrium ratios at high pressures.

For a real solution, the equilibrium ratios are no longer a function of the pressure and temperature alone, but also a function of the composition of the hydrocarbon mixture. This observation can be stated mathematically as

$$K_i = K(p,T,z_i)$$

Numerous methods have been proposed for predicting the equilibrium ratios of hydrocarbon mixtures. These correlations range from a simple mathematical expression to a complicated expression containing several compositional dependent variables. Some of these methods are next discussed.

Wilson's Correlation

Wilson (1968) proposed a simplified thermodynamic expression for estimating K-values. The proposed expression has the following form

$$K_i = \frac{p_{ci}}{p} \text{EXP}\left[5.37(1 + \omega_i)\left(1 - \frac{T_{ci}}{T}\right)\right] \qquad (5\text{-}17)$$

where p_{ci} = critical pressure of component i, psia
p = system pressure, psia
T_{ci} = critical temperature of component i, °R
T = system temperature, °R
ω_i = acentric factor of component i

The above relationship generates reasonable values for the equilibrium ratio when applied at low pressures.

Standing's Correlation

Standing (1979) derived a set of equations that fit the equilibrium ratio data of Katz and Hachmuth (1937) at pressures less than 1,000 psia and temperatures below 200°F.

The proposed form of the correlation is based on observations by Hoffmann et al. (1953), Brinkman and Sicking (1960), Kehn (1964), and Dykstra and Mueller (1965) that plots of $Log(K_ip)$ vs. F_i at a given pressure often form straight lines; with the component characterization factor F_i as defined by the following expression:

$$F_i = b_i \left[1/T_{bi} - 1/T\right] \tag{5-18}$$

with

$$b_i = \frac{Log(p_{ci}/14.7)}{[1/T_{bi} - 1/T_{ci}]} \tag{5-19}$$

where F_i = component characterization factor
T_{bi} = normal boiling point of component i, °R

The basic equation of the straight line relationship is given by

$$Log(K_ip) = a + c \, F_i$$

or

$$K_i = \frac{1}{p} \, 10^{a + c \, F_i} \tag{5-20}$$

where the coefficients a and c are the intercept and the slope of the line, respectively.

From a total of six isobar plots of $Log(K_ip)$ vs. F_i for 18 sets of equilibrium ratio values, Standing correlated the coefficients a and c with the pressure, to give

$$a = 1.2 + 0.00045 \, p + 15(10^{-8}) \, p^2 \tag{5-21}$$

$$c = 0.89 - 0.00017 \, p - 3.5(10^{-8}) \, p^2 \tag{5-22}$$

Standing pointed out that the predicted values of the equilibrium ratios of N_2, CO_2, H_2S, and C_1 through C_6 can be improved considerably by changing the correlating parameter b_i and the boiling point of these components. The author proposed the following modified values:

Component	b_i	T_{bi}, °R
N_2	470	109
CO_2	652	194
H_2S	1,136	331
C_1	300	94
C_2	1,145	303
C_3	1,799	416
i-C_4	2,037	471
n-C_4	2,153	491
i-C_5	2,368	542
n-C_5	2,480	557
C_6*	2,738	610

* Lumped Hexanes-fraction

When making flash calculations, the question of the equilibrium ratio to use for the lumped plus fraction always arises. One rule of thumb proposed by Katz and Hachmuth (1937) is that the K-value for C_{7+} can be taken as 15% of the K of C_7.

Standing offered an alternative approach for determining the K-value of the heptanes and heavier fractions. By imposing experimental equilibrium ratio values for C_{7+} on Equation 5-20, Standing calculated the corresponding characterization factors F_i for the plus fraction. The calculated F_i values were used to specify the pure normal paraffin hydrocarbon having the K-value of the C_{7+} fraction.

Standing suggested the following computational steps for determining the parameters b and T_b of the heptanes-plus fraction.

Step 1. Determine, from the following relationship, the number of carbon atoms n of the normal paraffin hydrocarbon having the K-value of the C_{7+} fraction,

$$n = 7.30 + 0.0075 \, (T - 460) + 0.0016 \, p \qquad (5\text{-}23)$$

Step 2. Calculate the correlating parameter b and the boiling point T_b from the following expression:

$$b = 1,013 + 324 \, n - 4.256 \, n^2 \qquad (5\text{-}24)$$

$$T_b = 301 + 59.85 \, n - 0.971 \, n^2 \qquad (5\text{-}25)$$

The above calculated values can then be used in Equation 5-18 to evaluate F_i.

Example 5-2. A hydrocarbon mixture with the following composition is flashed at 1,000 psia and 150°F.

Component	z_i
CO_2	0.009
N_2	0.003
C_1	0.535
C_2	0.115
C_3	0.088
i-C_4	0.023
n-C_4	0.023
i-C_5	0.015
n-C_5	0.015
C_6	0.015
C_{7+}	0.159

If the molecular weight and specific gravity of C_{7+} are 198.71 and 0.8527, respectively, calculate the equilibrium ratios by using

a. Wilson's Correlation
b. Standing's Correlation

Solution.

a. Solution by using Wilson's Correlation

Step 1. Calculate the critical pressure, critical temperature, and acentric factor of C_{7+} by using any of the characterization methods discussed in Chapter 2. From Example 2-6,

$$T_c = 1,294.1 \text{ °R}, \quad p_c = 263.67 \text{ psia}, \quad \omega = 0.5346$$

Step 2. Apply Equation 5-17, to give

Component	P_c, psia	T_c, °R	ω	K_i
CO_2	1,071	547.9	0.225	2.0923
N_2	493	227.6	0.040	16.343
C_1	667.8	343.37	0.0104	7.155
C_2	707.8	550.09	0.0986	1.263
C_3	616.3	666.01	0.1524	0.349
i-C_4	529.1	734.98	0.1848	0.144
n-C_4	550.7	765.65	0.2010	0.106
i-C_5	490.4	829.1	0.2223	0.046
n-C_5	488.6	845.7	0.2539	0.036
C_6	436.9	913.7	0.3007	0.013
C_{7+}	263.67	1,294.1	0.5346	$0.0255(10^{-3})$

b. Solution by Using Standing's Correlation

Step 1. Calculate the coefficients a and c from Equations 5-21 and 5-22 to give

$$a = 1.65$$
$$c = 0.685$$

Step 2. Calculate the number of carbon atoms n from Equation 5-23 to give

$$n = 10.025$$

Step 3. Determine the parameter b and the boiling point T_b for the hydrocarbon component with n carbon atoms by using Equations 5-24 and 5-25, to yield

$$b = 3833.369,$$
$$T_b = 803.41 \text{ °R}$$

Step 4. Apply Equation 5-20, to give

Component	b_i	T_{bi}	F_i Equation 5-18	K_i Equation 5-20
CO_2	652	194	2.292	1.660
N_2	470	109	3.541	11.901
C_1	300	94	2.700	3.159
C_2	1,145	303	1.902	0.897
C_3	1,799	416	1.375	0.391
$i\text{-}C_4$	2,037	471	0.985	0.211
$n\text{-}C_4$	2,153	491	0.855	0.172
$i\text{-}C_5$	2,368	542	0.487	0.096
$n\text{-}C_5$	2,480	557	0.387	0.082
C_6	2,738	610	0	0.045
C_{7+}	3,833.369	803.41	− 1.513	0.004

Convergence Pressure Method

Early high pressure phase-equilibria studies revealed that when a hydrocarbon mixture of a fixed overall composition is held at a constant temperature as the pressure increases, the equilibrium values of all components converge toward a common value of unity at certain pressure. This pressure is termed the convergence pressure P_k of the hydrocarbon mixture. The convergence pressure is essentially used to correlate the effect of the composition on equilibrium ratios.

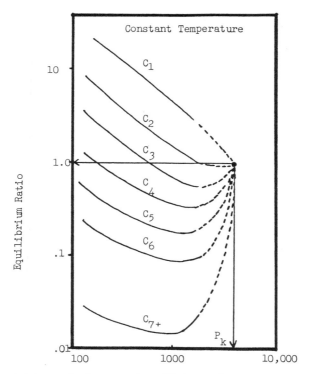

Figure 5-1. A schematic diagram of equilibrium ratios versus pressure relationship.

The concept of the convergence pressure can be better appreciated by examining Figure 5-1. The figure shows a schematic diagram of a typical set of equilibrium ratios plotted versus pressure on Log-Log paper for a hydrocarbon mixture held at a constant temperature. The illustration shows a tendency of the equilibrium ratios to converge isothermally to a value of $K_i = 1$ for all components at a specific pressure, i.e., convergence pressure. A different hydrocarbon mixture may exhibit a different convergence pressure.

The Natural Gas Processors Suppliers Association (NGPSA) correlated a considerable quantity of K-factor data as a function of temperature, pressure, component identity, and convergence pressure. These correlation charts were made available through the NGPSA's "Engineering Data Book" and are considered to be the most extensive set of published equilibrium ratios for hydrocarbons. They include the K-values for a number of convergence pressures, specifically 800, 1,000, 1,500, 2,000, 3,000, 5,000, and

10,000 psia. Equilibrium ratios for methane through decane for a convergence pressure of 5,000 psia are given in Appendix A.

Several investigators observed that for hydrocarbon mixtures with convergence pressures of 4,000 psia or greater, the values of the equilibrium ratio are essentially the same for hydrocarbon mixtures with system pressures less than 1,000 psia. This observation led to the conclusion that the overall composition of the hydrocarbon mixture has little effect on equilibrium ratios when the system pressure is less than 1,000 psia.

The problem with using the NGPSA equilibrium ratio graphical correlations is that the convergence pressure must be known before selecting the appropriate charts. Three of the methods of determining the convergence pressure are discussed below.

A. Hadden's Method. Hadden (1953) developed an iterative procedure for calculating the convergence pressure of the hydrocarbon mixture. The procedure, based on the "pseudo" or "equivalent" binary concept, uses the convergence pressure charts as shown in Figure 5-2. The method is summarized in the following steps:

Step 1. Estimate a value for the convergence pressure.

Step 2. From the appropriate equilibrium ratio charts, read the K-values of each component present in the mixture by entering the charts with the system pressure and temperature.

Step 3. Perform flash calculations using the calculated K-values and system composition.

Step 4. Identify the lightest hydrocarbon component that comprises at least 0.1 mole percent in the liquid phase.

Step 5. Convert the liquid mole fraction to weight fraction.

Step 6. Exclude the lighest hydrocarbon component, as identified in Step 4, and normalize the weight fractions of the remaining components.

Step 7. Calculate the weight average critical temperature and pressure of the lumped components (pseudo-conponent) from the following expressions:

$$T_c = \sum_{i=2} w_i^* T_{ci}$$

$$p_c = \sum_{i=2} w_i^* p_{ci}$$

where w_i^* is the normalized weight fraction of component i.

Step 8. Enter Figure 5-2 with the critical properties of the pseudo-component and trace the critical locus of the binary consisting of the light component and the pseudo-component.

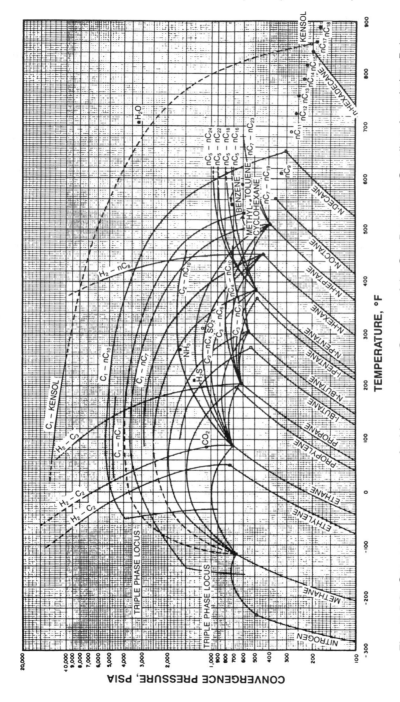

Figure 5-2. Convergence pressures for binary systems. Courtesy of the Gas Processors Suppliers Association. Published in the GPSA Engineering Data Book, Tenth Edition, 1987.

Step 9. Read the new convergence pressure (ordinate) from the point at which the locus crosses the temperature of interest.

Step 10. If the calculated new convergence pressure is not reasonably close to the assumed value, repeat Steps 2 through 9.

It should be noted that when the calculated new convergence pressure is between values for which charts are provided, interpolation between charts may be necessary. If the K-values do not change rapidly with the convergence pressure, i.e., $P_k \gg p$, then the set of charts nearest to the calculated P_k may be used.

B. Standing's Method. Standing (1977) suggested that the convergence pressure can be roughly correlated linearly with the molecular weight of the heptanes-plus fraction. Whitson and Torp (1981) expressed this relationship by the following equation:

$$P_k = 60 \, MW_{C_{7+}} - 4{,}200 \tag{5-26}$$

where $MW_{C_{7+}}$ is the molecular weight of the heptanes-plus fraction.

C. Rzasa's Method. Rzasa, Glass, and Opfell (1952) presented a simplified graphical correlation for predicting the convergence pressure of light hydrocarbon mixtures. They used the temperature and the product of the molecular weight and specific gravity of the heptane-plus fraction as correlating parameters. The graphical illustration of the proposed correlation is shown in Figure 5-3. The graphical correlation is expressed mathematically by the following equation:

$$P_k = -2{,}381.8542 + 46.341487 \, [MW \cdot \gamma]_{C_{7+}}$$

$$+ \sum_{i=1}^{3} a_i \left[[(MW \cdot \gamma)_{C_{7+}}/(T-460)] \right]^i \tag{5-27}$$

where $(MW)_{C_{7+}}$ = molecular weight of C_{7+}
$(\gamma)_{C_{7+}}$ = specific gravity of C_{7+}
$a_1 - a_3$ = coefficients of the correlation with the following values:
$a_1 = 6{,}124.3049$
$a_2 = -2{,}753.2538$
$a_3 = 415.42049$

The above mathematical expression can be used for determining the convergence pressure of hydrocarbon mixtures at a temperature in the range of 50°F to 300°F.

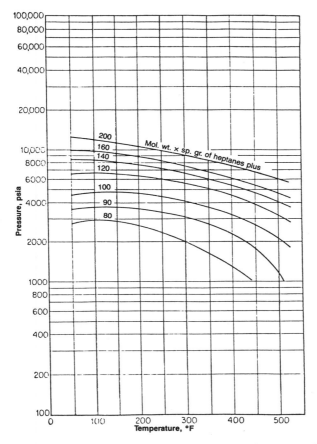

Figure 5-3. Rzasa's convergence pressure correlation. Courtesy of the American Institute of Chemical Engineers.

Whitson and Torp Correlation

Whitson and Torp (1981) reformulated Wilson's equation (Equation 5-17) to yield accurate results at higher pressures. Wilson's equation was modified by incorporating the convergence pressure into the correlation, to give:

$$K_i = \left[\frac{p_{ci}}{P_k} \right]^{A-1} \left[\frac{p_{ci}}{p} \right] EXP \left[5.37 \; A \; (1 + \omega_i) \left(1 - \frac{T_{ci}}{T} \right) \right] \tag{5-28}$$

with

$$A = 1 - \left[\frac{p - 14.7}{P_k - 14.7} \right]^{0.6}$$

where   p = system pressure, psia
       P_k = convergence pressure, psia
       T = system temperature, °R
       ω_i = acentric factor of component i

Example 5-3. Rework Example 5-2 and calculate the equilibrium ratios by using the Whitson and Torp Method.

Solution.

Step 1. Determine the convergence pressure from Equation 5-27 to give

P_k = 9,473.89

Step 2. Calculate the equilibrium ratios from Equation 5-28 to give

Component	K_i Equation 5-28
CO_2	3.087
N_2	14.202
C_1	7.691
C_2	2.122
C_3	0.817
i-C_4	0.422
n-C_4	0.337
i-C_5	0.182
n-C_5	0.152
C_6	0.073
C_{7+}	$0.693(10^{-3})$

Winn's Correlation

Winn (1954) presented a graphical correlation for determining the equilibrium ratios of hydrocarbon mixtures with a convergence pressure of 5,000 psia. The nomograph, as shown in Figure 5-4, consists of a temperature-pressure network, an equilibrium ratio scale, and a component identification scale. Determining the equilibrium ratios of a hydrocarbon mixture is done by simply reading the K-values scale where it is crossed by a straight line passing through the desired compo-

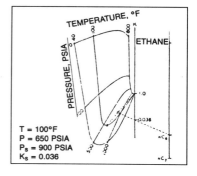

Simplification of Figure 5-4.

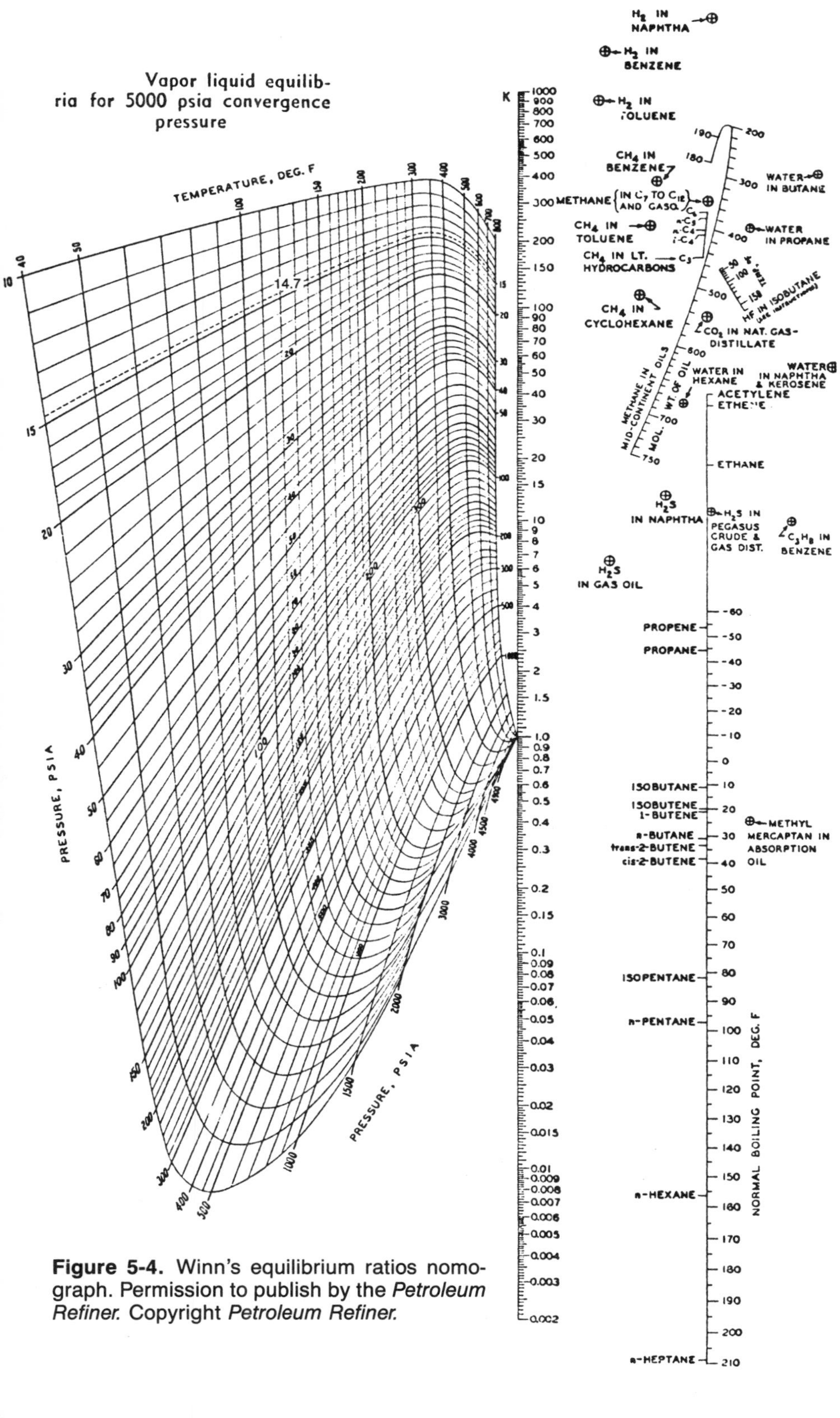

Figure 5-4. Winn's equilibrium ratios nomograph. Permission to publish by the *Petroleum Refiner.* Copyright *Petroleum Refiner.*

nent, and a point on the pressure-temperature network corresponding to the pressure and temperature of the system.

For hydrocarbon mixtures with a convergence pressure other than 5,000 psia, Winn provided an adjustment chart which is a coordinate plot showing the relationship between system pressure, convergence pressure, and "grid" pressure. This relationship is shown in Figure 5-5. The grid pressure is used in conjunction with Figure 5-4 to obtain the K-values at convergence pressures other than 5,000 psia. The adjustment procedure is summarized in the following steps:

Step 1. Locate the system pressure-temperature point on Figure 5-4.

Step 2. Draw a straight line connecting the point on Figure 5-4 to 1.0 on the equilibrium ratio scale.

Step 3. Locate and circle the point of intersection of this line with the pressure line corresponding to the grid pressure of Figure 5-5.

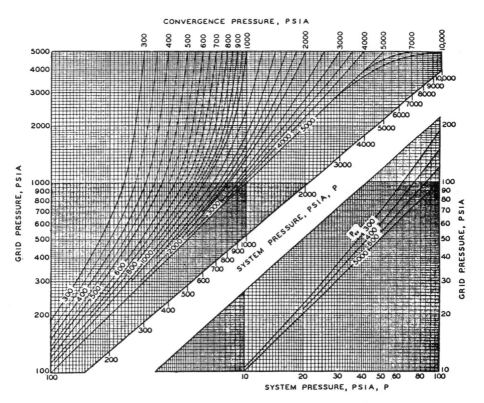

Figure 5-5. Grid pressure relationship. Permission to publish by the *Petroleum Refiner.* Copyright *Petroleum Refiner.*

This point is called grid point to distinguish it from system p-T point.

Step 4. Read the equilibrium ratio scale where it is crossed by a straight line passed through the grid point from Step 2 and the desired component.

Equations of State

In general, equations of state provide a more sophisticated and theoretically-based predictive technique for calculating equilibrium ratios. A detailed description of using equations of state to predict K-values is given in Chapter 6.

EQUILIBRIUM RATIOS FOR THE PLUS-FRACTION

The equilibrium ratios of the plus-fraction often behave in a manner different from the other components of a system. This is because the plus-fraction in itself is a mixture of components.

Several techniques have been proposed for estimating the K-value of the plus-fractions. Some of these techniques are presented below:

Campbell's Method

Campbell (1976) found that the plot of the log of K_i versus T_{ci}^2 for each component is a linear relationship for any hydrocarbon system. Campbell suggested that by drawing the best straight line through the points for propane through hexane components, the resulting line can be extrapolated to obtain the K-value of the plus-fraction. He pointed out that the plot of Log K_i versus $1/T_{bi}$ of each heavy fraction in the mixture is also a straight line relationship. The line can be extrapolated to obtain the equilibrium ratio of the plus-fraction from the reciprocal of its average boiling point. Campbell's suggestions are shown graphically in Figures 5-6 and 5-7.

Winn's Method

Winn (1954) proposed the following expression for determining the equilibrium ratio of heavy fractions with a boiling point above 210°F.

$$K_{C_+} = \frac{K_{C_7}}{(K_{C_2}/K_{C_7})^b} \tag{5-29}$$

where K_{C_+} = value of the plus-fraction

         K_{C_7} = K-value of n-heptane at system pressure, temperature, and convergence pressure

         K_{C_2} = K-value of ethane

         b = volatility exponent

Winn correlated, graphically, the volatility component b of the heavy fraction, with the atmosphere boiling point, as shown in Figure 5-8. This graphical correlation can be expressed mathematically by the following equation:

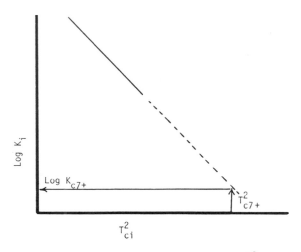

Figure 5-6. Schematic illustration of log K_i versus T_{ci}^2 relationship.

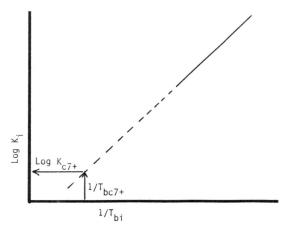

Figure 5-7. Schematic illustration of log K_i versus $1/T_{bi}$ relationship.

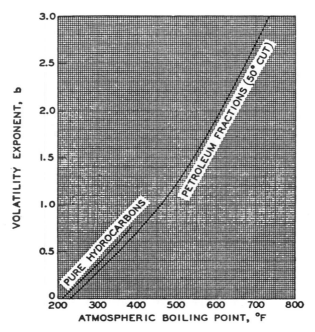

Figure 5-8. Volatility exponent. Permission to publish by the *Petroleum Refiner.* Copyright *Petroleum Refiner.*

$$b = a_1 + a_2 (T_b - 460) + a_3 (T_b - 460)^2 + a_4 (T_b - 460)^3$$
$$+ a_5/(T_b - 460) \qquad\qquad (5\text{-}30)$$

where T_b = boiling point, °R
    a_1–a_5 = coefficients with the following values:

$$a_1 = 1.6744337$$
$$a_2 = -3.4563079 \times 10^{-3}$$
$$a_3 = 6.1764103 \times 10^{-6}$$
$$a_4 = 2.4406839 \times 10^{-9}$$
$$a_5 = -2.9289623 \times 10^{2}$$

Katz's Method

Katz et al. (1957) suggested that a factor of 0.15 times the equilibrium ratio for the heptane component will give a reasonably close approximation to the equilibrium ratio for heptanes and heavier. This suggestion is expressed mathematically by the following equation:

$$K_{C_{7+}} = 0.15 \, K_{C_7} \qquad\qquad (5\text{-}31)$$

where $K_{C_{7+}}$ = equilibrium ratio of C_{7+}
$\quad\quad K_{C_7}$ = equilibrium ratio of the heptane fraction

Hoffmann's Method

Hoffmann, Crump, and Hocott (1953) proposed that the plot of the logarithms of the equilibrium ratio, times the system absolute pressure for the individual hydrocarbons (i.e., $K_i p$) so plotted, yields reasonably straight lines against the characterization factor F_i. The characterization factor is defined by the following relationship:

$$F_i = \left[\frac{\text{Log}(P_{ci}/14.7)}{1/T_{bi} - 1/T_{ci}} \right][1/T_{bi} - 1/T_{ci}] \tag{5-32}$$

where T_{bi} = boiling point of component i, °R
$\quad\quad T_{ci}$ = critical temperature of component i, °R
$\quad\quad T$ = system temperature, °R

The resulting straight line can be extrapolated to obtain the K-value of the plus-fraction by evaluating its characterization factor.

Example 5-4. A hydrocarbon system* exists at 201°F and 1,000 psia. The experimental K-values of the system, except the K-value of C_{7+}, are given below:

Component	K_i
C_1	4.769
C_2	1.515
C_3	0.650
i-C_4	0.360
n-C_4	0.295
i-C_5	0.165
n-C_5	0.142
C_6	0.0741
C_{7+}	—

* Data reported by Hoffman (1953)

The molecular weight and specific gravity of C_{7+} are 198.7 and 0.841. If the convergence pressure of the mixture is 5,000 psia, calculate the equilibrium ratio of C_{7+} by using

a. Standing's Correlation
b. Whitson and Torp Correlation

c. Campbell's Method
d. Winn's Equation
e. Katz's Equation
f. Hoffmann's Method

Solution. The heptanes-plus fraction should be first characterized in terms of its critical properties, acentric factor, and boiling point. Using the molecular weight and specific gravity of C_{7+} as input parameters in the Riazi and Daubert Correlation (Chapter 2, Equation 2-1), calculate T_b, T_c, and p_c, to yield

$T_b = 972.6°R$
$T_c = 1307.7°R$
$p_c = 257.3$ psia

Applying the above calculated properties in the Edmister acentric factor correlation (Chapter 2, Equation 2-21), gives

$\omega = 0.546$

a. Solution by using Standing's Correlation

Step 1. Calculate the parameters a and c by using Equations 5-21 and 5-22 to give:

$a = 1.8$
$c = 0.685$

Step 2. Calculate the characterization factor for C_{7+} by applying Equation 5-18 to give

$F_i = -2.2869$

Step 3. Calculate K_{C7+} from Equation 5-20:

$$K_{C7+} = \frac{1}{1,000} 10^{[1.8+0.685(-2.2869)]} = 0.00171$$

b. Solution by using Whitson and Torp Correlation

Step 1. Calculate the coefficient A in Equation 5-28:

$$A = 1 - \left(\frac{1,000 - 14.7}{5,000 - 14.7} \right)^{0.6} = 0.62197$$

Step 2. Calculate K_{C7+} from Equation 5-28 to give

$$K_{C7+} = 0.0051$$

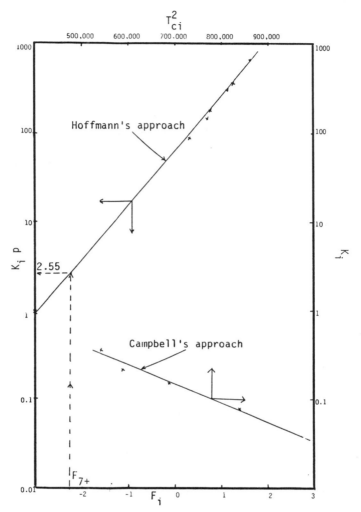

Figure 5-9. The calculation of $K_{C_{7+}}$ of example 6-4.

c. Solution by using Campbell's Method

Step 1. Plot $Log(K_i)$ vs. T_{ci}^2 as shown in Figure 5-9.

Step 2. Extrapolate the line connecting the above points to the reciprocal of T_C^2 of the C_{7+} to read:

$$K_{C_{7+}} = 0.032$$

d. Solution by using Winn's Correlation

Step 1. Estimate the volatility exponent b from Figure 5-8 or Equation 5-30, to give:

$$b = 1.25$$

Step 2. Determine the equilibrium ratios of ethane and heptane fractions from Appendix A, to give:

$$K_{C_2} = 1.52$$
$$K_{C_7} = 0.05$$

Step 3. Calculate K_{C_7+} from Equation 5-29:

$$K_{C_7+} = \frac{0.05}{(1.52/0.5)^{1.25}} = 0.0007$$

e. Solution by using Katz's Correlation.

Determine the equilibrium ratio of the heptanes-plus fraction from Equation 5-31:

$$K_{C_7+} = (0.15)(0.05) = 0.0075$$

f. Solution by using Hoffmann's Method

Step 1. Calculate $\text{Log}(K_i p)$ and the characterization factor F_i, as defined by Equation 5-32, for hydrocarbon fractions:

Component	F_i	$K_i p$
C_1	2.786	4,769
C_2	2.114	1,515
C_3	1.602	650
i-C_4	1.245	360
n-C_4	1.128	295
i-C_5	0.786	165
n-C_5	0.702	142
C_6	0.309	74.1
C_{7+}	-2.2869	—

Step 2. Plot the calculated $\text{Log}(K_i p)$ vs. F_i and connect the points with a straight line, as shown in Figure 5-9. Extrapolate the resulting line to $F_i = -2.2869$ and read the corresponding value of $[K_{C_7+} p]$, to give

$$K_{C_7+} \, p = 2.55$$

Step 3. Solve for K_{C_7+}:

$$K_{C_7+} = 2.55/1,000 = 0.00255$$

Summary of the Results

Method	$K_{C_{7+}}$
Standing	0.00171
Whitson-Torp	0.0051
Campbell	0.032
Winn	0.0075
Hoffmann	0.00255

APPLICATIONS OF THE EQUILIBRIUM RATIO IN PETROLEUM ENGINEERING

The vast amount of experimental and theoretical work that has been done on equilibrium ratio studies indicates their importance in solving phase equilibrium problems in reservoir and process engineering. Some of their practical applications are discussed next.

Determination of the Dew-Point Pressure

The dew-point pressure p_d of a hydrocarbon system is defined as the pressure at which an infinitesimal quantity of liquid is in equilibrium with a large quantity of gas. For a total of one pound-mole of a hydrocarbon mixture, i.e., $n = 1$, the following conditions are applied at the dew-point pressure

$$n_L = 0$$
$$n_v = 1$$

Under these conditions, the composition of the vapor phase y_i is equal to the overall composition z_i.

Applying the above constraints to Equation 5-14, yields

$$\sum_i \frac{z_i}{K_i} = 1 \tag{5-33}$$

where z_i is the total composition of the system under consideration.

The solution of Equation 5-33 for the dew-point pressure p_d involves a trial-and-error process. The process is summarized in the following steps:

Step 1.   Assume a trial value of p_d. A good starting value can be obtained by applying Wilson's Equation (Equation 5-17) for calculating K_i to Equation 5-33, to give

$$\sum_i \left[\frac{z_i}{\frac{p_{ci}}{p_d} \text{EXP} \left[5.37(1 + \omega_i) \left(1 - \frac{T_{ci}}{T} \right) \right]} \right] = 1$$

Solving for p_d, yields

$$\text{initial } p_d = 1/ \sum_i \left[z_i / \left[p_{ci} \text{ EXP} \left[5.37(1 + \omega_1) \left(1 - \frac{T_{ci}}{T} \right) \right] \right] \right] \quad (5\text{-}34)$$

Step 2. Using the assumed dew-point pressure, calculate the equilibrium ratio, K_i, for each component at the system temperature.

Step 3. Compute the summation of Equation 5-33.

Step 4. If the sum is less than one, steps 2 and 3 are repeated at a higher initial value of pressure; conversely, if the sum is greater than one, repeat the calculations with a lower initial value of p_d. The correct value of the dew-point pressure is obtained when the sum is equal to one.

Example 5-5. A natural gas reservoir at 250°F has the following composition:

Component	z_i
C_1	0.80
C_2	0.05
C_3	0.04
i-C_4	0.03
n-C_4	0.02
i-C_5	0.03
n-C_5	0.02
C_6	0.005
C_{7+}	0.005

If the molecular weight and specific gravity of C_{7+} are 140 and 0.8, calculate the dew-point pressure.

Solution.

Step 1. Calculate the convergence pressure of the mixture from Rzasa's correlation, i.e., Equation 5-26, to give

$$P_k = 5,000 \text{ psia}$$

Step 2. Determine an initial value for the dew-point pressure from Equation 5-34, to give

$$p_d = 207 \text{ psia}$$

Step 3. Solve for the dew-point pressure by applying the iterative procedure outlined previously, and using Equation 5-33, to give

Component	z_i	K_i at 207 psia	z_i/K_i	K_i at 300 psia	z_i/K_i	K_i at 222.3 psia	z_i/K_i
C_i	0.78	19	0.0411	13	0.06	18	0.0433
C_2	0.05	6	0.0083	4.4	0.0114	5.79	0.0086
C_3	0.04	3	0.0133	2.2	0.0182	2.85	0.0140
i-C_4	0.03	1.8	0.0167	1.35	0.0222	1.75	0.0171
n-C_4	0.02	1.45	0.0138	1.14	0.0175	1.4	0.0143
i-C_5	0.03	0.8	0.0375	0.64	0.0469	0.79	0.0380
n-C_5	0.02	0.72	0.0278	.55	0.0364	0.69	0.029
C_6	0.005	0.35	0.0143	0.275	0.0182	0.335	0.0149
C_{7+}	0.02	0.0255*	0.7843	0.02025*	0.9877	0.0243*	0.8230
			0.9571		1.2185		1.0022

* Equation 5-31

The dew-point pressure is therefore 222 psia at 250°F.

Determination of the Bubble-Point Pressure

At the bubble-point, the hydrocarbon system is essentially liquid, except for an infinitesimal amount of vapor. For a total of one pound-mole of the hydrocarbon mixture, the following conditions are applied at the bubble-point pressure

$$n_L = 1$$
$$n_v = 0$$

Obviously, under the above conditions, $x_i = z_i$. Applying the above constraints to Equation 5-15 yields

$$\sum_i (z_i K_i) = 1 \tag{5-35}$$

Following the procedure outlined in the dew-point pressure determination, Equation 5-35 is solved for the bubble-point pressure p_b by assuming various pressures and determining the pressure that will produce K-values satisfying Equation 5-35.

During the iterative process, if

$$\sum_i (z_i K_i) < 1 \rightarrow \text{the assumed pressure is high}$$

$$\sum_i (z_i K_i) > 1 \rightarrow \text{the assumed pressure is low}$$

Wilson's Equation can be used to give a good starting value for the iterative process:

$$\sum_i \left[z_i \frac{p_{ci}}{p_b} \text{EXP} \left[5.37 \ (1 + \omega) \left(1 - \frac{T_{ci}}{T} \right) \right] \right] = 1$$

or

$$p_b = \sum_i \left[z_i \ p_{ci} \ \text{EXP} \left[5.37 \ (1 + \omega_i) \left(1 - \frac{T_{ci}}{T} \right) \right] \right] \tag{5-36}$$

Example 5-6. A crude oil reservoir has a temperature of 200°F and a composition as given below. Calculate the bubble-point pressure of the oil.

Component	x_i
C1	0.42
C2	0.05
C3	0.05
i-C4	0.03
n-C4	0.02
i-C5	0.01
n-C5	0.01
C6	0.01
C7+	0.40*

* $(MW)_{C_{7+}} = 216.0$
$(\gamma)_{C_{7+}} = 0.8605$
$(T_b)_{C_{7+}} = 977°R$

Solution.

Step 1. Calculate the convergence pressure of the system by using Standing's correlation Equation 5-26.

$$P_k = (60)(216) - 4{,}200 = 8{,}760 \text{ psia}$$

Step 2. Calculate the critical pressure and temperature by the Riazi and Daubert equation to give

$$p_c = 230.4 \text{ psia}$$
$$T_c = 1{,}279.8°R$$

Step 3. Calculate the acentric factor by employing the Edmister correlation to yield

$$\omega = 0.653$$

Step 4. Estimate the bubble-point pressure from Equation 5-36 to give

$$P_b = 3,924 \text{ psia}$$

Step 5. Employing the iterative procedure outlined previously and using the Whitson and Torp equilibrium ratio correlation gives

Component	z_i	K_i at 3,924 psia	z_i/K_i	K_i at 3,950 psia	z_i/K_i	K_i at 4,329 psia	z_i/K_i
C_1	0.42	2.257	0.9479	2.242	0.9416	2.0430	0.8581
C_2	0.05	1.241	0.06205	1.237	0.0619	1.1910	0.0596
C_3	0.05	0.790	0.0395	0.7903	0.0395	0.793	0.0397
$i\text{-}C_4$	0.03	0.5774	0.0173	0.5786	0.0174	0.5977	0.0179
$n\text{-}C_4$	0.02	0.521	0.0104	0.5221	0.0104	0.5445	0.0109
$i\text{-}C_5$	0.01	0.3884	0.0039	0.3902	0.0039	0.418	0.0042
$n\text{-}C_5$	0.01	0.3575	0.0036	0.3593	0.0036	0.3878	0.0039
C_6	0.01	0.2530	0.0025	0.2549	0.0025	0.2840	0.0028
C_{7+}	0.40	0.0227	0.0091	0.0232	0.00928	0.032	0.0128
			1.09625		1.09008		1.0099

The calculated bubble-point pressure = 4,330 psia.

Separator Calculations

Produced reservoir fluids are complex mixtures of different physical characteristics. As a well stream flows from the high-temperature, high-pressure petroleum reservoir, it experiences pressure and temperature reductions. Gases evolve from the liquids and the well stream changes in character. The physical separation of these phases is by far the most common of all field-processing operations, and one of the most critical. The manner in which the hydrocarbon phases are separated at the surface influences the stock-tank oil recovery. The principal means of surface separation of gas and oil is the conventional stage separation.

Stage separation is a process in which gaseous and liquid hydrocarbons are flashed (separated) into vapor and liquid phases by two or more separators. These separators are usually operated in series at consecutively lower pressures. Each condition of pressure and temperature at which hydrocarbon phases are flashed is called a stage of separation. Examples of two- and three-stage separation processes are shown in Figure 5-10. Traditionally, the stock-tank is normally considered a separate stage of separation.

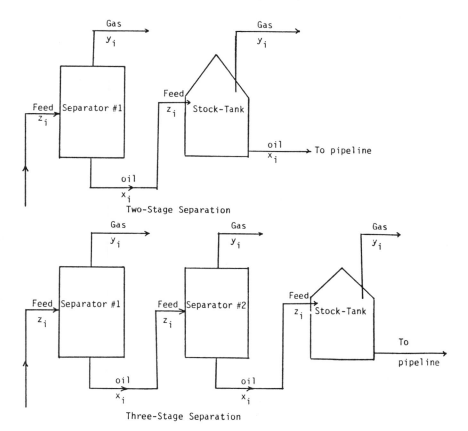

Figure 5-10. A schematic illustration of two- and three-stage separation processes.

Mechanically, there are two types of gas-oil separation:

- "Differential" separation
- "Flash" or "equilibrium" separation

To explain the various separation processes, it is convenient to define the composition of a hydrocarbon mixture by three groups of components. These groups are

- The very volatile components ("lights"), such as nitrogen, methane, and ethane.
- The components of intermediate volatility, i.e., intermediates, such as propane through hexane.
- The components of less volatility, or the "heavies," such as heptane and heavier components.

In the differential separation, the liberated gas (which is composed mainly of lighter components) is removed from contact with the oil as the pressure on the oil is reduced. As pointed out by Clark (1960), when the gas is separated in this manner, the maximum amount of heavy and intermediate components will remain in the liquid, there will be minimum shrinkage of the oil and, therefore, greater stock-tank oil recovery will occur. This is due to the fact that the gas liberated earlier at higher pressures is not present at lower pressures to attract the intermediate and heavy components and pull them into the gas phase.

In the flash (equilibrium) separation, the liberated gas remains in contact with oil until its instantaneous removal at the final separation pressure. A maximum proportion of intermediate and heavy components are attracted into the gas phase by this process and this results in a maximum oil shrinkage and, thus, a lower oil recovery.

In practice, the differential process is introduced first in field separation when gas or liquid is removed from the primary separator. In each subsequent stage of separation, the liquid initially undergoes a flash liberation followed by a differential process as actual separation occurs. As the number of stages increases, the differential aspect of the overall separation becomes greater.

The purpose of stage separation then is to reduce the pressure on the produced oil in steps so that more stock-tank oil recovery will result.

Separator calculations are basically performed to determine:

- Optimum separation conditions: separator pressure and temperature
- Compositions of the separated gas and oil phases
- Oil formation volume factor
- Producing gas-oil ratio
- API gravity of the stock-tank oil

Note that if the separator pressure is high, large amounts of light components will remain in the liquid phase at the separator and be lost along with other valuable components to the gas phase at the stock-tank. On the other hand, if the pressure is too low, large amounts of light components will be separated from the liquid and they will attract substantial quantities of intermediate and heavier components. An intermediate pressure, called "optimum separator pressure," should be selected to maximize the oil volume accumulation in the stock-tank. This optimum pressure will also yield (as shown schematically in Figure 5-11:

- A maximum in the stock-tank API gravity
- A minimum in the oil formation volume factor (i.e., less oil shrinkage)
- A minimum in the gas-oil ratio

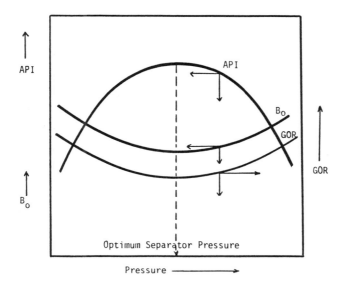

Figure 5-11. Effect of the separator pressure on API, B_o, and GOR.

The computational steps of the "separator calculations" are described below in conjunction with Figure 5-12, which schematically shows a bubble-point reservoir flowing into a surface separation unit consisting of n-stages operating at successively lower pressures.

Step 1. Calculate the volume of oil occupied by one pound-mole of the crude at the reservoir pressure and temperature. This volume, as denoted by V_o, is calculated by recalling and applying the equation which defines the number of moles, to give

$$n = \frac{m}{MW_a} = \frac{\rho_o V_o}{MW_a} = 1$$

Solving for the oil volume, gives

$$V_o = \frac{MW_a}{\rho_o} \tag{5-37}$$

where m = total weight of one lb-mole of the crude oil, lb/mole
V_o = volume of one lb-mole of the crude oil at reservoir conditions, ft³/mole
MW_a = apparent molecular weight
ρ_o = density of the reservoir oil, lb/ft³

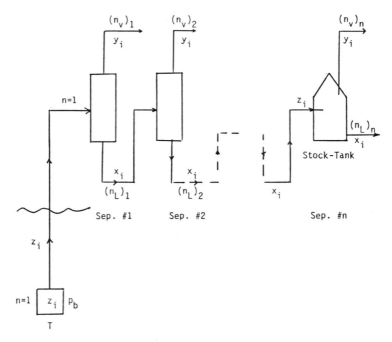

Figure 5-12. A schematic illustration of n-separation stages.

Step 2. Given the composition of the feed stream z_i to the first separator and the operating conditions of the separator, i.e., separator pressure and temperature, calculate the equilibrium ratios of the hydrocarbon mixture.

Step 3. Assuming a total of one mole of the feed entering the first separator and using the above calculated equilibrium ratios, perform flash calculations to obtain the compositions and quantities, in moles, of the gas and the liquid leaving the first separator. Designating these moles as $(n_L)_1$ and $(n_v)_1$, the actual number of moles of the gas and the liquid leaving the first separation stage are

$$[n_{v1}]_a = (n)(n_v)_1 = (1)(n_v)_1$$
$$[n_{L1}]_a = (n)(n_L)_1 = (1)(n_L)_1$$

where $[n_{v1}]_a$ = actual number of moles of vapor leaving the first separator
$[n_{L1}]_a$ = actual number of moles of liquid leaving the first separator

Step 4. Using the composition of the liquid leaving the first separator as the feed (i.e., z_i) for the second separator, calculate the equilibrium ratios of the hydrocarbon mixture at the prevailing pressure and temperature of the separator.

Step 5. Based on one mole of the feed, perform flash calculations to determine the compositions and quantities of the gas and liquid leaving the second separation stage. The actual number of moles of the two phases are then calculated from

$$[n_{v2}]_a = [n_{L1}]_a (n_v)_2 = (1) (n_L)_1 (n_v)_2$$
$$[n_{L2}]_a = [n_{L1}]_a (n_L)_2 = (1) (n_L)_1 (n_L)_2$$

where $[n_{v2}]_a$, $[n_{L2}]_a$ = actual moles of gas and liquid leaving separator two
$(n_v)_2$, $(n_L)_2$ = moles of gas and liquid as determined from flash calculations

Step 6. The previously outlined procedure is repeated for each separation stage, including the stock-tank stage, and the calculated moles and compositions are recorded. The total number of moles of gas given off in all stages are then calculated as

$$(n_v)_t = \sum_{i=1}^{n} (n_{va})_i = (n_v)_1 + (n_L)_1(n_v)_2$$

$$+ (n_L)_1(n_v)_2(n_v)_3 + \ldots + (n_L)_1 \ldots (n_L)_{n-1}(n_v)_n$$

In a more compacted form, the above expression can be written as

$$(n_v)_t = (n_v)_1 + \sum_{i=2}^{n} (n_v)_i \prod_{j=1}^{i-1} (n_L)_j \qquad (5\text{-}38)$$

where $(n_v)_t$ = total moles of gas given off in all stages, lb-mole/mole of feed
n = number of separation stages

Total moles of liquid remaining in the stock-tank can also be calculated as

$$(n_L)_{st} = \prod_{i=1}^{n} (n_L)_i \qquad (5\text{-}39)$$

where $(n_L)_{st}$ is the number of moles of liquid remaining in the stock-tank.

Step 7. Calculate the volume, in standard cubic feed, of all the liberated solution gas from

$$V_g = 379.4 \, (n_v)_t \qquad (5\text{-}40)$$

where V_g = total volume of the liberated solution gas scf/mole of feed.

Step 8. Determine the volume of stock-tank oil occupied by $(n_L)_{st}$ moles of liquid from

$$(V_o)_{st} = \frac{(n_L)_{st} \, (MW_a)_{st}}{(\rho_o)_{st}} \qquad (5\text{-}41)$$

where $(V_o)_{st}$ = volume of stock-tank oil, ft^3/mole of feed
$(MW_a)_{st}$ = apparent molecular weight of the stock-tank oil
$(\rho_o)_{st}$ = density of the stock-tank oil, lb/ft^3

Step 9. Calculate the specific gravity and the API gravity of the stock-tank oil by applying Equations 4-1 and 4-2, to give

$$\gamma_o = \frac{(\rho_o)_{st}}{62.4}$$

$$^\circ API = \frac{141.5}{\gamma_o} - 131.5$$

Step 10. Calculate the total gas-oil ratio

$$GOR = \frac{V_g}{(V_o)_{st}/5.615} = \frac{(5.615)(379.4)(n_v)_t}{(n_L)_{st} \, (MW)_{st}/(\rho_o)_{st}}$$

or

$$GOR = \frac{2{,}130.331 \, (n_v)_t \, (\rho_o)_{st}}{(n_L)_{st} \, (MW)_{st}} \, \text{scf/STB} \qquad (5\text{-}42)$$

where GOR = gas-oil ratio, scf/STB

Step 11. Calculate the oil formation volume factor from the relationship

$$B_o = \frac{V_o}{(V_o)_{st}}$$

Combining Equations 5-37 and 5-41 with the above expression gives

$$B_o = \frac{MW_a \, (\rho_o)_{st}}{\rho_o \, (n_L)_{st} \, (MW_a)_{st}} \qquad (5\text{-}43)$$

where B_o = oil formation volume factor, bbl/STB
MW_a = apparent molecular weight of the feed
ρ_o = density of crude oil at reservoir conditions, lb/ft³

The separator pressure can be optimized by calculating the API gravity, GOR, and B_o in the manner outlined above at different assumed pressures. The optimum pressure corresponds to a maximum in the API gravity and a minimum in gas-oil ratio and oil formation volume factor.

Example 5-7. A crude oil, with the composition as given below, exists at its bubble-point pressure of 1,708.7 psia and at a temperature of 131°F. The crude oil is flashed through three-stage separation facilities. The operating conditions of the three separators are

Separator	Pressure, psia	Temperature, °F
1	400	72
2	350	72
3	14.7	60

The composition of the crude oil is given below.

Component	z_i
CO_2	0.0008
N_2	0.0164
C_1	0.2840
C_2	0.0716
C_3	0.1048
$i\text{-}C_4$	0.0420
$n\text{-}C_4$	0.0420
$i\text{-}C_5$	0.0191
$n\text{-}C_5$	0.0191
C_6	0.0405
C_{7+}	0.3597

The molecular weight and specific gravity of C_{7+} are 252 and 0.8429. Calculate B_o, R_s, stock-tank density, and the API gravity of the hydrocarbon system.

Solution.

Step 1. Calculate the apparent molecular weight of the crude oil to give

$$MW_a = 113.5102$$

Step 2. Calculate the density of the bubble-point crude oil by using the Standing and Katz correlation to yield

$$\rho_o = 44.794 \text{ lb/ft}^3$$

Step 3. Flash the original composition through the first separator by generating the equilibrium ratios by using the Standing correlation (Equation 5-20) to give:

Component	z_i	K_i	x_i	y_i
CO_2	0.0008	3.509	0.0005	0.29791
N_2	0.0164	39.900	0.0014	0.0552
C_1	0.2840	8.885	0.089	0.7877
C_2	0.0716	1.3490	0.0652	0.088
C_3	0.1048	0.373	0.1270	0.0474
$i-C_4$	0.0420	0.161	0.0548	0.0088
$n-C_4$	0.042	0.120	0.0557	0.0067
$i-C_5$	0.0191	0.054	0.0259	0.0014
$n-C_5$	0.0191	0.043	0.0261	0.0011
C_6	0.0405	0.018	0.0558	0.0010
C_{7+}	0.3597	0.0021	0.4986	0.0010

with $n_L = 0.7209$ and $n_v = 0.29791$

Step 4. Use the calculated liquid composition as the feed for the second separator and flash the composition at the operating condition of the separator.

Component	z_i	K_i	x_i	y_i
CO_2	0.0005	3.944	0.0005	0.0018
N_2	0.0014	46.18	0.0008	0.0382
C_1	0.089	10.06	0.0786	0.7877
C_2	0.0652	1.499	0.0648	0.0971
C_3	0.1270	0.4082	0.1282	0.0523
$i-C_4$	0.0548	0.1744	0.0555	0.0097
$n-C_4$	0.0557	0.1291	0.0564	0.0072
$i-C_5$	0.0259	0.0581	0.0263	0.0015
$n-C_5$	0.0261	0.0456	0.0264	0.0012
C_6	0.0558	0.0194	0.0566	0.0011
C_{7+}	0.4980	0.00228	0.5061	0.0012

with $n_L = 0.9851$ and $n_v = 0.0149$

Step 5. Repeat the above calculation for the stock-tank stage, to give

Component	z_i	K_i	x_i	y_i
CO_2	0.0005	81.14	0000	0.0014
N_2	0.0008	1,159	0000	0.026
C_1	0.0784	229	0.0011	0.2455
C_2	0.0648	27.47	0.0069	0.1898
C_3	0.1282	6.411	0.0473	0.3030
i-C_4	0.0555	2.518	0.0375	0.0945
n-C_4	0.0564	1.805	0.0450	0.0812
i-C_5	0.0263	0.7504	0.0286	0.0214
n-C_5	0.0264	0.573	0.02306	0.0175
C_6	0.0566	0.2238	0.0750	0.0168
C_{7+}	0.5061	0.03613	0.7281	0.0263

with $n_L = 0.6837$ and $n_v = 0.3163$

Step 6. Calculate the density of the stock-tank oil by using the Standing correlation to give

$(\rho_o)_{st} = 50.920$

Step 7. Calculate the API gravity of the stock-tank oil.

$API = (141.5/0.816) - 131.5 = 41.9$

Step 8. Calculate the gas solubility from Equation 5-42 to give

$R_s = 571.46$ scf/STB

Step 9. Calculate B_o from Equation 5-43 to give

$B_o = 1.321$ bbl/STB

PROBLEMS

1. A hydrocarbon system has the following composition:

Component	z_i
C_1	0.30
C_2	0.10
C_3	0.05
i-C_4	0.03
n-C_4	0.03
i-C_5	0.02
n-C_5	0.02
C_6	0.05
C_{7+}	0.40

Given

$$\text{System pressure} = 2,100 \text{ psia}$$
$$\text{System temperature} = 150°\text{F}$$
$$\text{Specific gravity of } C_{7+} = 0.80$$
$$\text{Molecular weight of } C_{7+} = 140$$

Calculate the equilibrium ratios of the above system.

2. A well is producing oil and gas with the compositions given below at a gas-oil ratio of 500 scf/STB.

Component	x_i	y_i
C_1	0.35	0.60
C_2	0.08	0.10
C_3	0.07	0.10
n-C_4	0.06	0.07
n-C_5	0.05	0.05
C_6	0.05	0.05
C_{7+}	0.34	0.05

Given the following additional data:

$$\text{Current reservoir pressure} = 3,000 \text{ psia}$$
$$\text{Bubble-point pressure} = 2,800 \text{ psia}$$
$$\text{Reservoir temperature} = 120°\text{F}$$
$$\text{MW of } C_{7+} = 125$$
$$\text{Specific gravity of } C_{7+} = 0.823$$

Calculate the composition of the reservoir fluid.

3. The hydrocarbon mixture with the composition given below exists in a reservoir at 234°F and 3,500 psig.

Component	z_i
C_1	0.3805
C_2	0.0933
C_3	0.0885
C_4	0.0600
C_5	0.0378
C_6	0.0356
C_{7+}	0.3043

The C_{7+} has a molecular weight of 200 and a specific gravity of 0.8366.

Calculate:

a. The bubble-point pressure of the mixture.
b. The compositions of the two phases if the mixture is flashed at 500 psia and 150°F.
c. The density of the liquid phase.
d. The compositions of the two phases if the liquid from the first separator is further flashed at 14.7 psia and 60°F.
e. The oil formation volume factor at the bubble-point pressure.
f. The original gas solubility.
g. The oil viscosity at the bubble-point pressure.

4. A crude oil exists in a reservoir at its bubble-point pressure of 2,520 psig and a temperature of 180°F. The oil has the following composition:

Component	x_i
CO_2	0.0044
N_2	0.0045
C_1	0.3505
C_2	0.0464
C_3	0.0246
i-C_4	0.0683
n-C_4	0.0083
i-C_5	0.0080
n-C_5	0.0080
C_6	0.0546
C_{7+}	0.4824

The molecular weight and specific gravity of C_{7+} are 225 and 0.8364. The reservoir contains initially 122 MMbbl of oil. The surface facilities consist of two separation stages connecting in series. The first separation stage operating at 500 psig and 100°F. The second stage operating under standard conditions.

a. Characterize C_{7+} in terms of its critical properties, boiling point, and acentric factor.
b. Calculate the initial oil in place in STB.
c. Calculate the standard cubic feet of gas initially in solution.
d. Calculate the composition of the free gas and the composition of the remaining oil at 2,495 psig, assuming the overall composition of the system will remain constant.

REFERENCES

1. Brinkman, F. H. and Sicking, J. N., "Equilibrium Ratios for Reservoir Studies," *Trans. AIME*, 1960, Vol. 219, pp. 313–319.
2. Campbell, J. M., *Gas Conditioning and Processing*, Campbell Petroleum Series, Vol. 1, Norman, Oklahoma, 1976.
3. Clark, N., *Elements of Petroleum Reservoirs*, Society of Petroleum Engineers: Dallas, Texas, 1960.
4. Dykstra, H. and Mueller, T. D., "Calculation of Phase Composition and Properties for Lean-or Enriched-Gas Drive," *SPEJ*, Sept. 1965, pp. 239–246.
5. Hadden, J. T., "Convergence Pressure in Hydrocarbon Vapor-Liquid Equilibria," Chem. Eng. Progr. Symposium Ser., 1953, Vol. 49, No. 7, p. 53.
6. Hoffmann, A. E., Crump, J. S., and Hocott, R. C., "Equilibrium Constants for a Gas-Condensate System," *Trans. AIME*, 1953, Vol. 198, pp. 1–10.
7. Katz, D. L. and Hachmuth, K. H., "Vaporization Equilibrium Constants in a Crude Oil-Natural Gas System," *Ind. Eng. Chem.*, 1937, Vol. 29, p. 1072.
8. Katz, D., et al., *Handbook of Natural Gas Engineering*, McGraw-Hill Book Company, New York, 1959.
9. Kehn, D. M., "Rapid Analysis of Condensate Systems by Chromatography," *JPT*, April 1964, pp. 435–440.
10. Rzasa, M. J., Glass, E. D., and Opfell, J. B., "Prediction of Critical Properties and Equilibrium Vaporization Constants for Complex Hydrocarbon Systems," Chem. Eng. Progr. Symposium Ser., 1952, Vol. 48, No. 2, p. 28.
11. Standing, M. B., *Volumetric and Phase Behavior of Oil Field Hydrocarbon Systems*, 8th Printing, Society of Petroleum Engineers of AIME, Dallas, 1977.
12. Standing, M. B., "A Set of Equations for Computing Equilibrium Ratios of a Crude Oil/Natural Gas System at Pressures Below 1,000 psia," *JPT*, Sept 1979, pp. 1193–1195.
13. Whitson, C. H. and Torp, S. B., "Evaluating Constant Volume Depletion Data," paper SPE 10067, presented at the SPE 56th Annual Fall Technical Conference, San Antonio, Oct. 5–7, 1981.
14. Wilson, G., "A Modified Redlich-Kwong EOS, Application to General Physical Data Calculations," Paper 15C, presented at the Annual AIChE National Meeting held in Cleveland, Ohio, May 4–7, 1968.
15. Winn, F. W., "Simplified Nomographic Presentation, Hydrocarbon Vapor-Liquid Equilibria," Chem. Eng. Progr. Symposium Ser., 1954, Vol. 33, No. 6, pp. 131–135.

6
Equations of State

An Equation of State (EOS) is an analytical expression relating the pressure p to the temperature T and the volume V. A proper description of this PVT relationship for real hydrocarbon fluids is essential in determining the volumetric and phase behavior of petroleum reservoir fluids and in predicting the performance of surface separation facilities.

The best known and the simplest example of an equation of state is the ideal gas equation, expressed mathematically by the expression

$$p = \frac{RT}{V} \tag{6-1}$$

where V = gas volume in ft^3 per one mole of gas.

The above PVT relationship is only used to describe the volumetric behavior of real hydrocarbon gases at pressures close to the atmospheric pressure for which it was experimentally derived.

The extreme limitations of the applicability of Equation 6-1 prompted numerous attempts to develop an equation of state suitable for describing the behavior of real fluids at extended ranges of pressures and temperatures.

The main objective of this chapter is to review developments and advances in the field of empirical cubic equations of state and demonstrate their applications in petroleum engineering.

Van der Waals' EOS

In developing the ideal gas EOS, i.e., Equation 6-1, two assumptions were made:

First assumption: The volume of the gas molecules is insignificant compared to the volume of the container and distance between the molecules.
Second assumption: There are no attractive or repulsive forces between the molecules or the walls of the container.

Van der Waals (1873) attempted to eliminate these two assumptions in developing an empirical equation of state for real gases. In his attempt to eliminate the first assumption, Van der Waals pointed out that the gas molecules occupy a significant fraction of the volume at higher pressures and proposed that the volume of the molecules, as denoted by the parameter b, be subtracted from the actual molar volume V in Equation 6-1, to give

$$p = \frac{RT}{V - b} \tag{6-2}$$

where the parameter b is known as the co-volume and is considered to reflect the volume of molecules. The variable V represents the actual volume in ft^3 per one mole of gas.

To eliminate the second assumption, Van der Waals subtracted a corrective term, as denoted by a/V_2, from Equation 6-2 to account for the attractive forces between molecules. In a mathematical form, Van der Waals proposed the following expression:

$$p = \frac{RT}{V - b} - \frac{a}{V^2} \tag{6-3}$$

where p = system pressure, psia
       T = system temperature, °R
       R = gas constant, 10.73 psi-ft³/lb-mole, °R
       V = volume, ft³/mole

The parameters a and b are constants characterizing the molecular properties of the individual components. The symbol a is a measure of the intermolecular attractive forces between the molecules. Equation 6-3 shows the following important characteristics:

a. At low pressures and large volumes, the parameter b becomes negligible in comparison with V and the attractive forces term a/V_2 becomes insignificant, and therefore the Van der Waals equation reduces to the ideal gas equation (Equation 6-1).

b. At high pressure, i.e., $p \to \infty$, the volume V becomes very small and approaches the value b, which is the actual molecular volume.

The Van der Waals EOS or any other equation of state can be expressed in a more generalized form as follows:

$$p = p_{repulsive} - p_{attractive} \tag{6-4}$$

where the repulsive pressure term $p_{repulsive}$ is represented by the term RT/(V − b) and the attractive pressure term $p_{attractive}$ is described by a/V_2.

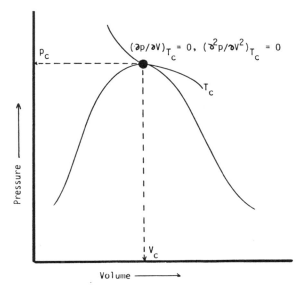

$(\partial p / \partial V)_{T_c} = 0, \; (\partial^2 p / \partial V^2)_{T_c} = 0$

Figure 6-1. An idealized pressure-volume relationship for a pure compound.

In determining the values of the two constants a and b for any pure substance, Van der Waals observed that the critical isotherm has a horizontal slope and an inflection point at the critical point, as shown in Figure 6-1. This observation can be expressed mathematically as follows:

$$\left[\frac{\partial p}{\partial V}\right]_{T_c} = 0, \quad \left[\frac{\partial^2 p}{\partial V^2}\right]_{T_c} = 0 \tag{6-5}$$

Differentiating Equation 6-3 with respect to the volume at the critical point results in

$$\left[\frac{\partial p}{\partial V}\right]_{T_c} = \frac{-R\,T}{(V_c - b)^2} + \frac{2a}{V_c^3} = 0 \tag{6-6}$$

$$\left[\frac{\partial^2 p}{\partial V^2}\right]_{T_c} = \frac{2R\,T_c}{(V_c - b)^3} - \frac{6a}{V_c^4} = 0 \tag{6-7}$$

Solving Equations 6-6 and 6-7 simultaneously for the parameters a and b gives

$$b = \left(\frac{1}{3}\right) V_c \tag{6-8}$$

$$a = \left(\frac{8}{9}\right) R\ T_c\ V_c \tag{6-9}$$

Notice that Equation 6-8 suggests that the volume of the molecules b is approximately 0.333 of the critical volume of the substance. Experimental studies reveal that b is in the range of 0.24–0.28 of the critical volume.

By applying Equation 6-3 to the critical point (i.e., by setting $T = T_c$, $p = p_c$, and $V = V_c$) and combining with Equations 6-8 and 6-9, we get:

$$p_c\ V_c = (0.375)\ R\ T_c \tag{6-10}$$

Equation 6-10 shows that regardless of the type of the substance, the Van der Waals EOS produces a universal critical gas compressibility factor Z_c of 0.375. Experimental studies show that Z_c values for substances range between 0.23 to 0.31.

Equation 6-10 can be combined with Equations 6-8 and 6-9 to give a more convenient expression for calculating the parameters a and b, to yield

$$a = \Omega_a \frac{R^2\ T_c^2}{p_c} \tag{6-11}$$

$$b = \Omega_b \frac{R\ T_c}{p_c} \tag{6-12}$$

where R = gas constant, 10.73 psia-ft^3/lb-mole-°R
p_c = critical pressure, psia
T_c = critical temperature, °R
Ω_a = 0.421875
Ω_b = 0.125

Equation 6-3 can also be expressed in a cubic form in terms of the volume V as follows:

$$V^3 - \left(b + \frac{RT}{p}\right) V^2 + \left(\frac{a}{p}\right) V - \left(\frac{ab}{p}\right) = 0 \tag{6-13}$$

Equation 6-13 is usually referred to as the Van der Waals two-parameter cubic equation of state. The term two-parameter refers to the parameters a and b. The term cubic equation of state implies an equation which, if expanded, would contain volume terms raised to the first, second, and third power.

Perhaps the most significant features of Equation 6-13 is that it describes the liquid-condensation phenomenon and the passage from the gas to the liquid phase as the gas is compressed. These important features of the Van der Waals EOS is discussed below in conjunction with Figure 6-2.

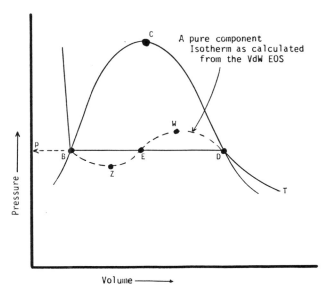

Figure 6-2. The volumetric behavior of a pure component as predicted by the Van Der Waal s EOS.

Consider a pure substance with a P-V behavior as shown in Figure 6-2. Assume that the substance is kept at a constant temperature T below its critical temperature. At this temperature, Equation 6-13 has three real roots (volumes) for each specified pressure p. A typical solution of Equation 6-13 at constant temperature T is shown graphically by the dashed isotherm: the constant temperature curve DWEZB, in Figure 6-2. The three values of V are the intersections B, E, and D on the horizontal line, corresponding to a fixed value of the pressure. This dashed calculated line (DWEZB) then appears to give a continuous transition from the gaseous phase to the liquid phase, but in reality, the transition is abrupt and discontinuous, with both liquid and vapor existing along the straight horizontal line DB. Examining the graphical solution of Equation 6-13 shows that the largest root (volume), as indicated by point D, corresponds to the volume of the saturated vapor while the smallest positive volume, as indicated by point B, corresponds to the volume of the saturated liquid. The third root, point E, has no physical meaning. It should be noted that these values become identical as the temperature approaches the critical temperature T_c of the substance.

Equation 6-13 can be expressed in a more practical form in terms of the compressibility factor Z. Replacing the molar volume V in Equation 6-12 with ZRT/p gives

$$Z^3 - (1 + B) Z^2 + AZ - AB = 0 \qquad (6\text{-}14)$$

where $A = \dfrac{ap}{R^2 T^2}$ $\qquad (6\text{-}15)$

$\qquad B = \dfrac{bp}{RT}$ $\qquad (6\text{-}16)$

Z = compressibility factor
p = system pressure, psia
T = system temperature, °R

Equation 6-14 yields one real root* in the one-phase region and three real roots in the two-phase region (where system pressure equals the vapor pressure of the substance). In the latter case, the largest root corresponds to the compressibility factor of the vapor phase Z^v, while the smallest positive root corresponds to that of the liquid Z^L.

An important practical application of Equation 6-14 is density calculations, as illustrated in the following example.

Example 6-1. A pure propane is held in a closed container at 100°F. Both gas and liquid are present. Calculate, by using the Van der Waals EOS, the density of the gas and liquid phases.

Solution.

Step 1. Determine the vapor pressure P_v of the propane from the Cox chart. This is the only pressure at which two phases can exist at the specified temperature.

$\qquad P_v = 185$ psia

Step 2. Calculate the parameters a and b from Equations 6-11 and 6-12, respectively.

$$a = 0.421875 \frac{(10.73)^2(666)^2}{616.3} = 34{,}957.4$$

---

* In some super critical regions, Equation 6-14 can yield three real roots for Z. From the three real roots, the largest root is the value of the compressibility with physical meaning.

$$b = 0.125 \frac{10.73 \ (666)}{616.3} = 1.4494$$

Step 3. Compute the coefficients A and B by applying Equations 6-15 and 6-16, respectively.

$$A = \frac{(34,957.4)(185)}{(10.73)^2(560)^2} = 0.179122$$

$$B = \frac{(1.4494)(185)}{(10.73)(560)} = 0.044625$$

Step 4. Substitute the values of A and B into Equation 6-13, to give

$$Z^3 - 1.044625Z^2 + 0.179122Z - 0.007993 = 0$$

Step 5. Solve the above third degree polynomial by extracting the largest and smallest roots of the polynomial by using the appropriate direct or iterative method, to give

$$Z^v = 0.72365$$
$$Z^L = 0.07534$$

Step 6. Solve for the density of the gas and liquid phases by using Equation 3-26:

$$\rho_g = \frac{(185)(44.0)}{(0.72365)(10.73)(560)} = 1.87 \ \text{lb/ft}^3$$

$$\rho_L = \frac{(185)(44)}{(0.07534)(10.73)(560)} = 17.98 \ \text{lb/ft}^3$$

The Van der Waals equation of state, despite its simplicity, provides a correct description, at least qualitatively, of the PVT behavior of substances in the liquid and gaseous states. Yet it is not accurate enough to be suitable for design purposes.

With the rapid development of computers the equation of state approach for the calculation of physical properties and phase equilibria proved to be a powerful tool, and much energy was devoted to the development of new and accurate equations of state. These equations, many of them a modification of the VdW equation of state, range in complexity from simple expressions containing two or three parameters to complicated forms containing more than fifty parameters. Although the complexity of any equation of state presents no computational problem, most authors prefer to retain the simplicity found in the Van der Waals cubic equation while improving its accuracy through modifications.

All equations of state are generally developed for pure fluids first, then extended to mixtures through the use of mixing rules. These mixing rules are simply means of calculating mixture parameters equivalent to those of pure substances.

Redlich-Kwong Equation of State

Redlich and Kwong (1948) demonstrated that by a simple adjustment, the Van der Waals' attractive pressure term (a/V_2) could considerably improve the prediction of the volumetric and physical properties of the vapor phase. The authors replaced the attractive pressure term with a generalized temperature dependence term. Their equation has the following form:

$$p = \frac{RT}{V - b} - \frac{a}{V(V + b)T^{0.5}} \tag{6-17}$$

where T is the system temperature in °R.

Redlich and Kwong, in their development of the equation, noted that as the system pressure becomes very large, i.e., $p \rightarrow \infty$, the molar volume V of the substance shrinks to about 26% of its critical volume regardless of the system temperature. Accordingly, they constructed Equation 6-17 to satisfy the following condition:

$$b = 0.26 \, V_c \tag{6-18}$$

Imposing the critical point conditions (as expressed by Equation 6-5) on Equation 6-17, and solving the resulting equations simultaneously, gives

$$a = \Omega_a \frac{R^2 \, T_c^{2.5}}{p_c} \tag{6-19}$$

$$b = \Omega_b \frac{R \, T_c}{p_c} \tag{6-20}$$

where $\Omega_a = 0.42747$
$\Omega_b = 0.08664$

Equating Equation 6-20 with 6-18 gives

$$p_c \, V_c = 0.333 \, R \, T_c \tag{6-21}$$

Equation 6-21 shows that the Redlich-Kwong EOS produces a universal critical compressibility factor (Z_c) of 0.333 for all substances. As indicated earlier, the critical gas compressibility ranges from 0.23 to 0.31 for most of the substances.

Replacing the molar volume V in Equation 6-17 with ZRT/p gives

$$Z^3 - Z^2 + (A - B - B^2) Z - AB = 0 \qquad (6\text{-}22)$$

where

$$A = \frac{ap}{R^2 \, T^{2.5}} \qquad (6\text{-}23)$$

$$B = \frac{bp}{RT} \qquad (6\text{-}24)$$

As in the VdW EOS, Equation 6-22 yields one real root in the one-phase region (gas phase region or liquid phase region), and three real roots in the two-phase region. In the latter case, the largest root corresponds to the compressibility factor of the gas phase Z^v while the smallest positive root corresponds to that of the liquid Z^L.

Example 6-2. Rework Example 6-1 by using the Redlich-Kwong equation of state.

Solution.

Step 1. Calculate the parameters a, b, A, and B.

$$a = 0.42747 \, \frac{(10.73)^2 (666)^{2.5}}{616.3} = 914{,}110.1$$

$$b = 0.08664 \, \frac{(10.73)(666)}{616.3} = 1.0046$$

$$A = \frac{(914{,}110.1)(185)}{(10.73)^2 (560)^{2.5}} = 0.197925$$

$$B = \frac{(1.0046)(185)}{(10.73)(560)} = 0.03093$$

Step 2. Substitute the parameters A and B into Equation 6-22 and extract the largest and the smallest root, to give

$$Z^3 - Z^2 + 0.1660384Z - 0.0061218 = 0$$

Largest root $Z^v = 0.802641$
Smallest root $Z^L = 0.0527377$

Step 3. Solve for the density of the liquid phase and gas phase.

$$\rho^L = \frac{(185)(44)}{(0.0527377)(10.73)(560)} = 25.7 \text{ lb/ft}^3$$

$$\rho^v = \frac{(185)(44)}{(0.802641)(10.73)(560)} = 1.688 \text{ lb/ft}^3$$

Redlich and Kwong extended the application of their equation to hydrocarbon liquid or gas mixtures by employing the following mixing rules:

$$a_m = \left[\sum_{i=1}^{n} x_i a_i^{0.5}\right]^2 \tag{6-25}$$

$$b_m = \sum_{i=1}^{n} [x_i b_i] \tag{6-26}$$

where n = number of components in the mixture
   a_i = Redlich-Kwong a parameter for the ith component as given by Equation 6-19
   b_i = Redlich-Kwong b parameter for the ith component as given by Equation 6-20
   a_m = mixture a parameter
   b_m = mixture b parameter
   x_i = mole fraction of component i in the liquid phase

To calculate a_m and b_m for a hydrocarbon gas mixture with a composition of y_i, use Equations 6-25 and 6-26 and replace x_i with y_i. The compressibility factor of the gas phase or the liquid is given by Equation 6-22 with the coefficients A and B as defined by Equations 6-23 and 6-24. The application of the Redlich and Kwong equation of state for hydrocarbon mixtures can be best illustrated through the following examples.

Example 6-3. Calculate the density of a crude oil with the following composition at 4,000 psia and 160°F. Use the Redlich-Kwong EOS.

Component	x_i	MW	p_c	T_c
C_1	0.45			
C_2	0.05			
C_3	0.05			
C_4	0.03			
C_5	0.01			
C_6	0.01			
C_{7+}	0.4	215	285	825

Solution.

Step 1. Determine the parameters a_i and b_i for each component by using Equations 6-19 and 6-20.

Component	a_i	b_i
C_1	161,044.3	0.4780514
C_2	493,582.7	0.7225732
C_3	914,314.8	1.004725
C_4	1,449,929	1.292629
C_5	2,095,431	1.609242
C_6	2,845,191	1.945712
C_{7+}	1.022348E7	4.191958

Step 2. Calculate the mixture parameters a_m and b_m from Equations 6-25 and 6-26, to give

$$a_m = 2,591,967$$
$$b_m = 2.0526$$

Step 3. Compute the coefficients A and B by using Equations 6-23 and 6-26, to produce

$$A = 9.406539$$
$$B = 1.234049$$

Step 4. Solve Equation 6-22 for the largest positive root, to yield

$$Z^L = 1.548126$$

Step 5. Calculate the apparent molecular weight of the crude oil

$$MW_a = 100.2547$$

Step 6. Solve for the density of the crude oil

$$\rho^L = \frac{(4,000)(100.2547)}{(10.73)(620)(1.548120)} = 38.93 \text{ lb/ft}^3$$

Notice, liquid density as calculated by Standing's correlation (Example 4-1) gives a value of 46.23 lb/ft^3.

Example 6-4. Calculate the density of a gas phase with the following composition at 4,000 psia and 160°F. Use the R-K EOS.

Component	y_i	MW	p_c	T_c
C_1	0.86			
C_2	0.05			
C_3	0.05			
C_4	0.02			
C_5	0.01			
C_6	0.005			
C_{7+}	0.005	215	285	825

Solution.

Step 1. Calculate a_m and b_m by using Equations 6-25 and 6-26, to give

$$a_m = 241,118$$
$$b_m = 0.5701225$$

Step 2. Calculate A and B, to yield

$$a = 0.8750$$
$$B = 0.3428$$

Step 3. Solve for Z^v, to give

$$Z^v = 0.907$$

Step 4. Calculate the density of the gas mixture.

$$\rho^v = \frac{(4,000)(20.89)}{(10.73)(620)(0.907)} = 13.85 \text{ lb/ft}^3$$

Soave-Redlich-Kwong Equation of State and Its Modifications

One of the most significant milestones in the development of cubic equations of state was the publication by Soave (1972) of a modification in the evaluation of the parameter a in the attractive pressure term of the Redlich-Kwong equation of state (Equation 6-17). Soave replaced the term $(a/T^{0.5})$ in Equation 6-17 with a more general temperature-dependent term as denoted by $(a\alpha)$, to give

$$p = \frac{RT}{V - b} - \frac{a\alpha}{V(V + b)} \tag{6-27}$$

where α is a dimensionless factor which becomes unity at $T = T_c$. At temperatures other than critical temperature, the parameter α is defined by the following expression:

$$\alpha = (1 + m(1 - T_r^{0.5}))^2 \tag{6-28}$$

The parameter m is correlated with the acentric factor, to give

$$m = 0.480 + 1.574\,\omega - 0.176\,\omega^2 \tag{6-29}$$

where T_r = reduced temperature T/T_c
  ω = acentric factor of the substance

For any pure component, the constants a and b in Equation 6-27 are found by imposing the classical Van der Waals' critical point constraints (Equation 6-5), on Equation 6-27 and solving the resulting equations, to give

$$a = \Omega_a \frac{R^2\,T_c^2}{p_c} \tag{6-30}$$

$$b = \Omega_b \frac{R\,T_c}{p_c} \tag{6-31}$$

where Ω_a and Ω_b are the Soave-Redlich-Kwong (SRK) dimensionless pure component parameters and have the following values:

$\Omega_a = 0.42747$
$\Omega_b = 0.08664$

Edmister and Lee (1984) showed that the two parameters a and b can be determined by a more convenient method. For the critical isotherm:

$$(V - V_c)^3 = V^3 - 3V_c\,V^2 + 3V_c^2\,V - V_c^3 = 0 \tag{6-32}$$

Equation 6-27 can also be put into a cubic form, to give

$$V^3 - \left[\frac{RT}{p}\right]V^2 + \left[\frac{a\alpha}{p} - \frac{bRT}{p} - b^2\right]V - \left[\frac{a\alpha b}{p}\right] = 0 \tag{6-33}$$

At the critical point, Equations 6-32 and 6-33 are identical and $\alpha = 1$. Equating the like terms:

$$3V_c = \frac{R\,T_c}{p_c} \tag{6-34}$$

$$3V_c^2 = \frac{a}{p_c} - \frac{bR\,T_c}{p_c} - b^2 \tag{6-35}$$

$$V_c^3 = \frac{ab}{p_c} \tag{6-36}$$

Solving the above equations for the parameters a and b yields expressions for the parameters as given by Equations 6-30 and 6-31.

Equation 6-34 indicated that the SRK equation of state gives a universal critical gas compressibility factor of 0.333. Combining Equation 6-34 with 6-31, gives

$$b = 0.26 \, V_v$$

Introducing the compressibility factor, Z, into Equation 6-33 by replacing the molar volume V in the equation with (ZRT/p) and rearranging, gives

$$Z^3 - Z^2 + (A - B - B^2)Z - AB = 0 \tag{6-37}$$

with

$$A = \frac{(a\alpha)p}{(RT)^2} \tag{6-38}$$

$$B = \frac{bp}{RT} \tag{6-39}$$

where p = system pressure, psia
       T = system temperature, °R
       R = 10.730 psia ft^3/lb-mole °R

Example 6-5. Rework Example 6-1 and solve for the density of the two phases by using the SRK EOS.

Solution.

Step 1. Determine the critical pressure, critical temperature, and acentric factor from Table 1-1, to give

$$T_c = 666.01°R$$
$$p_c = 616.3 \text{ psia}$$
$$\omega = 0.1524$$

Step 2. Calculate the reduced temperature

$$T_r = 560/666.01 = 0.8408$$

Step 3. Calculate the parameter m by applying Equation 6-29, to yield

$$m = 0.7051$$

Step 4. Solve for the parameter α by using Equation 6-28, to give

$$\alpha = 1.120518$$

Step 5. Compute the coefficients a and b by applying Equations 6-30 and 6-31, to yield

$$a = 35,427.6$$
$$b = 1.00471$$

Step 6. Calculate the coefficients A and B from Equations 6-38 and 6-39, to produce

$$A = 0.203365$$
$$B = 0.034658$$

Step 7. Solve Equation 6-37 for Z^L and Z^v, to yield

$$Z^L = 0.06729$$
$$Z^v = 0.80212$$

Step 8. Calculate the gas and liquid density, to give

$$\rho^v = 1.6887 \text{ lb/ft}^3$$
$$\rho^L = 20.13 \text{ lb/ft}^3$$

To use Equation 6-37 with mixtures, mixing rules are required to determine the terms a and b for the mixtures. Soave adopted the following mixing rules:

$$(a\alpha)_m = \sum_i \sum_j [x_i x_j (a_i a_j \alpha_i \alpha_j)^{0.5}(k_{ij} - 1)] \qquad (6\text{-}40)$$

$$b_m = \sum_i [x_i \, b_i] \qquad (6\text{-}41)$$

with

$$A = \frac{(a\alpha)_m \, p}{(RT)^2} \qquad (6\text{-}42)$$

and

$$B = \frac{b_m \, p}{RT} \qquad (6\text{-}43)$$

The parameter k_{ij} is an empirically determined correction factor called the binary interaction coefficient, characterizing the binary formed by component i and component j in the hydrocarbon mixture.

These binary interaction coefficients are used to model the intermolecular interaction through empirical adjustment of the $(a\alpha)_m$ term as represented mathematically by Equation 6-41. They are dependent on the difference in molecular size of components in a binary system and they are characterized by the following properties, as summarized by Slot-Petersen (1987):

- The interaction between hydrocarbon components increases as the relative difference between their molecular weights increases:

$$k_{i,j+1} > k_{i,j}$$

- Hydrocarbon components with the same molecular weight have a binary interaction coefficient of zero:

$$k_{i,i} = 0$$

- The binary interaction coefficient matrix is symmetric:

$$k_{i,j} = k_{j,i}$$

Slot-Petersen (1987) and Vidal and Daubert (1978) presented a theoretical background to the meaning of the interaction coefficient and techniques for determining their values.

Groboski and Daubert (1978) and Soave (1972) suggested that no binary interaction coefficients are required for hydrocarbon systems. However, with non-hydrocarbons present, binary interaction parameters can greatly improve the volumetric and phase behavior predictions of the mixture by the SRK EOS.

In solving Equation 6-32 for the compressibility factor of the liquid phase, the composition of the liquid x_i is used to calculate the coefficients A and B of Equations 6-42 and 6-43 through the use of the mixing rules as described by Equations 6-40 and 6-41. For determining the compressibility factor of the gas phase Z_v, the foregoing outlined procedure is used but with composition of the gas phase y_i replacing x_i.

Example 6-6. A two-phase hydrocarbon system exists in equilibrium at 4,000 psia and 160°F. The system has the following composition:

Component	x_i	y_i
C_1	0.45	0.86
C_2	0.05	0.05
C_3	0.05	0.05
C_4	0.03	0.02
C_5	0.01	0.01
C_6	0.01	0.005
C_{7+}	0.40	0.0005

The heptanes-plus fraction has the following properties:

MW = 215
p_c = 285 psia
T_c = 700°F
ω = 0.52

Assuming k_{ij} = 0, calculate the density of each phase by using the SRK EOS.

Solution.

Step 1. Calculate the parameters α, a, and b by applying Equations 6-28, 6-30, and 6-31.

Component	α_i	a_i	b_i
C_1	0.6869	8,689.3	0.4780
C_2	0.9248	21,040.8	0.7225
C_3	1.0502	35,422.1	1.0046
C_4	1.1616	52,390.3	1.2925
C_5	1.2639	72,041.7	1.6091
C_6	1.3547	94,108.4	1.9455
C_{7+}	1.7859	232,367.9	3.7838

Step 2. Calculate the mixture parameters $(a\alpha)_m$ and b_m for the gas phase and liquid phase by applying Equations 6-40 and 6-41, to give

For the gas phase:

$(a\alpha)_m$ = 9,219.3
b_m = 0.5680

For the liquid phase:

$(a\alpha)_m$ = 104,362.9
b_m = 1.8893

Step 3. Calculate the coefficients A and B for each phase by applying Equations 6-42 and 6-43, to yield

For the gas phase:

A = 0.8332
B = 0.3415

For the liquid phase:

A = 9.4324
B = 1.136

Step 4. Solve Equation 6-37 for the compressibility factor of the gas phase, to produce

$$Z^v = 0.9267$$

Step 5. Solve Equation 6-37 for the compressibility factor of the liquid phase, to produce

$$Z^L = 1.4121$$

Step 6. Calculate the apparent molecular weight of the gas phase and liquid phase from their composition, to yield

For the gas phase:

$$MW_a = 20.89$$

For the liquid phase:

$$MW_a = 100.25$$

Step 7. Calculate the density of each phase.

$$\rho^v = \frac{(4,000)(20.89)}{(10.73)(620)(0.9267)} = 13.556 \text{ lb/ft}^3$$

$$\rho^L = \frac{(4,000)(100.25)}{(10.73)(620)(1.4121)} = 42.68 \text{ lb/ft}^3$$

It is appropriate at this time to introduce and define the concept of the fugacity and the fugacity coefficient of the component.

The fugacity f is a measure of the molar Gibbs energy of a real gas. It is evident from the definition that the fugacity has the units of pressure, in fact, the fugacity may be looked upon as a vapor pressure modified to represent correctly the escaping tendency of the molecules from one phase into the other. In a mathematical form, the fugacity of a component is defined by the following expression:

$$f = p \text{ EXP} \left[\int_o^P \left(\frac{Z-1}{p} \right) dp \right] \tag{6-44}$$

where f = fugacity, psia
       p = pressure, psia
       Z = compressibility factor

The ratio of the fugacity to the pressure, i.e., f/p, is called the fugacity coefficient Φ and is calculated from Equation 6-44.

$$\Phi = \text{EXP} \left[\int_{o}^{p} \left(\frac{Z - 1}{p} \right) dp \right]$$

Soave applied the above generalized thermodynamic relationship to Equation 6-27 to determine the fugacity coefficient of a pure component, to give

$$\text{Ln}(\Phi) = Z - 1 - \text{Ln}(Z - B) - \frac{A}{B} \text{Ln} \left[\frac{Z + B}{Z} \right] \tag{6-45}$$

In practical petroleum engineering applications we are concerned with the phase behavior of the hydrocarbon liquid mixture, which at a specified pressure and temperature, is in equilibrium with a hydrocarbon gas mixture at the same pressure and temperature.

The component fugacity in each phase is introduced to develop a criterion for thermodynamic equilibria. Physically, the fugacity of a component in one phase with respect to the fugacity of the component in a second phase is a measure of the potential for transfer of the component between phases. The phase with the lower component fugacity accepts the component from the second phase. Equal fugacities of a component in the two phases results in a zero net transfer. A zero transfer for all components implies a hydrocarbon system that is in thermodynamic equilibrium. Therefore, the condition of the thermodynamic equilibrium can be expressed mathematically by

$$f_i^v = f_i^L \qquad 1 \le i \le n \tag{6-46}$$

where f_i^v = fugacity of component i in the gas phase, psi
f_i^L = fugacity of component i in the liquid phase, psi
n = number of components in the system

The fugacity coefficient of component i in a hydrocarbon liquid mixture or hydrocarbon gas mixture is a function of the system pressure, mole fraction, and the fugacity of the component, and is defined by the following expressions:

$$\Phi_i^v = \frac{f_i^v}{y_i p} \tag{6-47}$$

$$\Phi_i^L = \frac{f_i^L}{y_i p} \tag{6-48}$$

where Φ_i^v = fugacity coefficient of component i in the vapor phase
Φ_i^L = fugacity coefficient of component i in the liquid phase

It is clear that at equilibrium ($f_i^L = f_i^v$), the equilibrium ratio K_i as previously defined by Equation 6-1 can be determined from the following relationship:

$$K_i = \frac{[f_i^L/(x_ip)]}{[f_i^v/(y_ip)]} = \frac{\Phi_i^L}{\Phi_i^v} \tag{6-49}$$

Reid and Sherwood (1977) defined the fugacity coefficient of component i in a hydrocarbon mixture by the following generalized thermodynamic relationship:

$$Ln(\Phi_i) = \left(\frac{-1}{RT}\right)\left[\int_\infty^v \left(\frac{p}{n_i} - \frac{RT}{V}\right)dV\right] - Ln(Z) \tag{6-50}$$

where V = total volume of n moles of the mixture
n_i = number of moles of component i
Z = compressibility factor of the hydrocarbon mixture

By combining the above thermodynamic definition of the fugacity with the SRK EOS (Equation 6-37), Soave proposed the following expression for the fugacity coefficient of component i in the liquid phase:

$$Ln(\Phi_i^L) = \left[\frac{b_i(Z^L - 1)}{b_m}\right] - Ln(Z^L - B)$$

$$- \left|\frac{A}{B}\right|\left[\frac{2\psi_i}{\psi} - \frac{b_i}{b_m}\right] Ln\left[1 + \frac{B}{Z^L}\right] \tag{6-51}$$

where $\psi_i = \sum_j [x_j(a_ia_j\alpha_i\alpha_j)^{.5}(1 - k_{ij})]$ \qquad (6-52)

$$\psi = \sum_i \sum_j [x_ix_j(a_ia_j\alpha_i\alpha_j)^{.5}(1 - k_{ij})] \tag{6-53}$$

Equation 6-52 is also used to determine Φ_i^v by using the composition of the gas phase y_i in calculating A, B, Z^v, and other composition-dependent terms.

Modifications of the SRK EOS

Groboski and Daubert (1978) proposed a new expression for calculating the parameter m of Equation 6-29 to improve the pure component vapor pressure predictions by the SRK equation of state. The proposed relationship was originated from analyzing extensive experimental data for pure hydrocarbons. The relationship has the following form:

$$m = 0.48508 + 1.55171 \, \omega - 0.15613 \, \omega^2 \tag{6-54}$$

Sim and Daubert (1980) pointed out that because the coefficients of Equation 6-54 were determined by analyzing vapor pressure data of low molecular weight hydrocarbons it is unlikely that Equation 6-54 will suffice for high molecular weight petroleum fractions. Realizing that the acentric factors for the heavy petroleum fractions are calculated from an equation such as the Edmister correlation or the Lee and Kesler correlation (as discussed in Chapter 2), the authors proposed the following expressions for determining the parameter m:

- If the acentric factor is determined by using the Edmister correlation, then:

$$m = 0.431 + 1.57 \, \omega_i - 0.161 \, \omega_i^2 \tag{6-55}$$

- If the acentric factor is determined by using the Lee and Kesler correlation, then:

$$m = 0.315 + 1.60 \, \omega_i - 0.166 \, \omega_i^2 \tag{6-56}$$

Elliot and Daubert (1985) stated that the optimal binary interaction coefficient would minimize the error in the representation of all the thermodynamic properties of a mixture. Properties of particular interest in phase equilibrium calculations include bubble-point pressure, dew-point pressure, and equilibrium ratios. The authors, in their evaluation of optimal interaction coefficients of asymmetric mixtures*, proposed the following set of expressions for calculating k_{ij}:

- For N_2 Systems:

$$k_{ij} = 0.107089 + 2.9776 \, k_{ij}^{\infty} \tag{6-57}$$

- For CO_2 Systems:

$$k_{ij} = 0.08058 - 0.77215 \, k_{ij}^{\infty} - 1.8407 \, (k_{ij}^{\infty})^2 \tag{6-58}$$

- For H_2S Systems:

$$k_{ij} = 0.07654 + 0.017921 \, k_{ij}^{\infty} \tag{6-59}$$

---

* An asymmetric mixture is defined as one in which two of the components are considerably different in their chemical behavior. Mixtures of methane with hydrocarbons of ten or more carbon atoms can be considered asymmetric. Mixtures containing gases such as nitrogen or hydrogen are asymmetric.

- For Methane Systems with compounds of 10 carbons or more:

$$k_{ij} = 0.17985 + 2.6958 \; k_{ij}^{\infty} + 10.853 \; (k_{ij}^{\infty})^2 \tag{6-60}$$

where, for the above expression:

$$k_{ij}^{\infty} = [- (\epsilon_i - \epsilon_j)^2]/(2\epsilon_i\epsilon_j) \tag{6-61}$$

and

$$\epsilon_i = [a_i \; Ln(2)]^{.5}/b_i \tag{6-62}$$

The parameters a_i and b_i of Equation 6-62 are defined previously by Equations 6-30 and 6-31.

The major drawback in the SRK EOS is that the critical compressibility factor takes on the unrealistic universal critical compressibility of 0.333 for all substances. Consequently, the molar volumes are typically overestimated, i.e., densities are underestimated.

Peneloux et al. (1982) developed a procedure for improving the volumetric predictions of the SRK EOS by introducing a volume correction parameter c_i into the equation. The third parameter does not change the vapor-liquid equilibrium conditions determined by the unmodified SRK equation, but modifies the liquid and gas volumes by effecting the following translation along the volume axis. They proposed the following:

$$V_{corr}^{L} = V^{L} - \sum_i (x_i c_i) \tag{6-63}$$

$$V_{corr}^{v} = V^{v} - \sum_i (y_i c_i) \tag{6-64}$$

where V^{L} = liquid molar volume = $Z^{L} \; RT/p$, ft^3/mole
V^{v} = gas molar volume = $Z^{v} \; RT/p$, ft^3/mole
V_{corr}^{L} = corrected liquid molar volume, ft^3/mole
V_{corr}^{v} = corrected gas molar volume, ft^3/mole
x_i = mole fraction of component i in the liquid phase
y_i = mole fraction of component i in the gas phase

The authors proposed six different schemes for calculating the correction factor c_i for each component. For petroleum fluids and heavy hydrocarbons, Peneloux and coworkers suggested that the best correlating parameter for c_i is the Rackett compressibility factor Z_{RA}. The correction factor is defined mathematically by the following relationship:

$$c_i = 4.43797878(0.29441 - Z_{RA})T_{ci}/p_{ci} \tag{6-65}$$

where c_i = correction factor for component i, ft^3/lb-mole
   T_{ci} = critical temperature of component i, °R
   p_{ci} = critical pressure of component i, psia

Z_{RA} is a unique constant for each compound. The values of Z_{RA} are in general not much different from those of the critical compressibility factors Z_c. If their values are not available, Peneloux et al. proposed the following correlation for calculating c_i:

$$c_i = 4.43797878(0.00261 + 0.0928\ \omega_i)T_{ci}/p_{ci} \qquad (6\text{-}66)$$

where ω_i = acentric factor of component i

Example 6-7. Rework Example 6-6 by incorporating the Peneloux volume correction approach in the solution.

Solution.

Step 1. Calculate the correction factor c_i by using Equation 6-66.

Component	c_i	x_i	$c_i x_i$	y_i	$c_i y_i$
C_1	0.00839	0.45	0.003776	0.86	0.00722
C_2	0.03807	0.05	0.001903	0.05	0.00190
C_3	0.07729	0.05	0.003861	0.05	0.00386
C_4	0.1265	0.03	0.00379	0.02	0.00253
C_5	0.19897	0.01	0.001989	0.01	0.00198
C_6	0.2791	0.01	0.00279	0.005	0.00139
C_{7+}	0.91881	0.40	0.36752	0.005	0.00459
			0.38564		0.02349

Step 2. Calculate the uncorrected volume of the gas and liquid phase by using the compressibility factors as calculated in Example 6-6.

$$V^v = \frac{(10.73)(620)(0.9267)}{4,000} = 1.54119 \ \text{ft}^3/\text{mole}$$

$$V^L = \frac{(10.73)(620)(1.4121)}{4,000} = 2.3485 \ \text{ft}^3/\text{mole}$$

Step 3. Calculate the corrected gas and liquid volumes by applying Equations 6-63 and 6-64.

$$V^L_{corr} = 2.3485 - 0.38564 = 1.962927 \ \text{ft}^3/\text{mole}$$
$$V^v_{corr} = 1.54119 - 0.02349 = 1.5177 \ \text{ft}^3/\text{mole}$$

Step 4. Calculate the corrected compressibility factors.

$$Z_{corr}^v = \frac{(4,000)(1.5177)}{(10.73)(620)} = 0.91254$$

$$Z_{corr}^L = \frac{(4,000)(1.962927)}{(10.73)(620)} = 1.18025$$

Step 5. Determine the corrected densities of both phases.

$$\rho^v = \frac{(4,000)(20.89)}{(10.73)(620)(0.91254)} = 13.767 \text{ lb/ft}^3$$

$$\rho^L = \frac{(4,000)(100.25)}{(10.73)(620)(1.18025)} = 51.07 \text{ lb/ft}^3$$

Peng-Robinson Equation of State and Its Modifications

Peng and Robinson (1975) conducted a comprehensive study to evaluate the use of the SRK equation of state for predicting the behavior of naturally occurring hydrocarbon systems. They showed the need for an improvement in the ability of the equation of state to predict liquid densities and other fluid properties particularly in the vicinity of the critical region. As a basis for creating an improved model, Peng and Robinson (P-R) proposed the following expression:

$$p = \frac{RT}{V - b} - \frac{a\alpha}{(V + b)^2 - cb^2}$$

where a, b, and α have the same significance as they have in the SRK model, and c is a whole number optimized by analyzing the Z_c and b/V_c values obtained from the equation. It is generally accepted that Z_c should be close to 0.28 and that b/V_c should be approximately 0.26. An optimized value of c = 2 gave $Z_c = 0.307$ and $(b/V_c) = 0.253$. Based on this value of c, Peng and Robinson proposed the following equation of state:

$$p = \frac{RT}{V - b} - \frac{a\alpha}{V(V + b) + b(V - b)} \tag{6-67}$$

Imposing the classical critical point conditions (Equation 6-5) on Equation 6-67 and solving for the parameters a and b, yields

$$a = \Omega_a \frac{R^2 T_c^2}{p_c} \tag{6-68}$$

$$b = \Omega_b \frac{R\,T_c}{p_c} \tag{6-69}$$

where $\Omega_a = 0.45724$
$\Omega_b = 0.07780$

This equation predicts a universal critical gas compressibility factor of 0.307 compared to 0.333 for the SRK model. Peng and Robinson also adopted Soave's approach for calculating the parameter α:

$$\alpha = (1 + m(1 - T_r^{.5}))^2 \tag{6-70}$$

where $m = 0.3746 + 1.5423\,\omega - 0.2699\,\omega^2$

This was later expanded by the investigators (1978) to give the following relationship:

$$m = 0.379642 + 1.48503\,\omega - 0.1644\,\omega^2 + 0.016667\,\omega^3 \tag{6-71}$$

Rearranging Equation 6-67 into the compressibility factor form, gives

$$Z^3 + (B - 1)\,Z^2 + (A - 3B^2 - 2B)Z - (AB - B^2 - B^3) = 0 \tag{6-72}$$

where A and B are given by Equations 6-38 and 6-39 for pure component and by Equations 6-42 and 6-43 for mixtures.

Example 6-8. Using the composition given in Example 6-6, calculate the density of the gas phase and liquid phase by using the Peng-Robinson EOS. Assume $k_{ij} = 0$.

Solution.

Step 1. Calculate the mixture parameters $(a\alpha)_m$ and b_m for the gas and liquid phase, to give

For the gas phase

$(a\alpha)_m = 10{,}423.54$
$b_m = 0.862528$

For the liquid phase

$(a\alpha)_m = 107{,}325.4$
$b_m = 1.696543$

Step 2. Calculate the coefficients A and B, to give

For the gas phase

A = 0.94209
B = 0.30669

For the liquid phase

A = 9.700183
B = 1.020078

Step 3. Solve Equation 6-72 for the compressibility factor of the gas phase and the liquid phase, to give

Z^v = 0.8625
Z^L = 1.2645

Step 4. Calculate the density of both phases.

$$\rho^v = \frac{(4,000)(20.89)}{(10.73)(620)(0.8625)} = 14.566 \text{ lb/ft}^3$$

$$\rho^L = \frac{(4,000)(100.25)}{(10.73)(620)(1.2645)} = 47.67 \text{ lb/ft}^3$$

Applying the thermodynamic relationship, as given by Equation 6-45, to Equation 6-68, yields the following expression for the fugacity of a pure component:

$$\ln\left(\frac{f}{p}\right) = Z - 1 - \ln(Z - B) - \left[\frac{A}{2.82843B}\right] \ln\left[\frac{Z + 2.414B}{Z - 0.414B}\right] \tag{6-73}$$

The fugacity coefficient of component i in a hydrocarbon liquid mixture is calculated from the following expression:

$$\ln(\Phi_i^L) = b_i (Z^L - 1)/b_m - \ln (Z^L - B)$$

$$- \frac{A}{2.82843B} \left[\frac{2\psi_i}{\psi} - \frac{b_i}{b_m}\right] \ln\left[\frac{Z^L + 2.414B}{Z^L - 0.414B}\right] \tag{6-74}$$

where the mixture parameters b_m, B, A, ψ_i, and ψ are defined previously.

Equation 6-74 can be used to determine the fugacity coefficient of any component in the gas phase by using the composition of the gas phase in calculating the composition-dependent terms of the equation.

Modifications of the PR Equation of State

To obtain a good representation of the phase equilibria in multicomponent mixtures containing N_2, CO_2, and CH_4, Nikos et al. (1986) proposed a generalized correlation for evaluating the binary interaction coefficient, k_{ij}, of the PR EOS as a function of the pressure, temperature, and acentric factor of the hydrocarbon. These generalized correlations were originated with all the binary experimental data available in the literature. The authors proposed the following generalized form for k_{ij}:

$$k_{ij} = \delta_2\, T_{rj}^2 + \delta_1\, T_{rj} + \delta_0 \tag{6-76}$$

where i refers to the principal components N_2, CO_2, or CH_4; and j refers to the other hydrocarbon component of the binary. The acentric factor dependent coefficients δ_0, δ_1, and δ_2 are determined for each set of binaries by applying the following expressions:

For Nitrogen-Hydrocarbons:

$$\delta_0 = 0.1751787 - 0.7043\, \text{Log}(\omega_j) - 0.862066\, [\text{Log}(\omega_j)]^2 \tag{6-76}$$

$$\delta_1 = -0.584474 + 1.328\, \text{Log}\,(\omega_j) + 2.035767\, [\text{Log}\,(\omega_j)]^2 \tag{6-77}$$

and

$$\delta_2 = 2.257079 + 7.869765\, \text{Log}(\omega_j) + 13.50466\, [\text{Log}(\omega_j)]^2 + 8.3864\, [\text{Log}(\omega_j)]^3 \tag{6-78}$$

They also suggested the following pressure correction:

$$k_{ij}' = k_{ij}\, (1.04 - 4.2 \times 10^{-5}p) \tag{6-79}$$

where p is the pressure in pounds per square inch.

For Methane-Hydrocarbons:

$$\delta_0 = -0.01664 - 0.37283\, \text{Log}(\omega_j) + 1.31757\, [\text{Log}(\omega_j)]^2 \tag{6-80}$$

$$\delta_1 = 0.48147 + 3.35342\, \text{Log}\,(\omega_j) - 1.0783\, [\text{Log}(\omega_j)]^2 \tag{6-81}$$

and

$$\delta_2 = -0.4114 - 3.5072\, \text{Log}(\omega_j) - 0.78798\, [\text{Log}(\omega_j)]^2 \tag{6-82}$$

For CO_2-Hydrocarbons:

$$\delta_0 = 0.4025636 + 0.1748927 \, \text{Log} \, (\omega_j) \qquad (6\text{-}83)$$

$$\delta_1 = -0.94812 - 0.6009864 \, \text{Log} \, (\omega_j) \qquad (6\text{-}84)$$

and

$$\delta_2 = 0.741843368 + 0.441775 \, \text{Log} \, (\omega_j) \qquad (6\text{-}85)$$

For the CO_2 interaction parameters, the following pressure correction is suggested:

$$k_{ij}' = k_{ij} \, (1.044269 - 4.375 \times 10^{-5} p) \qquad (6\text{-}86)$$

The applicability of the above correlations should be extended with caution for cases where T_{rj} far exceeds the value of 1.0.

Stryjek and Vera (1986) proposed an improvement in the reproduction of vapor pressures of pure component by the PR EOS in the reduced temperature range from 0.7 to 1.0, by replacing the m term in Equation 6-70 with the following expression:

$$m_0 = 0.378893 + 1.4897153 - 0.17131848 \, \omega^2 + 0.0196554 \, \omega^3 \qquad (6\text{-}87)$$

To reproduce vapor pressures at reduced temperatures below 0.7, Stryjek and Vera further modified the m parameter in the PR equation by introducing an adjustable parameter m_1 characteristic of each compound to Equation 6-70. They proposed the following generalized relationship for the parameter m:

$$m = m_0 + m_1 \, (1 + T_r^{.5})(0.7 - T_r) \qquad (6\text{-}88)$$

where T_r = reduced temperature of the pure component
m_0 = defined by Equation 6-87
m_1 = adjustable parameter

For all components with a reduced temperature above 0.7, Stryjek and Vera recommended setting $m_1 = 0$. For components with a reduced temperature greater than 0.7, the optimum values of m_1 for compounds of industrial interest are given in Table 6-1.

Due to the totally empirical nature of m_1, Stryjek and Vera could not find a generalized correlation for the parameter m_1 in terms of pure component parameters. They pointed out that the values of m_1 given in Table 6-1 should be used without changes.

Table 6-1
Parameter "m_1" of Pure Compounds

Compound	m_1	Compound	m_1
Nitrogen	0.01996	Nonane	0.04104
Carbon Dioxide	0.04285	Decane	0.04510
Water	− 0.06635	Undecane	0.02919
Methane	− 0.00159	Dodecane	0.05426
Ethane	0.02669	Tridecane	0.04157
Propane	0.03136	Tetradecane	0.02686
Butane	0.03443	Pentadecane	0.01892
Pentane	0.03946	Hexadecane	0.02665
Hexane	0.05104	Heptadecane	0.04048
Heptane	0.04648	Octadecane	0.08291
Octane	0.04464		

Jhaveri and Youngren (1984) found, when applying the unmodified Peng-Robinson equation of state to reservoir fluids, that the error associated with the equation in the prediction of gas phase Z-factors ranged from 3–5% and the error in the liquid density predictions ranged from 6–12%. Following the procedure proposed by Peneloux and co-workers (see the SRK EOS), Jhaveri and Youngren introduced the volume correction parameter c_i to the PR EOS. This third parameter has the same units as the second parameter b_i of the unmodified PR equation and is defined by the following relationship:

$$c_i = S_i \, b_i \tag{6-89}$$

where S_i = dimensionless parameter and is called the shift parameter
b_i = Peng-Robinson co-volume as given by Equation 6-69

The volume correction parameter c_i does not change the vapor-liquid equilibrium conditions, i.e., equilibrium ratios K_i. The corrected hydrocarbon phase volumes are given by the following expressions:

$$V_{corr}^L = V^L - \sum_{i=1} (x_i c_i)$$

$$V_{corr}^v = V^v - \sum_{i=1} (y_i c_i)$$

where V^L, V^v = volumes of the liquid phase and gas phase as calculated by unmodified PR EOS, ft^3/mole.
V_{corr}^L, V_{corr}^v = corrected volumes of the liquid and gas phase

Table 6-2
Shift Parameter For Hydrocarbons

Component	S_i
C_1	-0.1540
C_2	-0.1002
C_3	-0.08501
i-C_4	-0.07935
n-C_4	-0.06413
i-C_5	-0.04350
n-C_5	-0.04183
n-C_6	-0.01478

Table 6-2 gives values of the shift parameters for the well-defined lighter hydrocarbons as presented by the authors.

For the undefined components, i.e., C_{7+}, the authors proposed the following Equation for calculating the shift parameter:

$$S_i = 1 - \frac{d}{(MW)_i^e}$$

where   MW = molecular weight of component i
        d, e = positive correlation coefficients

Jhaveri and Youngren outlined the procedure of calculating the coefficients e and d. They also proposed that in the absence of the experimental information needed for calculating e and d, the power coefficient e can be set equal to 0.2051 and the coefficient d is adjusted to match the C_{7+} density. The values of d range from 2.2 to 3.2.

To use the Peng and Robinson equation of state to predict the phase and volumetric behavior of mixtures, one must be able to provide the critical pressure, the critical temperature, and the acentric factor for each component in the mixture. For pure compounds, the required properties are well-defined and known. Nearly all naturally occurring petroleum fluids contain a quantity of heavy fractions that are not well-defined. These heavy fractions are often lumped together as a heptanes-plus fraction. The problem of how to adequately characterize the C_{7+} fractions in terms of their critical properties and acentric factors has been long recognized in the petroleum industry. Changing the characterization of C_{7+} fractions present in even small amounts can have a profound effect on the PVT properties and the phase equilibria of a hydrocarbon system as predicted by the Peng and Robinson equation of state.

The usual approach for such situations is to "tune" the parameters in the EOS in an attempt to improve the accuracy of prediction. During the tuning process, the critical properties of the heptanes-plus fraction and the binary interaction coefficients are adjusted to obtain a reasonable match with experimental data available on the hydrocarbon mixture.

Recognizing that the inadequacy of the predictive capability of the PR EOS lies with the improper procedure for calculating the parameters a, b, and α of the equation for the C_{7+} fraction. Ahmed (1988) devised an approach for determining these parameters from the readily measured physical properties of C_{7+}: molecular weight and specific gravity. The approach is based on generating 49 density values for the C_{7+} by applying the Riazi and Daubert correlation. These values were subsequently subjected to 10 temperature and 10 pressure values in the range of 60°F–300°F and 14.7–7,000 psia, respectively. The Peng and Robinson EOS was then applied to match the 4,900 generated density values by optimizing the parameters a, b, and α using a non-linear regression model. The optimized parameters for the heptanes-plus fraction are given by the following expressions:

For the parameter α of C_{7+}:

$$\alpha = \left[1 + m \left[1 - \left[\frac{520}{T} \right]^{0.5} \right] \right]^2 \tag{6-90}$$

with m as defined by

$$m = \left[\frac{(MW/\gamma)_{C_{7+}}}{a_1 + a_2 (MW/\gamma)_{C_{7+}}} \right] + a_3 \, MW_{C_{7+}} + a_4 \, MW_{C_{7+}}^2 + a_5 / MW_{C_{7+}}$$

$$+ a_6 \, \gamma_{C_{7+}} + a_7 \, \gamma_{C_{7+}}^2 + a_8 / \gamma_{C_{7+}} \tag{6-91}$$

where $MW_{C_{7+}}$ = molecular weight of C_{7+}
$\gamma_{C_{7+}}$ = specific gravity of C_{7+}
$a_1 - a_8$ = coefficients as given in Table 6-3

Table 6-3
Coefficients $a_1 - a_8$ for Equations 6-91 and 6-92

Coefficient	m	a	b
a_1	− 36.91776	− 2.433525E7	− 6.8453198
a_2	− 5.2393763E-2	8.3201587E3	1.730243E-2
a_3	1.7316235E-2	− 18.444102	− 6.2055064E-6
a_4	− 1.3743308E-5	3.6003101E-2	9.0910383E-9
a_5	12.718844	3.4992796E7	13.378898
a_6	10.246122	2.838756E7	7.9492922
a_7	− 7.6697942	− 1.1325365E7	− 3.1779077
a_8	− 2.6078099	6.418828E6	1.7190311

For the parameters a and b of C_{7+}:

The following generalized correlation is proposed:

$$a \text{ or } b = a_1 + a_2(MW/\gamma)_{C_{7+}} + a_3(MW/\gamma)^2_{C_{7+}} + a_4(MW/\gamma)^3_{C_{7+}} \\ + a_5\,\gamma_{C_{7+}}/MW_{C_{7+}} + a_6\,\gamma_{C_{7+}} + a_7\,\gamma^2_{C_{7+}} + a_8/\gamma_{C_{7+}} \qquad (6\text{-}92)$$

The coefficients a_1–a_8 for the parameters a and b are given in Table 6-3.

The coefficients a, b, and m for carbon dioxide, nitrogen, methane, and ethane were also optimized by using this approach. The optimized values are:

Component	Coefficient		
	a	b	m In Equation 5-90
CO_2	1.499914×10^4	0.41503575	-0.73605717
N_2	4.5693589×10^3	0.4682582	-0.97962859
C_1	7.709708×10^3	0.46749727	-0.549765
C_2	2.416260×10^4	0.6690577	-0.6952108

Heyen Equation of State

All equations of state described previously contain two parameters, a and b, which are fixed at the critical point by the Van der Waals critical point conditions as described by Equation 6-5. Away from the critical point, a temperature dependence, α, is included in the attractive pressure term. The resulting critical compressibility predicted by these equations is constant for all components. To overcome this drawback, Heyen (1983) proposed a modification of the Peng-Robinson equation of state where a third parameter c is added to reproduce the experimental critical compressibility factor. Heyen also introduced temperature dependence into the parameter b to match observed saturated liquid volume. Heyen proposed the following three-parameter cubic equation of state:

$$p = \frac{RT}{V - b} - \frac{a}{V^2 + (b + c)V - bc} \qquad (6\text{-}93)$$

with

$$a = \Omega_a \frac{R^2\,T_c^2}{p_c}\,[EXP\,(k(1 - T_r^n))] \qquad (6\text{-}94)$$

$$b = \Omega_b \frac{R T_c}{p_c} \left[1 + m \left[\frac{1 - \text{EXP} \left[\Theta(T_r - 1) \right]}{1 + \text{EXP} \left[\Theta(T_r - 1) \right]} \right] \right] \tag{6-95}$$

$$c = \Omega_c \frac{R T_c}{p_c} \tag{6-96}$$

Heyen identified the parameters k, n, m, and Θ for any component by forcing Equation 6-93 to match experimental vapor pressures and saturated liquid volumes. This led to the following correlations for the above parameters with the acentric factor as the correlating parameter:

$$k = 0.49164 + 0.43882 \, \omega - 0.08821 \, \omega^2 \tag{6-97}$$

$$n = 1.637 + 1.389 \, \omega \tag{6-98}$$

$$\theta = 7.2562 + 14.153 \, \omega + 1.33137 \, \omega^2 \tag{6-99}$$

$$m = 0.23333 - 0.06737 \, \omega + 0.49110 \, \omega^2 \tag{6-100}$$

Imposing the Van der Waals critical point conditions on Equation 6-93, results in

$$\Omega_c = 1 - 3 Z_c \tag{6-101}$$

$$\Omega_b^3 + (2 - 3Z_c) \, \Omega_b^2 + 3 \, Z_c^2 \, \Omega_b - Z_c^3 = 0 \tag{6-102}$$

$$\Omega_a = 3 \, Z_c^2 + 2 \, \Omega_c \, \Omega_b + \Omega_b + \Omega_c + \Omega_b^2 \tag{6-103}$$

where Ω_b is the smallest positive real root of Equation 6-102. This cubic equation is readily solved by the Newton-Raphson method with an initial value for Ω_b as given by

$$\Omega_b = 0.32429 \, Z_c - 0.022005$$

Introducing the compressibility factor Z into Equation 6-93 results in

$$Z^3 + (C - 1) \, Z^2 + (A - 2BC - B - C - B^2)Z$$
$$+ (BC + B^2C - AB) = 0 \tag{6-104}$$

where

$$A = \frac{ap}{(RT)^2} \tag{6-105}$$

$$B = \frac{bp}{(RT)} \tag{6-106}$$

$$C = \frac{cp}{RT} \tag{6-107}$$

For mixtures, the following classical quadratic mixing rules are used to calculate a_m (to replace a in Equation 6-105), b_m (to replace b in Equation 6-106), and c_m (to replace c in Equation 6-107).

$$a_m = \sum_i \sum_j [x_i x_j (a_i a_j)^{0.5}(1 - k_{ij})] \tag{6-108}$$

$$b_m = \sum_i [x_i \, b_i] \tag{6-109}$$

$$c_m = \sum_i [x_i \, c_i] \tag{6-110}$$

The expression for the fugacity coefficient of component i in a hydrocarbon phase is given by

$$\ln(\Phi_i) = -\ln(Z - B) + \left[\frac{b_i}{Z - B} - \left[\frac{\psi_i}{RTd}\right]\right] \ln\left|\frac{Q + d}{Q - d}\right|$$

$$+ 0.5 \, A \left[\frac{b_i + c_i}{Q^2 - d^2}\right] + 0.125 \, A \, [c_i(3B + C) + b_i(3C + B)]$$

$$\left\{ \ln\left|\frac{Q + D}{Q - d}\right| - \frac{2Qd}{Q^2 - d^2}\right\} \left[\frac{p}{RT}\right] \tag{6-111}$$

where $\psi_i = \Sigma_j [x_j(a_i a_j)^{0.5}(1 - k_{ij})]$
$Q = Z + 0.5 \, (B + C)$
$d = [BC + 0.25 \, (B + C)^2]^{0.5}$
T = temperature, °R
p = pressure, psia
R = universal gas constant, 10.73 psia ft^3/lb-mole °R

Kubic Equation of State

Kubic (1982) presented a modification of the Martin (1979) equation of state for calculating vapor-liquid equilibria. The modified equation predicts a variable critical compressibility factor for accurate description of real fluids. Kubic proposed the following three-parameter cubic equation:

$$p = \frac{RT}{v - b} - \frac{a}{(v + c)^2} \qquad (6\text{-}112)$$

with

$$a = \Omega_a \frac{R^2 T_c^2}{p_c} (\alpha^0 + \omega'\alpha') \qquad (6\text{-}113)$$

$$b = \Omega_b \frac{R T_c}{p_c} \qquad (6\text{-}114)$$

$$c = \Omega_c \frac{R T_c}{p_c} \qquad (6\text{-}115)$$

where $\Omega_a = 0.421875$

$\alpha^0 = 0.7895 - 0.1514\, T_r + 0.3314/T_r + 0.029/T_r^2 + 0.0015/T_r^7$

$\alpha' = -0.237\, T_r - 0.7846/T_r + 1.0026/T_r^2 + 0.019/T_r^7$

$\omega' = 0.000756 + 0.90984\, \omega + 0.16226\, \omega^2 + 0.14549\, \omega^3$

$\Omega_b = 0.082 - 0.0713\, \omega'$

$\Omega_c = 0.043\, \alpha^0 + 0.0713\, \omega'\, \gamma'$

$\gamma^0 = 4.275051 - 8.878889/T_r + 8.508932/T_r^2$
$\qquad - 3.481408/T_r^3 + 0.576312/T_r^4$

$\gamma' = 12.856404 - 34.744125/T_r + 37.433095/T_r^2$
$\qquad - 18.059421/T_r^3 + 3.514050/T_r^4$

In terms of the compressibility factor, Equation 6-112 can be written as follows:

$$\begin{aligned} Z^3 + (2C - B - 1)Z^2 + (C^2 - 2CB - 2C + A)Z \\ - (BC^2 + C^2 + AB) = 0 \end{aligned} \qquad (6\text{-}116)$$

with

$$A = \frac{ap}{R^2 T^2}$$

$$B = \frac{bp}{RT}$$

$$C = \frac{cP}{RT}$$

For mixtures, the classical mixing rules as given by Equations 6-108 through 6-110 are suggested for determining a_m, b_m, and c_m. The fugacity coefficient is calculated from the following expression:

$$\ln(\Phi_i) = -\ln (Z - B) + \left[\frac{p}{RT}\right]\left[\frac{b_i}{Z - B}\right] - \left[\frac{p}{RT}\right]^2\left[\frac{2\psi_i}{(Z + C)}\right]$$

$$+ \left[\frac{p}{RT}\right]\left[\frac{A\ c_i}{(Z + C)^2}\right] \tag{6-117}$$

where $\psi_i = \sum_j [x_j(a_ia_j)^{0.5}(1 - k_{ij})]$

Patel and Teja Equation of State and Its Modifications

Patel and Teja (1982) proposed the following three-parameter cubic equation:

$$p = \frac{RT}{v - b} - \frac{a}{V^2 + (b + c)V - bc} \tag{6-118}$$

The above equation has the same form as that proposed by Heyen, i.e., Equation 6-93, where a is a function of the temperature, and b and c are constants characteristic of each component. Equation 6-118 was constrained to satisfy the following conditions

$$\frac{\partial p}{\partial V}_{T_c} = 0$$

$$\frac{\partial^2 p}{\partial V^2}_{T_c} = 0$$

$$\frac{p_cV_c}{R\ T_c} = \xi_c \tag{6-119}$$

Patel and Teja pointed out that the third parameter c in the equation allows the empirical parameter ξ_c to be chosen freely. Acceptable prediction of both low and high pressure densities requires that, in general, ξ_c be greater than the experimental critical compressibility. Application of Equation 6-119 to Equation 6-118 yields:

$$a = \Omega_a \frac{R^2\ T_c^2}{p_c} [1 + m(1 - T_r^{0.5})]^2 \tag{6-120}$$

$$b = \Omega_b \frac{R\,T_c}{p_c} \tag{6-121}$$

$$c = \Omega_c \frac{R\,T_c}{p_c} \tag{6-122}$$

where $\Omega_c = 1 - 3\,\xi_c$ (6-123)

$$\Omega_a = 3\,\xi_c^2 + 3(1 - 2\,\xi_c)\,\Omega_b + \Omega_b^2 + (1 - 3\,\xi_c) \tag{6-124}$$

and Ω_b is the smallest positive root of the following equation:

$$\Omega_b^3 + (2 - 3\,\xi_c)\,\Omega_b^2 + 3\,\xi_c^2\,\Omega_b - \xi_c^3 = 0 \tag{6-125}$$

Equation 6-125 can be solved for Ω_b by using the Newton-Raphson iterative method with an initial value for Ω_b as given by

$$\Omega_b = 0.32429\;Z_c - 0.002005$$

For nonpolar fluids, the parameters m and ξ_c are related to the acentric factor by the following relationships:

$$m = 0.452413 + 1.30982\;\omega - 0.295937\;\omega^2 \tag{6-126}$$

$$\xi_c = 0.329032 - 0.0767992\;\omega + 0.0211947\;\omega^2 \tag{6-127}$$

In terms of Z, Equation 6-118 can be rearranged to produce

$$Z^3 + (C - 1)Z^2 + (A - 2BC - B - C - B^2)Z$$

$$+ (BC + B^2C - AB) = 0 \tag{6-128}$$

where, for mixtures

$$A = \frac{a_m p}{(RT)^2} \tag{6-129}$$

$$B = \frac{b_m p}{RT} \tag{6-130}$$

$$C = \frac{c_m p}{RT} \tag{6-131}$$

with

$$a_m = \sum_i \sum_j \left[x_i x_j (a_i a_j)^{0.5} (1 - k_{ij}) \right] \tag{6-132}$$

$$b_m = \sum_i [x_i \, b_i] \tag{6-133}$$

$$c_m = \sum_i [x_i \, c_i] \tag{6-134}$$

For the Patel and Teja EOS, the fugacity coefficient is given by Equation 6-111.

Willman and Teja (1986) proposed that for undefined components such as C_{7+}, the parameters m and ξ_c can be correlated in terms of the boiling point T_b and specific gravity γ. They proposed the following relationships:

$$\xi_c = a_0 + a_1/T_b + a_2/\gamma + a_3/(T_b\gamma) \tag{6-135}$$

$$m = b_0 + b_1 \, T_b + b_2 \, \gamma + b_3 \, T_b\gamma \tag{6-136}$$

where T_b = boiling point of the plus-fraction, °R
γ = specific gravity of the plus-fraction

The constants a_0–a_3 and b_0–b_3 are given in Table 6-4.

Table 6-4
Constants of Equations 6-135 and 6-136

Coefficient	"PNA" Liquids	Aromatic Oil
a_0	0.2315	0.3119
a_1	83.86668	-9.23886
a_2	5.391×10^{-5}	-0.0707
a_3	-26.61606	69.05934
b_0	-1.0334	-1.9885
b_1	$3.32705556 \times 10^{-3}$	0.002972222
b_2	1.2123	0.7613
b_3	-0.00261111	$-1.8077778 \times 10^{-3}$

Valderrama and Cisternas (1986) used the experimental critical compressibility factor as a third input parameter instead of the parameter ξ_c in the PT EOS. The terms Ω_a, Ω_b, Ω_c, and m are correlated with Z_c by the following polynomials:

$$\Omega_a = 0.69368018 - 1.0634424\ Z_c + 0.68289995\ Z_c^2$$
$$- 0.21044403\ Z_c^3 + 0.003752658\ Z_c^4 \tag{6-137}$$

$$\Omega_b = 0.025987178 + 0.180754784\ Z_c + 0.061258949\ Z_c^2 \tag{6-138}$$

$$\Omega_c = 0.577500514 - 1.898414283\ Z_c \tag{6-139}$$

$$m = -6.608 + 70.43\ Z_c - 159.0\ Z_c^2 \tag{6-140}$$

Schmidt and Wenzel Equation of State

Schmidt and Wenzel (1980) proposed an attractive pressure term that introduces the acentric factor ω as a third parameter. The SW EOS has the following form:

$$p = \frac{RT}{V - b} - \frac{a}{V^3 + (1 + 3\omega)bV - 3\ \omega b^2} \tag{6-141}$$

with

$$a = \Omega_a \left| \frac{R^2\ T_c^2}{p_c} \right| \alpha \tag{6-142}$$

$$b = \Omega_b \left| \frac{R\ T_c}{p_c} \right| \tag{6-143}$$

where
$$\Omega_a = (1 - \xi_c\ [1 - \beta_c])^3 \tag{6-144}$$
$$\Omega_b = \beta_c\ \xi_c \tag{6-145}$$

The β_c parameter is given by the smallest positive root of the following equation:

$$(6\omega + 1)\beta_c^3 + 3\ \beta_c^2 + 3\beta_c - 1 = 0 \tag{6-146}$$

and

$$\xi_c = \frac{1}{3(1 + \beta_c\omega)} \tag{6-147}$$

Equation 6-146 can be solved for β_c by employing the Newton-Raphson iterative method with a starting value of

$$\beta_c = 0.25989 - 0.0217\ \omega + 0.00375\ \omega^2 \tag{6-148}$$

The dimensionless factor α in Equation 6-142 is related to the reduced temperature and the acentric factor by the following expressions:

- For $T_r > 1.0$, then

$$\alpha = 1 - (0.4774 + 1.328\ \omega)\ \ln(T_r) \qquad (6\text{-}149)$$

- For $T \le 1.0$, then

$$\alpha = [1 + m\ (1 - T_r^{0.5})]^2 \qquad (6\text{-}150)$$

The parameter m in the above expression is correlated with the acentric factor and reduced temperature as follows:

let

$$m_0 = 0.465 + 1.347\ \omega - 0.528\ \omega^2 \qquad \text{For } \omega \le 0.3671 \qquad (6\text{-}151)$$

$$m_0 = 0.5361 + 0.9593\ \omega \qquad \text{For } \omega > 0.3671 \qquad (6\text{-}152)$$

$$m_1 = m_0 + (1/70)(5\ T_r - 3\ m_0 - 1)^2 \qquad (6\text{-}153)$$

$$m_2 = m_0 + 0.71\ (T_r - 0.779)^2 \qquad (6\text{-}154)$$

Then

- For $\omega \le 0.40$

$$m = m_1$$

- For $\omega \ge 0.55$

$$m = m_2$$

- For $0.4 < \omega < 0.55$

$$m = [(0.55 - \omega)\ m_1 + (\omega - 0.40)\ m_2]/0.15$$

Equation 6-141 can be rearranged in terms of Z to give

$$Z^3 + (UB - B - 1)\ Z^2 + (WB^2 - UB^2 - UB + A)\ Z - (WB^3 + WB^2 + AB) = 0 \qquad (6\text{-}155)$$

For hydrocarbon mixtures, the coefficients of Equation 6-155 are defined by the following

$$A = \frac{a_m p}{R^2\ T^2}$$

$$B = \frac{b_m p}{R\,T}$$

$$U = 1 + 3\,\omega_m \tag{6-156}$$

$$W = -\,3\,\omega_m \tag{6-157}$$

Schmidt and Wenzel adopted the following mixing rules for calculating the mixture parameters a_m, b_m, and ω_m:

$$a_m = \sum_{i=1} \sum_{j=1} [x_i x_j (a_i a_j)^{0.5}(1 - k_{ij})]$$

$$b_m = \sum_i [x_i\, b_i]$$

$$\omega_m = \left[\sum_i (\omega_i\, x_i\, b_i^{0.7})\right] \Big/ \left[\sum_i (x_i\, b_i^{0.7})\right] \tag{6-158}$$

Finally, the authors developed the following expression for calculating the equilibrium ratios (K_i factors):

$$RT\,\ln(K_i) = F(y_i) - F(x_i) \tag{6-159}$$

where $F(x_i) = RT\,\ln\,[(Z - B)RT/p] - \dfrac{b_i p}{Z - B}$

$$+ \frac{1}{D}\left[2\psi_i - \frac{a_m \beta_i}{D}\right] \ln\left(\frac{G}{H}\right)$$

$$+ \left[\frac{E_i + \beta_i}{G} - \frac{E_i - \beta_i}{H}\right]\left(\frac{a_m}{2D}\right) \tag{6-160}$$

with

$$E_i = U\, b_i + \left[\frac{3R_i(\omega_i - \omega_m)b_m}{x_i}\right]$$

$$R_i = x_i\, a_i^{0.5} \Big/ \left[\sum_j (x_j\, a_j^{0.5})\right]$$

$$\beta_i = \bar{s}\, b_i + [9R_i\,(\omega_i - \omega_m)(1 + \omega_m)b_m]/(x_i \bar{s})$$

$$\bar{s} = (U^2 + 4\,U - 4)^{0.5}$$

$$D = \bar{s} \, b_m$$

$$G = \left[\frac{ZRT}{p}\right] + \left[\frac{(U + \bar{s})b_m}{2}\right]$$

$$H = \left[\frac{ZRT}{p}\right] + \left[\frac{(U - \bar{s})b_m}{2}\right]$$

Equation 6-160 is also used to determine $F(y_i)$ by replacing x_i with y_i and by calculating all the composition-dependent parameters using the gas composition.

Yu-Lu Equation of State

Yu and Lu (1987) proposed the following three-parameter cubic EOS:

$$p = \frac{RT}{v - b} - \frac{a\alpha}{v(v + c) + b(3v + c)} \tag{6-161}$$

where the parameter α is the only temperature-dependent coefficient in the above expression.

Yu and Lu expressed the parameters a, b, and c in terms of the acentric factor and critical properties as follows:

$$a = \Omega_a \, \frac{R^2 \, T_c^2}{p_c} \tag{6-162}$$

$$b = \Omega_b \, \frac{R \, T_c}{p_c} \tag{6-163}$$

$$c = \Omega_c \, \frac{R \, T_c}{p_c} \tag{6-164}$$

with $\Omega_c = (3 + u) \, \Omega_b$ ⠀⠀⠀⠀⠀⠀⠀⠀⠀⠀⠀⠀⠀⠀⠀⠀⠀⠀(6-165)

where $\Omega_a = 0.46863 - 0.0378304 \, \omega + 0.00751969 \, \omega^2$ ⠀⠀(6-166)
⠀⠀⠀⠀$\Omega_b = 0.0892828 - 0.0340903 \, \omega - 0.00518289 \, \omega^2$ ⠀(6-167)
⠀⠀⠀⠀$u = 1.70083 + 0.648463 \, \omega + 0.895926 \, \omega^2$ ⠀⠀⠀⠀(6-168)

The coefficient α is expressed in terms of the reduced temperature T_r and acentric factor as follows:

$$\text{Log}(\alpha) = M \, [A_0 + A_1 \, T_r + A_2 \, T_r^2] \, (1 - T_r) \tag{6-169}$$

Two acentric factor ranges are used in the application of Equation 6-169.

- For $\omega \leq 0.49$

$$M = 0.406846 + 1.87907\ \omega - 0.792636\ \omega^2 + 0.737519\ \omega^3 \quad (6\text{-}170)$$

with

$$A_0 = 0.536843 \qquad A_1 = -0.39244 \qquad A_2 = 0.26507$$

- For $0.49 < \omega \leq 1.0$

$$M = 0.581981 - 0.171416\ \omega + 1.84441\ \omega^2 - 1.19047\ \omega^3 \quad (6\text{-}171)$$

with

$$A_0 = 0.79355 \qquad A_1 = -0.53409 \qquad A_2 = 0.37273$$

In the application of Equation 6-169 to the $T_r > 1$ conditions, the term $(A_0 + A_1\ T_r + A_2\ T_r^2)$ reduces to $(A_0 + A_1 + A_2)$ with the term M and the coefficients A_0–A_2 defined by Equation 6-171.

In terms of the compressibility factor, Equation 6-161 is written as

$$Z^3 + [C + 2B - 1]Z^2 + [A - 3B^2 - C - 3B]Z$$
$$+ [AB - AC + B^2C] = 0 \quad (6\text{-}172)$$

For mixtures, the coefficients A, B, and C are defined as

$$A = \frac{(a\alpha)_m p}{R^2\ T^2} \quad (6\text{-}173)$$

$$B = \frac{b_m p}{RT} \quad (6\text{-}174)$$

$$C = \frac{c_m p}{RT} \quad (6\text{-}175)$$

with

$$(a\alpha)_m = \sum_i \sum_j x_i x_j (a_i a_j \alpha_i \alpha_j)^{0.5}(1 - k_{ij}) \quad (6\text{-}176)$$

$$b_m = \sum_i x_i\ b_i \quad (6\text{-}177)$$

$$c_m = \sum_i x_i\ c_i \quad (6\text{-}178)$$

In these equations, x_i is replaced by y_i for vapor mixtures parameters.

The expressions for calculating the fugacity coefficient of component i in the vapor phase is given by the following equation:

$$Ln(\Phi_i^y) = \frac{RT}{p}\left[\frac{b_i}{Z - B}\right] - Ln(Z - B)$$

$$+ \left[\frac{pA}{RTE^2}\right]\left[\frac{(b_iC + Bc_i)(2Z + 3B + C) - (3b_i + c_i)(3BZ + CZ + 2BC)}{Z^2 + (3B + C)Z + BC}\right]$$

$$+ \left[\frac{p}{RT}\right]\left[\frac{2\psi_i + [2b_ic_m + 2b_mc_i - (3b_m + c_m)(3b_i + c_i)]Ap/E^2}{RTE}\right]$$

$$Ln\left[\frac{2Z + (3B + C) - E}{2Z + (3B + C) + E}\right] \qquad (6\text{-}179)$$

where $E = [8B^2 + (B + C)^2]^{0.5}$

THE GENERALIZED FORM OF EQUATIONS OF STATE

Schmidt and Wenzel (1980) have shown that almost all cubic equations of state can be expressed in a generalized form by the following four-constant equation of state:

$$p = \frac{RT}{V - b} - \frac{a}{V^2 + ubV + wb^2} \qquad (6\text{-}180)$$

When the parameters u and w are assigned certain values, Equation 6-180 is reduced to a specific equation of state. The relationship between u and w for a number of cubic equations of state, as presented by Yu et al. (1986), is given below:

Type of EOS	u	w
Van der Waals	0	0
Redlich-Kwong	1	0
Soave-Redlich-Kwong	1	0
Peng-Robinson	2	-1
Heyen	$1 - w$	$f(\omega, b)$
Kubic	$f(\omega)$	$u^2/4$
Patel-Teja	$1 - w$	$f(\omega)$
Schmidt-Wenzel	$1 - w$	$f(\omega)$
Yu-Lu	$f(\omega)$	$u - 3$

Equation 6-180 can be expressed in terms of the compressibility coefficient as

$$Z^3 + (UB - B - 1)Z^2 + (WB^2 - UB^2 - UB + A)Z$$
$$- (WB^3 + WB^2 + AB) = 0 \qquad (6\text{-}181)$$

In terms of the fugacity coefficient, Equation 6-180 can be expressed as

$$Ln(\Phi_i) = \frac{b_i}{b_m} (Z - 1) - Ln(Z - B) + \left[\frac{2\psi_i}{\psi} - \frac{b_i}{b_m}\right]$$

$$\left[\frac{A}{(U^2 + 4W)^{.5}B}\right] Ln \left[\frac{2Z + B[U - (U^2 - 4W)^{0.5}]}{2Z + B[U + (U^2 - 4W)^{0.5}]}\right] \qquad (6\text{-}182)$$

APPLICATIONS OF THE EQUATION OF STATE IN PETROLEUM ENGINEERING

The Determination of the Equilibrium Ratios

A flow diagram is presented in Figure 6-3 to illustrate the procedure of determining equilibrium ratios of a hydrocarbon mixture. For this type of calculation, the system temperature T, the system pressure p, and the overall composition of the mixture z_i must be known. The procedure is summarized in the following steps in conjunction with Figure 6-3.

Step 1. Assume an initial value of the equilibrium ratio for each component in the mixture at the specified system pressure and temperature. Wilson's equation (see Chapter 5, Equation 5-17) can provide the starting values for K_i's.

$$K_i^A = \frac{P_{ci}}{p} \text{EXP} \left[5.37 \ (1 + \omega_i)(1 - T_{ci}/T)\right]$$

where K_i^A = assumed equilibrium ratio of component i

Step 2. Using the overall composition and the assumed K-values, perform flash calculations, as outlined in Chapter 5.

Step 3. Using the calculated composition of the liquid phase x_i, determine the fugacity coefficient Φ_i^L for each component in the liquid phase.

Step 4. Repeat Step 3 by using the calculated composition of the gas phase y_i to determine Φ_i^y.

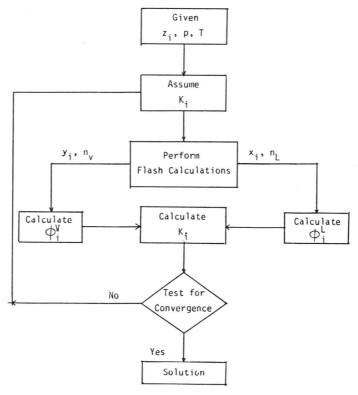

Figure 6-3. Flow diagram of the equilibrium ratio determination by an equation of state.

Step 5. Calculate the new set of equilibrium ratios by applying Equation 6-49, or

$$K_i = \frac{\Phi_i^L}{\Phi_i^y}$$

Step 6. Check for the solution by applying the following constraint:

$$\sum_{i=1}^{n} [K_i/K_i^A - 1]^2 \le \epsilon$$

where ϵ = preset error tolerance, e.g., 0.0001
           n = number of components in the system

If the above conditions are satisfied, then the solution has been reached. If not, Steps 1 through 6 are repeated by using the calculated equilibrium ratios as initial values.

Determination of the Dew-Point Pressure

A saturated vapor exists for a given temperature at the pressure at which an infinitesimal amount of liquid first appears. This pressure is referred to as the dew-point pressure p_d. The dew-point pressure of a mixture is described mathematically by the following conditions (see Chapter 5).

$$y_i = z_i \quad 1 \le i \le n \tag{6-183}$$
$$n_v = 1$$

$$\sum_{i=1}^{n} \left[\frac{z_i}{K_i} \right] = 1 \tag{6-184}$$

Applying the definition of K_i in terms of the fugacity coefficient to Equation 6-184, gives

$$\sum_{i=1}^{n} \left[\frac{z_i \, \Phi_i^y}{\Phi_i^L} \right] = \sum \left[\frac{z_i \, f_i^y}{z_i \, p_d \, \Phi_i^L} \right] = 1$$

or

$$p_d = \sum_{i=1}^{n} \left[\frac{f_i^y}{\Phi_i^L} \right]$$

The above equation is arranged to give

$$f(p_d) = \sum_{i=1}^{n} \left[\frac{f_i^y}{\Phi_i^L} \right] - p_d = 0 \tag{6-185}$$

where p_d = dew-point pressure, psia
   f_i^y = fugacity of component i in the vapor phase, psia
   Φ_i^L = fugacity coefficient of component i in the liquid phase

Equation 6-185 can be solved for the dew-point pressure by using the Newton-Raphson iterative method. To use the iterative method, the derivative of Equation 6-185 with respect to the pressure is required. This derivative is given by the following expression:

$$\frac{\partial f}{\partial p_d} = \sum_{i=1}^{n} \left[\frac{\Phi_i^L \, (\partial f_i^y / \partial p_d) - f_i^y \, (\partial \Phi_i^L / \partial p_d)}{(\Phi_i^L)^2} \right] - 1 \tag{6-186}$$

These derivatives may be determined numerically as follows:

$$\frac{\partial f_i^v}{\partial p_d} = \left[\frac{f_i^v(p_d + \Delta p_d) - f_i^v(p_d - \Delta p_d)}{2\Delta p_d}\right] \qquad (6\text{-}187)$$

and

$$\frac{\partial \Phi_i^L}{\partial p_d} = \left[\frac{\Phi_i^L(p_d + \Delta p_d) - \Phi_i^L(p_d - \Delta p_d)}{2\Delta p_d}\right] \qquad (6\text{-}188)$$

where
Δp_d = pressure increment, 5 psia, for example
$f_i^v(p_d + \Delta p_d)$ = fugacity of component i at $(p_d + \Delta p_d)$
$f_i^v(p_d - \Delta p_d)$ = fugacity of component i at $(p_d - \Delta p_d)$
$\Phi_i^L(p_d + \Delta p_d)$ = fugacity coefficient of component i at $(p_d + \Delta p_d)$
$\Phi_i^L(p_d - \Delta p_d)$ = fugacity coefficient of component i at $(p_d - \Delta p_d)$
Φ_i^L = fugacity coefficient of component i at p_d

The computational procedure of determining p_d is summarized in the following steps.

Step 1. Assume an initial value for the dew-point pressure p_d^A.

Step 2. Using the assumed value of p_d, calculate a set of equilibrium ratios for the mixture by using Wilson's correlation.

Step 3. Calculate the composition of the liquid phase, i.e., composition of the droplets of liquid, by applying the mathematical definition of K_i, to give

$$x_i = \frac{z_i}{K_i}$$

It should be noted that $y_i = z_i$.

Step 4. Using the composition of the gas phase z_i and the liquid phase x_i, calculate f_i^v and Φ_i^L at p_d, $p_d + \Delta p_d$, and $p_d - \Delta p_d$.

Step 5. Evaluate Equations 6-185 and 6-186.

Step 6. Using the values of the function $f(p_d)$ and the derivative $\partial f/\partial p_d$ as determined in Step 5, calculate a new dew-point pressure by applying the Newton-Raphson formula:

$$p_d = p_d^A - f(p_d)/[\partial f/\partial p_d] \qquad (6\text{-}189)$$

Step 7. The calculated value of p_d is checked numerically against the assumed value by applying the following condition:

$$|p_d - p_d^A| \leq 5$$

If the above condition is met, then the correct dew-point pressure p_d has been found. Otherwise, Steps 2 through 6 are repeated by using the calculated p_d as the new value for the next iteration.

Determination of the Bubble-Point Pressure

The bubble-point pressure p_b is defined as the pressure at which the first bubble of gas is formed. Accordingly, the bubble-point pressure is defined mathematically, as shown in Chapter 5, by the following equations:

$$x_i = z_i; \quad 1 \le i \le n \tag{6-189}$$

$$n_L = 1.0$$

$$\sum_{i=1}^{n} [z_i \, K_i] = 1 \tag{6-190}$$

Introducing the concept of the fugacity coefficient into Equation 6-190, gives

$$p_b = \sum_{i=1}^{n} \left[\frac{f_i^L}{\Phi_i^v} \right]$$

or

$$f(p_b) = \sum_{i=1}^{n} \left[\frac{f_i^L}{\Phi_i^v} \right] - p_b = 0 \tag{6-191}$$

Differentiating the above function with respect to the bubble-point pressure, yields

$$\frac{\partial f}{\partial p_b} = \sum_{i=1}^{n} \left[\frac{\Phi_i^v \, \partial f_i^L / \partial p_b - f_i^L \, \partial \Phi_i^v / \partial p_b}{(\Phi_i^v)^2} \right] - 1 \tag{6-192}$$

The iteration sequence for calculation of p_b is similar to that of the dew-point pressure.

Determination of the Mixture Critical Properties

Chueh and Prausnitz (1967) correlated the true critical temperature of the hydrocarbon mixture with the surface fraction Θ, composition of the mixture (x_i or y_i), and the critical temperature of the individual components present in the mixture. Their correlation has the following form:

$$T_{cm} = \sum_j (\Theta_j \, T_{cj}) + \sum_i \sum_j (\Theta_i \, \Theta_j \, \tau_{ij}) \qquad (6\text{-}193)$$

with the surface fraction Θ defined by the following expression:

$$\Theta_j = \frac{x_j \, (V_{cj})^{2/3}}{\sum_i [x_i \, (V_{ci})^{2/3}]} \qquad (6\text{-}194)$$

where T_{cm} = mixture critical temperature, °R
$\quad\quad\quad V_{ci}$ = critical volume of component i, ft^3/mole
$\quad\quad\quad \tau_{ij}$ = binary interaction coefficient

Using the extensive binary system data studied by Spencer et al. (1973), Reid et al. (1987) proposed the following 4th degree polynomial for determining τ_{ij}:

$$\psi_T = A + B \, \delta_T + C \, \delta_T^2 + D \, \delta_T^3 + E \, \delta_T^4 \qquad (6\text{-}195)$$

where $\psi_T = \dfrac{2 \, \tau_{ij}}{T_{ci} + T_{cj}} \qquad (6\text{-}196)$

$$\delta_T = \left| \frac{T_{ci} - T_{cj}}{T_{ci} + T_{cj}} \right| \qquad (6\text{-}197)$$

with $\tau_{ij} = 0$

The coefficients A–E for different binary systems are given below:

Binary	A	B	C	D	E
Containing H_2S	− 0.0479	− 5.725	70.974	− 161.319	0
Containing CO_2	− 0.0953	2.185	− 33.985	179.068	− 264.522
Containing aromatics	− 0.0219	1.227	− 24.277	147.673	− 259.433
All other systems	− 0.0076	0.287	− 1.343	5.443	− 3.038

Equation 6-195 is valid in the range of $0 \le \delta_T \le 0.5$.

Chueh and Prausnitz proposed also a similar correlation for determining the critical volume of a mixture V_{cm}. The correlation has the following form:

$$V_{cm} = \sum_j (\Theta_j V_{cj}) + \sum_i \sum_j (\Theta_i \Theta_j \nu_{ij}) \tag{6-198}$$

with the surface fraction Θ_j as defined by Equation 6-194. The interaction parameter ν_{ij} can be estimated by the following relationship:

$$\psi = A + B \delta_v + C \delta_v^2 + D \delta_v^3 + E \delta_v^4 \tag{6-199}$$

where
$$\psi_\nu = \frac{2 \nu_{ij}}{V_{ci} + V_{cj}} \tag{6-200}$$

$$\delta_v = \left| \frac{V_{ci}^{2/3} - V_{cj}^{2/3}}{V_{ci}^{2/3} + V_{cj}^{2/3}} \right| \tag{6-201}$$

with $\nu_{ij} = 0$

The coefficients of Equation 6-199, as given below, were determined from the experimental binary systems provided by Spencer and co-workers.

Binary	A	B	C	D	E
Aromatic-Aromatic	0	0	0	0	0
Paraffin-Aromatic	0.0753	− 3.332	2.220	0	0
Systems with CO_2					
or H_2S	− 0.4957	17.1185	− 168.56	587.05	− 698.89
All other systems	0.1397	− 2.9672	1.8337	− 1.536	0

Chueh and Prausnitz correlated the critical pressure of the mixture p_{cm} with the mixture critical temperature T_{cm} and critical volume V_{cm} by modifying the Redlich-Kwong equation of state. The modified R-K EOS is described by the following expressions:

$$p_{cm} = \frac{R \, T_{cm}}{V_{cm} - b_m} - \frac{a_m}{T_{cm}^{0.5} V_{cm}(V_{cm} + b_m)} \tag{6-203}$$

with

$$b_m = \sum_j \left[\frac{x_j \, \Omega_{bj}^* \, R \, T_{cj}}{p_{cj}} \right]$$

$$a_m = \sum_i \sum_j [x_i x_j a_{ij}]$$

$$\Omega_{bj}^* = 0.0867 - 0.0125 \, \omega_j + 0.011 \, \omega_j^2$$

$$a_{ii} = \frac{\Omega_{ai}^* \; R^2 \; T_{ci}^{2.5}}{p_{ci}}$$

$$a_{ij} = \frac{(\Omega_{ai}^* + \Omega_{aj}^*) \; R \; T_{cij}^{1.5} \; (V_{ci} + V_{cj})}{4 \, [0.291 - 0.04 \, (\omega_i + \omega_j)]}$$

$$T_{cij} = (1 - k_{ij})(T_{ci} \; T_{cj})^{0.5}$$

$$\Omega_{aj}^* = \left[\frac{R \; T_{cj}}{V_{cj} - b_j} - p_{cj}\right]\left[\frac{p_{cj} \; V_{cj} \; (V_{cj} + b_j)}{(R \; T_{cj})^2}\right]$$

where k_{ij} is the binary interaction coefficient. The co-volume b_j is defined by Equation 6-20.

Spencer et al. (1975) reported a large deviation between the mixture critical pressure as calculated by Equation 6-202 and the experimental value for systems containing methane.

Example 6-9. Calculate the critical temperature and volume of the following hydrocarbon mixture:

Component	x_i
C_3	0.40
C_4	0.30
C_5	0.30

Solution.

Calculation of the critical temperature

Step 1. From the individual critical properties of the individual components, calculate the surface fraction:

Component	x_i	T_c °R	V_c, ft³/mole	Θ_i	$\Theta_i \, T_{ci}$	$\Theta_i \, V_{ci}$
C_3	0.40	666	3.25	0.3503	233.3	1.1385
C_4	0.30	765.6	4.08	0.3057	234.0	1.2473
C_5	0.30	845.7	4.87	0.3440	290.9	1.6753
					758.2	4.0611

Step 2. Solve for τ_{ij} by applying Equations 6-195 through 6-197 and by using $A = -0.0076$, $B = 0.287$, $C = -1.343$, $D = 5.443$, and $E = -3.038$.

i	j	δ_T	ψ_T	τ_{ij}
C_3	C_4	0.0696	0.0076	5.44
C_3	C_5	0.1189	0.0161	12.17
C_4	C_5	0.0497	0.0040	3.22

Step 3. Calculate the critical temperature of the mixture from Equation 6-193.

$$T_{cm} = 758.2 + (2)(0.3503)(0.3057)(5.44)$$
$$+ (2)(0.3503)(0.344)(12.17)$$
$$+ (2)(0.3057)(0.344)(3.22) = 763°R$$

Calculation of the critical volume

Step 1. Solve for ν_{ij} by applying Equations 6-199 through 6-201 and by using A = 0.1397, B = -2.9672, C = 1.8337, D = -1.536, and E = 0.

i	j	δ_v	ψ_v	ν_{ij}
C_3	C_4	0.0757	-0.0751	-0.2752
C_3	C_5	0.1340	-0.2287	-0.9285
C_4	C_5	0.0589	-0.0290	-0.1298

Step 2. Calculate the critical volume of the mixture from Equation 6-198.

$$V_{cm} = 4.0611 + (2)(0.3503)(0.3057)(-0.2752)$$
$$+ (2)(0.3503)(0.344)(-0.9285)$$
$$+ (2)(0.3057)(0.344)(-0.1298) = 3.751 \text{ ft}^3/\text{mole}$$

THREE PHASE EQUILIBRIUM CALCULATIONS

Two and three phase equilibria occur frequently during the processing of hydrocarbon and related systems. Peng and Robinson (1976) proposed a three phase equilibrium calculation scheme of systems which exhibit a water-rich liquid phase, a hydrocarbon-rich liquid phase, and a vapor phase.

In applying the principle of mass-conservation to one mole of a water-hydrocarbon in a three-phase state of thermodynamic equilibrium at a fixed temperature T and pressure p, gives

$$n_L + n_w + n_v = 1 \tag{6-203}$$

$$n_L x_i + n_w x_{wi} + n_v y_i = z_i \tag{6-204}$$

$$\sum_i^n x_i = \sum_{i=1}^n x_{wi} = \sum_{i=1}^n y_i = \sum_{i=1}^n z_i = 1 \tag{6-205}$$

where n_L, n_w, n_v = number of moles of the hydrocarbon-rich liquid, the water-rich liquid, and the vapor, respectively.

x_i, x_{wi}, y_i = mole fraction of component i in the hydrocarbon-rich liquid, the water rich liquid, and the vapor, respectively.

The equilibrium relations between the compositions of each phase are defined by the following expressions:

$$K_i = \frac{y_i}{x_i} = \frac{\Phi_i^L}{\Phi_i^y} \qquad (6\text{-}206)$$

and

$$K_{wi} = \frac{y_i}{x_{wi}} = \frac{\Phi_i^w}{\Phi_i^y} \qquad (6\text{-}207)$$

where K_i = equilibrium ratio of component i between the vapor and hydrocarbon-rich liquid

K_{wi} = equilibrium ratio of component i between the vapor and water-rich liquid

Φ_i^L = fugacity coefficient of component i in the hydrocarbon-rich liquid

Φ_i^y = fugacity coefficient of component i in the vapor phase

Φ_i^w = fugacity coefficient of component i in the water-rich liquid

Combining Equations 6-203 through 6-207 gives the following conventional non-linear equations:

$$\sum_{i=1} x_i = \sum_{i=1} \left[\frac{z_i}{n_L(1 - K_i) + n_w\,(K_i/K_{wi} - K_i) + K_i} \right] = 1 \qquad (6\text{-}208)$$

$$\sum_{i=1} x_{wi} = \sum_{i=1} \left[\frac{z_i\,K_i/K_{wi}}{n_L(1 - K_i) + n_w\,(K_i/K_{wi} - K_i) + K_i} \right] = 1 \qquad (6\text{-}209)$$

$$\sum_{i=1} y_i = \sum_{i=1} \left[\frac{z_i\,K_i}{n_L(1 - K_i) + n_w\,(K_i/K_{wi} - K_i) + K_i} \right] = 1 \qquad (6\text{-}210)$$

Assuming that the equilibrium ratios between phases can be calculated, the above equations are combined to solve for the unknown n_L and n_v, and hence x_i, x_{wi}, and y_i. It is the nature of the specific equilibrium calculation that determines the appropriate combination of Equations 6-208 through 6-210. The combination of the above three expressions can then be used to determine the phase and volumetric properties of the three-phase system.

There are essentially three types of phase behavior calculations for the three-phase system:

- Bubble-point prediction
- Dew-point prediction
- Flash calculation

Peng and Robinson (1981) proposed the following combination schemes of Equations 6-208 through 6-210:

- For the bubble-point pressure determination

$$\sum_i x_i - \sum_i x_{wi} = 0 \qquad \left[\sum_i y_i\right] - 1 = 0$$

Substituting Equations 6-208 through 6-210 in the above relationships, gives

$$f(n_L, n_w) = \sum_i \left[\frac{z_i \,(1 - K_i/K_{wi})}{n_L(1 - K_i) + n_w \,(K_i/K_{wi} - K_i) + K_i}\right] = 0 \qquad (6\text{-}211)$$

and

$$g(n_L, n_w) = \sum_i \left[\frac{z_i \,K_i}{n_L(1 - K_i) + n_w \,(K_i/K_{wi} - K_i) + K_i}\right] - 1 = 0 \quad (6\text{-}212)$$

- For the dew-point pressure

$$\sum_i x_{wi} - \sum_i y_i = 0 \qquad \left[\sum_i x_i\right] - 1 = 0$$

Combining with Equations 6-208 through 6-210, yields

$$f(n_L, n_w) = \sum_i \left[\frac{z_i \,K_i \,(1/K_{wi} - 1)}{n_L(1 - K_i) + n_w \,(K_i/K_{wi} - K_i) + K_i}\right] = 0 \qquad (6\text{-}213)$$

and

$$g(n_L, n_w) = \sum_i \left[\frac{z_i}{n_L(1 - K_i) + n_w \,(K_i/K_{wi} - K_i) + K_i}\right] - 1 = 0 \quad (6\text{-}214)$$

- For flash calculations

$$\sum_i x_i - \sum_i y_i = 0 \qquad \left[\sum_i x_{wi}\right] - 1 = 0$$

or

$$f(n_L, n_w) = \sum_i \left[\frac{z_i (1 - K_i)}{n_L(1 - K_i) + n_w (K_i/K_{wi} - K_i) + K_i} \right] = 0 \qquad (6\text{-}215)$$

and

$$g(n_L, n_w) = \sum_i \left[\frac{z_i K_i/K_{wi}}{n_L(1 - K_i) + n_w (K_i/K_{wi} - K_i) + K_i} \right] - 1.0 = 0 \qquad (6\text{-}216)$$

It should be noted that in performing any of the above property predictions, we always have two unknown variables, i.e., n_L and n_w, and between them, two equations. Providing that the equilibrium ratios and the overall composition are known, the equations can be solved simultaneously by using the appropriate iterative technique, e.g., the Newton-Raphson method. The application of this iterative technique for solving Equations 6-215 and 6-216 is summarized in the following steps:

Step 1. Assume initial values for the unknown variables n_L and n_w.

Step 2. Calculate new values of n_L and n_v by solving the following two linear equations:

$$\begin{bmatrix} n_L \\ n_w \end{bmatrix}^{new} = \begin{bmatrix} n_L \\ n_w \end{bmatrix} - \begin{bmatrix} \partial f/\partial n_L & \partial f/\partial n_w \\ \partial g/\partial n_L & \partial g/\partial n_w \end{bmatrix}^{-1} \begin{bmatrix} f(n_L, n_w) \\ g(n_L, n_w) \end{bmatrix}$$

where $f(n_L, n_w)$ = value of the function $f(n_L, n_w)$ as expressed by Equation 6-215

$g(n_L, n_w)$ = value of the function $g(n_L, n_w)$ as expressed by Equation 6-216

The first derivative of the above functions with respect to n_L and n_w are given by the following expressions:

$$(\partial f/\partial n_L) = \sum_{i=1} \left[\frac{- z_i (1 - K_i)^2}{[n_L (1 - K_i) + n_w (K_i/K_{wi} - K_i) + K_i]^2} \right]$$

$$(\partial f/\partial n_w) = \sum_{i=1} \left[\frac{- z_i (1 - K_i)(K_i/K_{wi} - K_i)}{[n_L (1 - K_i) + n_w (K_i/K_{wi} - K_i) + K_i]^2} \right]$$

$$(\partial f/\partial n_L) = \sum_{i=1} \left[\frac{- z_i (K_i/K_{wi})(1 - K_i)}{[n_L (1 - K_i) + n_w (K_i/K_{wi} - K_i) + K_i]^2} \right]$$

$$(\partial f/\partial n_w) = \sum_{i=1} \left[\frac{- z_i (K_i/K_{wi})(K_i/K_{wi} - K_i)}{[n_L (1 - K_i) + n_w (K_i/K_{wi} - K_i) + K_i]^2} \right]$$

Step 3. The new calculated values of n_L and n_w are then compared with the initial values. If no changes in the values are observed, then the correct values of n_L and n_w have been obtained. Otherwise, the above steps are repeated with the new calculated values used as initial values.

Peng and Robinson (1980) proposed two modifications when using their equation of state for three-phase equilibrium calculations. The first modification concerns the use of the parameter α as expressed by Equation 6-70 for the water compound. Peng and Robinson suggested that when the reduced temperature of this compound is less than 0.85, the following equation is applied:

$$\alpha = [1.0085677 + 0.82154 \ (1 - T_r^{0.5})] \tag{6-217}$$

where T_r is the reduced temperature of the water component $T/(T_c)_{H_2O}$. At reduced temperature for the water compound greater than or equal to 0.85, Equation 6-70 still applies.

The second modification concerns the application of Equation 6-40 for the water-rich liquid phase. A temperature-dependent binary interaction coefficient was introduced into the equation, to give:

$$(a\alpha)_m = \sum_i \sum_j [x_{wi} \ x_{wj} \ (a_i a_j \alpha_i \alpha_j)^{0.5}(1 - \tau_{ij})] \tag{6-218}$$

where τ_{ij} is a temperature-dependent binary interaction coefficient. Peng and Robinson proposed graphical correlations for determining this parameter for each aqueous binary pair. Lim et al. (1984) expressed these graphical correlations mathematically by the following generalized equation:

$$\tau_{ij} = a_1 \left[\frac{T}{T_{ci}}\right]^2 \left[\frac{p_{ci}}{p_{cj}}\right]^2 + a_2 \left[\frac{T}{T_{ci}}\right]\left[\frac{p_{ci}}{p_{cj}}\right] + a_3 \tag{6-219}$$

where T = system temperature, °R
  T_{ci} = critical temperature of the component of interest, °R
  p_{ci} = critical pressure of the component of interest, psia
  p_{cj} = critical pressure of the water compound, psia

The coefficients a_1, a_2, and a_3 of the above polynomial are given below for selected binaries.

Component i	a_1	a_2	a_3
C_1	0	1.659	-0.761
C_2	0	2.109	-0.607
C_3	-18.032	9.441	-1.208
n-C_4	0	2.800	-0.488
n-C_6	49.472	-5.783	-0.152

For selected non-hydrocarbon components, values of interaction parameters are given by the following expressions:

For N_2-H_2O binary

$$\tau_{ij} = 0.402 \, (T/T_{ci}) - 1.586 \tag{6-220}$$

where τ_{ij} = binary parameter between nitrogen and the water compound
        T_{ci} = critical temperature of nitrogen, °R
and

For CO_2-H_2O binary, the parameter is given by

$$\tau_{ij} = -0.074 \left[\frac{T}{T_{ci}}\right]^2 + 0.478 \left[\frac{T}{T_{ci}}\right] - 0.503 \tag{6-221}$$

where T_{ci} is the critical temperature of CO_2.

In the course of making phase equilibrium calculations, it is always desirable to provide initial values for the equilibrium ratios so the iterative procedure can proceed as reliably and rapidly as possible. Peng and Robinson adopted Wilson's equilibrium ratio correlation to provide initial K-values for the hydrocarbon-vapor phase

$$K_i = p_{ci}/p \; EXP \, [5.3727 \, (1 + \omega_i)(1 - T_{ci}/T)]$$

While for the water-vapor phase, Peng and Robinson proposed the following expression:

$$K_{wi} = 10^6 \, [p_{ci} \, T/(T_{ci} \, p)]$$

PROBLEMS

1. A pure n-butane exists in the two phase region at 120°F. Calculate the density of the coexisting phase by using the following equations of state:

a. Van der Waals
b. Redlich-Kwong
c. Soave-Redlich-Kwong
d. Peng-Robinson

2. A crude oil system with the following composition exists at its bubble-point pressure of 3,250 psia and 155°F.

Component	x_i
C_1	0.42
C_2	0.08
C_3	0.06
C_4	0.02
C_5	0.01
C_6	0.04
C_{7+}	0.37

If the molecular weight and specific gravity of the heptanes-plus fraction are 225 and 0.823, calculate the density of the crude by using

a. Standing-Katz density correlation
b. Alani-Kennedy density correlation
c. Peng-Robinson EOS
d. Schmidt-Wenzel EOS

3. Calculate the true critical temperature, volume, and pressure of the following hydrocarbon mixture.

Component	Mole Fraction
C_5	0.30
C_6	0.30
C_7	0.40

REFERENCES

1. Ahmed, T., "A Practical Modification of the Peng-Robinson Equation of State," Paper SPE 18532, Presented at the SPE Eastern Regional Meeting, Charleston, West Virginia, Nov. 1–4, 1988.
2. Chueh, P. and Prausnitz, J., "Vapor-Liquid Equilibria at High Pressures: Calculation of Critical Temperatures, Volumes, and Pressures of Nonpolar Mixtures," *AIChE Journal*, 1967, Vol. 13, No. 6, pp. 1107–1112.
3. Edmister, W. and Lee, B., *Applied Hydrocarbon Thermodynamics*, Volume 1, Second Edition, Gulf Publishing Company: Houston, 1986, p. 52.

4. Elliot, J. and Daubert, T., "Revised Procedure for Phase Equilibrium Calculations with Soave Equation of State," *Ind. Eng. Chem. Process Des. Dev.*, 1985, Vol. 23, pp. 743–748.

5. Gibbons, R. and Laughton, A., "An Equation of State for Polar and Non-Polar Substances and Mixtures," *J. Chem. Soc.*, 1984, Vol. 80, pp. 1019–1038.

6. Graboski, M. S. and Daubert, T. E., "A Modified Soave Equation of State for Phase Equilibrium Calculations 1. Hydrocarbon System," *Ind. Eng. Chem. Process Des. Dev.*, 1978, Vol. 17, pp. 443–448.

7. Heyen, G., "A Cubic Equation of State with Extended Range of Application," Proc., Second World Congress Chem. Eng., Montreal, October 4–9, 1983.

8. Jhaveri, B. S. and Youngren, G. K., "Three-Parameter Modification of the Peng-Robinson Equation of State to Improve Volumetric Predictions," Paper SPE 13118, presented at the 1984 SPE Annual Technical Conference, Houston, September 16–19.

9. Kubic, W. L. J., "A Modification of the Martin Equation of State for Calculating Vapor-Liquid Equilibria," *Fluid Phase Equilibria*, 1982, Vol. 9, pp. 79–97.

10. Lim, D., et al., "Calculation of Liquid Dropout for Systems Containing Water," Paper SPE 13094, presented at the 59th Annual Technical Conference of the SPE, held in Houston, TX, September 16–19, 1984.

11. Nikos, V., et al., "Phase Behavior of Systems Comprising North Sea Reservoir Fluids and Injection Gases," *JPT*, November 1986, pp. 1221–1233.

12. Patel, N. and Teja, A., "A New Equation of State for Fluids and Fluid Mixtures," *Chem. Eng. Sci.*, 1982, Vol. 37, No. 3, pp. 463–473.

13. Peneloux, A., Rauzy, E., and Freze, R., "A Consistent Correlation for Redlich-Kwong-Soave Volumes," *Fluid Phase Equilibria*, 1982, Vol. 8, pp 7–23.

14. Peng, D. and Robinson, D., "A New Two Constant Equation of State," *Ind. & Eng. Chem. Fund.*, 1976, Vol. 15, No. 1, pp. 59–64.

15. Peng, D. and Robinson, D., "Two and Three Phase Equilibrium Calculations for Systems Containing Water," *Canadian J. Chem. Eng.*, 1976, Vol. 54, pp. 595–598.

16. Peng, D. and Robinson, D., "Two and Three Phase Equilibrium Calculations for Coal Gasification and Related Processes," ACS Symposium Series, No. 133, Thermodynamics of Aqueous Systems with Industrial Applications, 1980.

17. Redlich, O. and Kwong, J., "On The Thermodynamics of Solutions. An Equation of State. Fugacities of Gaseous Solutions," Chemical Reviews, Vol. 44, 1949, pp. 233–247.

18. Schmidt, G. and Wenzel, H., "A Modified Van der Waals Type Equation of State," *Chem. Eng. Sci.*, 1980, Vol. 135, pp. 1503–1512.
19. Slot-Petersen, C., "A Systematic and Consistent Approach to Determine Binary Interaction Coefficients for the Peng-Robinson Equation of State," Paper SPE 16941, presented at the 62nd Annual Technical Conference of the SPE, held in Dallas, TX, September 27–30, 1987.
20. Soave, G., "Equilibrium Constants from a Modified Redlich-Kwong Equation of State," *Chem. Eng. Sci.*, 1972, Vol. 27, pp. 1197–1203.
21. Spencer, C., Daubert, T., and Danner, R., "A Critical Review of Correlations for the Critical Properties of Defined Mixtures," *AIChE Journal*, 1973, Vol. 19, No. 3, pp. 522–527.
22. Stryjek, R. and Vera, J. H., "PRSV: An Improvement Peng-Robinson Equation of State for Pure Compounds and Mixtures," *Canadian J. Chem. Eng.*, April 1986, Vol. 64, pp. 323–333.
23. Valderrama, J. and Cisternas, L., "A Cubic Equation of State for Polar and Other Complex Mixtures," *Fluid Phase Equilibria*, 1986, Vol. 29, pp. 431–438.
24. Van der Waals, J. D., *On the Continuity of the Liquid and Gaseous State*, Ph.D. Dissertation, Sigthoff, Leiden, 1873.
25. Vidal, J. and Daubert, T., "Equations of State—Reworking the Old Forms," *Chem. Eng. Sci*, 1978, Vol. 33, pp. 787–791.
26. Willman, B. T. and Teja, B. T., "Continuous Thermodynamics of Phase Equilibria Using a Multivariable Distribution Function and An Equation of State," *AIChE Journal*, December 1986, Vol. 32, No. 12, pp. 2067–2078.

7
Splitting and Lumping Schemes of Petroleum Fractions

The hydrocarbon plus fractions that comprise a significant portion of naturally occurring hydrocarbon fluids create major problems when predicting the thermodynamic properties and the volumetric behavior of these fluids by equations of state. These problems arise due to the difficulty of properly characterizing the plus fractions (heavy ends) in terms of their critical properties and acentric factors.

Whitson (1980) and Maddox-Erbar (1982 and 1984), among others, have shown the distinct effect of the heavy fractions characterization procedure on PVT relationships prediction by equations of state. Usually, these undefined plus fractions, commonly known as the C_{7+} fractions, contain an indefinite number of components with a carbon number higher than six. Molecular weight and specific gravity of the C_{7+} fraction may be the only measured data available.

In the absence of detailed analytical data for the plus fraction in a hydrocarbon mixture, erroneous predictions and conclusions can result if the plus fraction is used directly as a single component in the mixture phase behavior calculations. Numerous authors have indicated that these errors can be substantially reduced by "splitting" or "breaking down" the plus fraction into a manageable number of fractions (pseudo-components) for equation of state calculations.

The problem, then, is how to adequately split a C_{7+} fraction into a number of pseudo-components characterized by:

- Mole fractions
- Molecular weights
- Specific gravities

These characterization properties, when properly combined, should match the measured plus fraction properties, i.e., $(MW)_{7+}$ and $(\gamma)_{7+}$.

348

SPLITTING SCHEMES

Splitting schemes refer to the procedures of dividing the heptanes-plus fraction into hydrocarbon groups with a single carbon number (C_7, C_8, C_9, etc.) and are described by the same physical properties used for pure components.

Several authors have proposed different schemes for extending the molar distribution behavior of C_{7+}, i.e., the molecular weight and specific gravity. In general, the proposed schemes are based on the observation that lighter systems such as condensates usually exhibit exponential molar distribution, while heavier systems often show left-skewed distributions. This behavior is shown schematically in Figure 7-1.

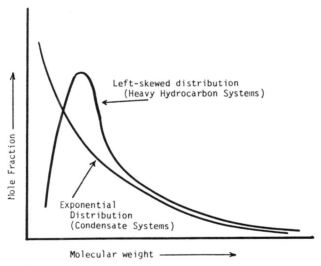

Figure 7-1. The exponential and left-skewed distribution functions.

Three important requirements should be satisfied when applying any of the proposed splitting models. These requirements are

- The sum of the mole fractions of the individual pseudo-components is equal to the mole fraction of C_{7+}.
- The sum of the products of the mole fraction and the molecular weight of the individual pseudo-components is equal to the product of the mole fraction and molecular weight of C_{7+}.
- The sum of the product of the mole fraction and molecular weight divided by the specific gravity of each individual component is equal to that of C_{7+}.

The above requirements can be expressed mathematically by the following relationships:

$$\sum_{n=7}^{N+} z_n = z_{7+} \tag{7-1}$$

$$\sum_{n=7}^{N+} [z_n \, MW_n] = z_{7+} \, MW_{7+} \tag{7-2}$$

$$\sum_{n=7}^{N+} \frac{z_n \, MW_n}{\gamma_n} = \frac{z_{7+} \, MW_{7+}}{\gamma_{7+}} \tag{7-3}$$

where

z_{7+} = mole fraction of C_{7+}
n = number of carbon atoms
$N+$ = last hydrocarbon group in the C_{7+} with n carbon atoms, e.g., 20 +
z_n = mole fraction of pseudo-component with n carbon atoms
MW_{7+}, γ_{7+} = measured molecular weight and specific gravity of C_{7+}
MW_n, γ_n = molecular weight and specific gravity of the pseudo-component with n carbon atoms

Several splitting schemes have been proposed recently. These schemes, as discussed below, are used to predict the compositional distribution of the heavy plus fraction.

Katz's Method

Katz (1983) presented an easy-to-use graphical correlation for breaking down into pseudo-components the C_{7+} fraction present in condensate systems. The method was originated by studying the compositional behavior of six condensate systems using detailed extended analyses. On semi-log scale, the mole percent of each constituent of the C_{7+} fraction versus the carbon number in the fraction was plotted. The resulting relationship can be conveniently expressed mathematically by the following expression:

$$z_n = 1.38205 \, z_{7+} \, e^{-0.25903n} \tag{7-4}$$

where z_{7+} = mole fraction of C_{7+} in the condensate system

   n = number of carbon atoms of the pseudo-component

   z_n = mole fraction of the pseudo-component with number of carbon atoms of n

Equation 7-4 is repeatedly applied until Equation 7-1 is satisfied. The molecular weight and specific gravity of the last pseudo-component can be calculated from Equations 7-2 and 7-3, respectively.

The computational procedure of Katz's method is best explained through the following example.

Example 7-1. A naturally occurring condensate gas system has the following composition:

Component	z_i	Component	z_i
C_1	0.9135	i-C_5	0.0015
C_2	0.0403	n-C_5	0.0019
C_3	0.0153	C_6	0.0039
i-C_4	0.0039	C_{7+}	0.0154
n-C_4	0.0043		

Molecular weight and specific gravity of C_{7+} are 141.25 and 0.797, respectively.

a. Using Katz's splitting scheme, extend the compositional distribution of C_{7+} to the pseudo-fraction C_{16+}.
b. Calculate MW, γ, T_b, p_c, T_c, and ω of C_{16+}.

Solution.

a. Calculation of the molar distribution using Katz's Method
Applying Equation 7-4 with z_{7+} = 0.0154 gives

	Experimental	**Equation 7-4**
n	z_n	z_n
7	0.00361	0.00347
8	0.00285	0.00268
9	0.00222	0.00207
10	0.00158	0.001596
11	0.00121	0.00123
12	0.00097	0.00095
13	0.00083	0.00073
14	0.00069	0.000566
15	0.00050	0.000437
16 +	0.00094	0.001671*

* This value is obtained by applying Equation 7-1, i.e.,

$$0.0154 - \sum_{n=7}^{15} z_n = 0.001671$$

b. Characterization of C_{16+}

Step 1. Calculate the molecular weight and specific gravity of C_{16+} by solving Equations 7-2 and 7-3 for these properties:

$$MW_{16+} = z_{7+} MW_{7+} - \sum_{n=7}^{15} (z_n \cdot MW_n)/z_{16+}$$

and

$$\gamma_{16+} = z_{16+} MW_{16+}/(z_{7+}MW_{7+}/\gamma_{7+}) - \sum_{n=7}^{15} (z_n MW_n/\gamma_n)$$

where MW_n, γ_n = molecular weight and specific gravity of the hydrocarbon group with n carbon atoms, as determined from Table 2-10, Chapter 2.

The calculations are performed using values from the following chart.

n	z_n	MW (Table 2-2)	z_n MW	γ (Table 2-2)	$z_n \cdot MW/\gamma_n$
7	0.00347	96	0.33312	0.727	0.4582
8	0.00268	107	0.28676	0.749	0.3829
9	0.00207	121	0.25047	0.768	0.3261
10	0.001596	134	0.213864	0.782	0.27348
11	0.00123	147	0.18081	0.793	0.22801
12	0.00095	161	0.15295	0.804	0.19024
13	0.00073	175	0.12775	0.815	0.15675
14	0.000566	190	0.10754	0.826	0.13019
15	0.000437	206	0.09002	0.836	0.10768
16 +	0.001671	—	—	—	—
			1.743284		2.25355

$$MW_{16+} = \frac{(0.0154)(141.25) - 1.743284}{0.001671} = 258.5$$

$$\gamma_{16+} = (0.001671)(258.5)/\frac{(0.0154)(141.25)}{(0.797)} - 2.25355 = 0.908$$

Step 2. Calculate the boiling points, critical pressure, and critical temperature of C_{16+} by using the Riazi-Daubert correlation (Equation 2-1, Chapter 2) to give:

$$T_b = 1,136°R$$
$$p_c = 215 \text{ psia}$$
$$T_c = 1,473°R$$

Step 3. Calculate the acentric factor of C_{16+} by applying the Edmister correlation, to give:

$$\omega = 0.684$$

Lohrenz's Method

Lohrenz et al. (1964) proposed that the heptanes plus fraction could be divided into pseudo-components with carbon number ranges from 7 to 40. They mathematically stated that the mole fraction z_n is related to its number of carbon atoms n and the mole fraction of the hexane fraction z_6 by the expression:

$$z_n = z_6 \, e^{A(n-6)^2 + B(n-6)} \tag{7-5}$$

The constants A and B are determined such that the constraints given by Equations 7-1 through 7-3 are satisfied.

The use of Equation 7-5 assumes that the individual C_{7+} components are distributed through the hexane mole fraction, and tail off to an extremely small quantity of heavy hydrocarbons.

Example 7-2. Rework Example 7-1 by using the Lohrenz splitting scheme and assuming that a partial molar distribu-tion of C_{7+} is available. The composition is given below:

Component	z_i	Component	z_i
C_1	0.9135	C_6	0.0039
C_2	0.0403	C_7	0.00361
C_3	0.0153	C_8	0.00285
i-C_4	0.0039	C_9	0.00222
n-C_4	0.0043	C_{10}	0.00158
i-C_5	0.0015	C_{11+}	0.00514
n-C_5	0.0019		

Solution.

Step 1. Determine the coefficients A and B of Equation 7-5 by the least-squares fit to the mole fractions C_6 through C_{10}, to give:

$$A = -0.043046307$$
$$B = -0.057954278$$

Step 2. Solve for the mole fraction of C_{10} through C_{15} by applying Equation 7-5 and setting $z_6 = 0.0039$.

Component	Experimental z_n	Equation 7-5 z_n
C_7	0.00361	0.00361
C_8	0.00285	0.00285
C_9	0.00222	0.00222
C_{10}	0.00158	0.00158
C_{11}	0.00121	0.00100
C_{12}	0.00097	0.00058
C_{13}	0.00083	0.00032
C_{14}	0.00069	0.00016
C_{15}	0.00050	0.00007
C_{16+}	0.00094	0.00301*

* Obtained by applying Equation 7-1

Step 3. Calculate the molecular weight and specific gravity of C_{16+} by applying Equations 7-2 and 7-3, to give:

$$(MW)_{16+} = 233.3$$
$$(\gamma)_{16+} = 0.943$$

Step 4. Solve for T_b, p_c, T_c, and ω by applying the Riazi-Daubert and Edmister correlations, to give:

$$T_b = 1,103°R$$
$$p_c = 251 \text{ psia}$$
$$T_c = 1,467°R$$
$$\omega = 0.600$$

Pedersen's Method

Pedersen et al. (1982) proposed that for naturally occurring hydrocarbon mixtures, an exponential relationship exists between the mole fraction of a component and the corresponding carbon number. They expressed this relationship mathematically in the following form:

$$z_n = e^{(n-A)/B} \tag{7-6}$$

where A and B are constants.

For condensates and volatile oils, Pedersen and coworkers suggested that A and B can be determined by a least squares fit to the molar distribution of the lighter fractions. Equation 7-6 can then be used to calculate the molar content of each of the heavier fractions by extrapolation. The classical constraints as given by Equations 7-1 through 7-3 are also imposed.

Example 7-3. Rework Example 7-2 by using the Pedersen splitting correlation.

Solution.

Step 1. Calculate the coefficients A and B by the least-squares fit to the molar distribution of C_6 through C_{10}, to give

$$A = -14.404639$$
$$B = -3.8125739$$

Step 2. Solve for the mole fraction of C_{10} through C_{15} by applying Equation 7-6:

Component	Experimental z_n	Calculated z_n
C_7	0.00361	0.00361
C_8	0.00285	0.00285
C_9	0.00222	0.00222
C_{10}	0.00158	0.00166
C_{11}	0.00121	0.00128
C_{12}	0.00097	0.00098
C_{13}	0.00083	0.00076
C_{14}	0.00069	0.00058
C_{15}	0.00050	0.00045
C_{16+}	0.00094	0.00101*

* From Equation 7-1

Whitson's Method

Whitson (1980) proposed that a probability function called the Three Parameter Gamma Function can be used to model the molar distribution. Unlike the previous models, the gamma function has the flexibility to describe a wider range of distributions by adjusting its variance, which is left as an adjustable parameter. Whitson expressed this function in the following form:

$$z_n = \frac{z_{7+}}{\Gamma(\alpha)} \sum_{j=0}^{\infty} \frac{Y_{n+1}^{\alpha+j} e^{-Y_{n+1}} - Y_n^{\alpha+j} e^{-Y_n}}{(\alpha + j)!} \tag{7-7}$$

where z_{7+} = mole fraction of the heptanes-plus fraction
        z_n = mole fraction of the pseudo-component with a number of carbon atoms of n
        $Y_n = MW_n - \eta/\beta$
        α = adjustable parameter
     $\Gamma(\alpha)$ = gamma function
        η = the lowest molecular weight expected to occur in the pseudo-component state. A good approximation of η is given by:
             $\eta = 14 n - 6$
        $\beta = MW_{7+} - \eta/\alpha$

The summation in Equation 7-7 can be discontinued when $\Sigma_{j+1} - \Sigma_j < 10^{-6}$.

Using Equation 7-7, the heptanes plus fraction with a mole fraction of z_{7+} can be divided into several pseudo-components each with composition z_n and molecular weight MW_n. The shape of the curve (exponential or left-skewed) representing the molar distribution depends on the value of the adjustable parameter α.

Whitson proposed that the parameter α can be optimized by minimizing the $E(\alpha)$ which is defined as the sum of the squares of differences in measured (from partial extend analysis) and calculated compositions. Mathematically, the objective function is expressed as follows:

$$E(\alpha) = \sum_{i=n}^{N+} (z_i^{cal} - z_i^{exp})^2 \qquad (7\text{-}8)$$

where z_i^{cal} = calculated mole fraction of pseudo-component i
       z_i^{exp} = experimental mole fraction of component i
      $N+$ = last hydrocarbon group in the C_{7+} fraction

Whitson suggested that reasonable limits for α are 0.5 to 3.0. Figure 7-2 illustrates the proposed model for several values of the parameter α. For $\alpha = 1$, the distribution is exponential. Values equaling less than one give accelerated exponential distributions, while values greater than one yield left-skewed distributions.

Ahmed's Method

Ahmed et al. (1985) devised a simplified method for splitting the C_{7+} fraction into pseudo-components. The method originated from studying the molar behavior of thirty-four condensate and crude oil systems through detailed laboratory compositional analyses of the heavy fractions. The only re-

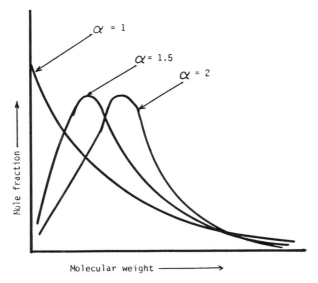

Figure 7-2. Illustration of different probability functions.

quired input data for the proposed method are the molecular weight and the total mole fraction of the heptanes-plus fraction.

The splitting scheme is based on calculating the mole fraction z_n at a progressively higher number of carbon atoms. The extraction process continues until the sum of the mole fraction of the pseudo-components equals the total mole fraction of the heptanes-plus (z_{7+}).

The authors proposed the following expressions for estimating z_n:

$$z_n = z_{n+} \left[\frac{MW_{(n+1)+} - MW_{n+}}{MW_{(n+1)+} - MW_n} \right] \tag{7-9}$$

where z_n = mole fraction of the pseudo-component with a number of carbon atoms of n (z_7, z_8, z_9, etc.)

MW_n = molecular weight of the hydrocarbon group with n carbon atoms as given in Table 2-10

MW_{n+} = molecular weight of the n + fraction as calculated by the following expression

$$MW_{n+} = MW_{7+} + S(n - 7) \tag{7-10}$$

where n is the number of carbon atoms and S is the coefficient of Equation 7-10 with values:

Number of Carbon Atoms	Condensate Systems	Crude Oil Systems
n ≤ 8	15.5	16.5
n > 8	17.0	20.1

The authors also proposed the following equation for calculating the specific gravity of the n + fraction:

$$\gamma_{n+} = \gamma_{7+} \left[1 + a\, e^{-bn} \left(\frac{MW_{n+}}{MW_{7+}} - 1 \right) \right] \tag{7-11}$$

where γ_{n+} = specific gravity of C_{n+}
γ_{7+} = specific gravity of C_{7+}

The coefficients a and b are given for each type of hydrocarbon system:

Coefficients of Equation 7-11	Condensate Systems	Crude Oil Systems
a	0.06773	0.247308
b	0.008405	0.063241

The step-wise calculation sequences of the proposed correlation are summarized in the following steps:

Step 1. According to the type of the hydrocarbon system under investigation (condensate or crude oil), select the appropriate values for the coefficient S.

Step 2. Knowing the molecular weight of C_{7+} fraction (MW_{7+}), calculate the molecular weight of the octanes-plus fraction (MW_{8+}) by applying Equation 7-10.

Step 3. Calculate the mole fraction of the heptane fraction (z_7) by using Equation 7-9.

Step 4. Steps 2 and 3 are repeatedly applied for each component in the system (C_8, C_9, etc.) until the sum of the calculated mole fractions is equal to the mole fraction of C_{7+} of the system.

The splitting scheme is best explained through the following example.

Example 7-4. Rework Example 7-1 by using Ahmed's splitting method.

Solution.

Step 1. Calculate the molecular weight of C_{8+} by applying Equation 7-10.

$$MW_{8+} = 141.25 + 15.5 \ (8 - 7) = 156.75$$

Step 2. Solve for the mole fraction of heptane (z_7) by applying Equation 7-9.

$$z_7 = z_{7+}[(MW_{8+} - MW_{7+})/(MW_{8+} - MW_7)]$$
$$z_7 = 0.0154 \ [(156.75 - 141.25)/(156.75 - 96)] = 0.00393$$

Step 3. Calculate the molecular weight of C_{9+} from Equation 7-10.

$$MW_{9+} = 141.25 + 15.5 \ (9 - 7) = 172.25$$

Step 4. Determine the mole fraction of C_8 from Equation 7-9.

$$z_8 = z_{8+} \ [(MW_{9+} - MW_{8+})/(MW_{9+} - MW_8)]$$
$$z_8 = (0.0154 - 0.00393)[(172.5 - 156.75)/(172.5 - 107)]$$
$$= 0.00276$$

Step 5. This extracting method is repeated as outlined in the above steps, to give

Component	n	MW_{n+} Equation 7-10	MW_n Table 2-10	z_n Equation 7-9
C_7	7	141.25	96	0.00393
C_8	8	156.75	107	0.00276
C_9	9	175.25	121	0.00200
C_{10}	10	192.25	134	0.00144
C_{11}	11	209.25	147	0.00106
C_{12}	12	226.25	161	0.0008
C_{13}	13	243.25	175	0.00061
C_{14}	14	260.25	190	0.00048
C_{15}	15	277.25	206	0.00038
C_{16+}	16 +	294.25	222	0.00159*

* Calculated from Equation 7-1

Step 6. The boiling point, critical properties, and the acentric factor of C_{16+} are then determined by using the appropriate methods, to give

$$MW = 222$$
$$\gamma = 0.856$$
$$T_b = 1{,}174.6°R$$
$$p_c = 175.9 \text{ psia}$$
$$T_c = 1{,}449.3°R$$
$$\omega = 0.742$$

LUMPING SCHEMES

Equation of state calculations are frequently burdened by the large number of components necessary to describe the hydrocarbon mixture for accurate phase behavior modeling. Often, the problem is either lumping together the many experimental determined fractions, or modeling the hydrocarbon system when the only experimental data available for the C_{7+} fraction are the molecular weight and specific gravity.

Generally, with a sufficiently large number of pseudo-components used in characterizing the heavy fraction of a hydrocarbon mixture, a satisfactory prediction of the PVT behavior by the equation of state can be obtained. However, in compositional models, the cost and computing time can increase significantly with the increased number of components in the system. Therefore, there are strict limitations on the maximum number of components that can be used in compositional models and the original components have to be lumped into a smaller number of new pseudo-components.

The term "lumping" or "pseudoization" then denotes the reduction in the number of components used in equation of state calculations for reservoir fluids. This reduction is accomplished by employing the concept of the pseudo-component. The pseudo-component denotes a group of pure components lumped together and represented by a single component.

There are several problems associated with "regrouping" the original components into a smaller number without losing the predicting power of the equation of state. These problems include:

- How to select the groups of pure components to be represented by one pseudo-component each.
- What mixing rules should be used for determining the EOS constants (p_c, T_c, and ω) for the new lumped pseudo-components.

There are several unique published techniques which can be used to address the above lumping problems; notably the methods proposed by

- Lee et al. (1979)
- Whitson (1981)
- Mehra (1982)
- Montel-Gouel (1984)
- Schlijper (1984)
- Behrens-Sandler (1986)
- Gonzalez-Colonomos-Rusinek (1986)

Several of these techniques are presented in the following discussion.

Whitson's Lumping Scheme

Whitson (1980) proposed a regrouping scheme whereby the compositional distribution of the C_{7+} fraction is reduced to only a few Multiple-Carbon-Number (MCN) groups. Whitson suggested that the number of MCN groups necessary to describe the plus fraction is given by the following empirical rule:

$$N_g = Int \, [1 + 3.3 \, Log \, (N - n)] \tag{7-12}$$

where N_g = number of MCN groups
  Int = Integer
  N = number of carbon atoms of the last component in the hydrocarbon system
  n = number of carbon atoms of the first component in the plus fraction

The integer function requires that the real expression evaluated inside the brackets be rounded to the nearest integer. Whitson pointed out that for black-oil systems, the calculated value of N_g can be reduced by one.

The molecular weights separating each MCN group are calculated from the following expression:

$$(MW)_I = (MW)_n \left[Exp \left[\left(\frac{1}{N_g} \right) Ln \left(\frac{(MW)_N}{(MW)_n} \right) \right] \right]^I \tag{7-13}$$

where $(MW)_N$ = molecular weight of the last reported component in the extended analysis of the plus fraction
  $(MW)_n$ = molecular weight of the first hydrocarbon group in the extended analysis of the plus fraction
  $I = 1, 2, ..., N_g$

Molecular weight of hydrocarbon groups (molecular weight of C_7-group, C_8-group, etc.) falling within the boundaries of these values are included in the Ith MCN group.

Example 7-5 illustrates the use of Equations 7-12 and 7-13.

Example 7-5. Given the following compositional analysis of the C_{7+} fraction in a condensate system, determine the appropriate number of pseudo-components forming the C_{7+}.

Component	z_i
C_7	0.0034
C_8	0.00268
C_9	0.00207
C_{10}	0.001596
C_{11}	0.00123
C_{12}	0.00095
C_{13}	0.00073
C_{14}	0.000566
C_{15}	0.000437
C_{16+}	0.001671

$MW_{16+} = 259$

Solution.

Step 1. Determine the molecular weight of each component in the system from Table 2-10.

Component	z_i	MW
C_7	0.00347	96
C_8	0.00268	107
C_9	0.00207	121
C_{10}	0.001596	134
C_{11}	0.00123	147
C_{12}	0.00095	161
C_{13}	0.00073	175
C_{14}	0.000566	190
C_{15}	0.000437	206
C_{16+}	0.001671	259*

Step 2. Calculate the number of pseudo-components from Equation 7-12.

$$N_g = \text{Int } [1 + 3.3 \text{ Log } (16 - 7)]$$
$$N_g = \text{Int } [4.15]$$
$$N_g = 4$$

Step 3. Determine the molecular weights separating the hydrocarbon groups by applying Equation 7-13.

$$(MW)_I = 96 \left[\text{Exp} \left[\left(\frac{1}{4} \right) \text{Ln} \left(\frac{259}{96} \right) \right] \right]^I$$

$$(MW)_I = 96 \, [1.282]^I$$

or

I	$(MW)_I$
1	123
2	158
3	202
4	259

- First Pseudo-Component
  The first pseudo-component includes all components with molecular weight in the range of 96 to 123. This group then includes C_7, C_8, and C_9.
- Second Pseudo-Component
  The second pseudo-component contains all components with a molecular weight higher than 123 to a molecular weight of 158. This group includes C_{10} and C_{11}.
- Third Pseudo-Component
  The third pseudo-component includes components with a molecular weight higher than 158 to a molecular weight of 202. Therefore, this groups includes C_{12}, C_{13}, and C_{14}.

- Fourth Pseudo-Component
  This pseudo-component includes all the remaining components, i.e., C_{15} and C_{16+}.

Group I	Component	z_i	z_I
1	C_7	0.00347	
	C_8	0.00268	0.00822
	C_9	0.00207	
2	C_{10}	0.001596	0.002826
	C_{11}	0.00123	
3	C_{12}	0.00095	
	C_{13}	0.00073	0.002246
	C_{14}	0.000566	
4	C_{15}	0.000437	0.002108
	C_{16+}	0.001671	

It is convenient at this stage to present the mixing rules which can be employed to characterize the pseudo-component in terms of its pseudo-physical and pseudo-critical properties. Since there are numerous ways to mix the properties of the individual components, all giving different properties for the pseudo-components, the choice of a correct mixing rule is as important as the lumping scheme. Some of these mixing rules are given next.

Hong's Mixing Rules. Hong (1982) concluded that the weight fraction average w_i is the best mixing parameter in characterizing the C_{7+} fractions by the following mixing rules:

- Pseudo-critical pressure $p_{cL} = \sum\limits^{L} w_i \, p_{ci}$

- Pseudo-critical temperature $T_{cL} = \sum\limits^{L} w_i \, T_{ci}$

- Pseudo-critical volume $V_{cL} = \sum\limits^{L} w_i \, V_{ci}$

- Pseudo-acentric factor $\omega_L = \sum\limits^{L} w_i \, \omega_i$

- Pseudo-molecular weight $MW_L = \sum\limits^{L} w_i \, MW_i$

- Binary interaction coefficient $K_{kL} = 1 - \sum\limits_{i}^{L} \sum\limits_{j}^{L} w_i \, w_j \, (1 - k_{ij})$

  with

$$w_i = \frac{z_i \, MW_i}{\sum\limits^{L} z_i \, MW_i}$$

where w_i = average weight fraction
K_{kL} = binary interaction coefficient between the kth component and the lumped fraction

The subscript L in the above relationships denotes the lumped fraction.

Lee's Mixing Rules. Lee et al. (1979), in their proposed regrouping model, employed Kay's mixing rules as the characterizing approach for determining the properties of the lumped fractions. Defining the normalized mole fraction of the component i in the lumped fraction as

$$\phi_i = z_i / \sum\limits^{L} z_i$$

the following rules are proposed:

- $MW_L = \sum\limits^{L} \phi_i \, MW_i$ (7-14)

- $\gamma_L = MW_L / \sum\limits^{L} [\phi_i \, MW_i/\gamma_i]$ (7-15)

- $V_{cL} = \sum\limits^{L} [\phi_i \, MW_i \, V_{ci}/MW_L]$ (7-16)

- $p_{cL} = \sum\limits^{L} [\phi_i \, p_{ci}]$ (7-17)

- $T_{cL} = \sum\limits^{L} [\phi_i \, T_{ci}]$ (7-18)

- $\omega_L = \sum\limits^{L} [\phi_i \, \omega_i]$ (7-19)

Example 7-6. Using Lee's mixing rules, determine the physical and critical properties of the four pseudo-components in Example 7-5.

Solution.

Step 1. Assign the appropriate physical and critical properties to each component.

Group	Compo- nent	z_i	z_I	MW_i	γ_i	V_{ci}	p_{ci}	T_{ci}	ω_i
1	C_7	0.00347		96*	0.727*	0.06289*	453*	985*	0.280*
	C_8	0.00268	0.00822	107	0.749	0.06264	419	1,036	0.312
	C_9	0.00207		121	0.768	0.06258	383	1,058	0.348
2	C_{10}	0.001596	0.002826	134	0.782	0.06273	351	1,128	0.385
	C_{11}	0.00123		147	0.793	0.06291	325	1,166	0.419
3	C_{12}	0.00095		161	0.804	0.06306	302	1,203	0.454
	C_{13}	0.00073	0.002246	175	0.815	0.06311	286	1,236	0.484
	C_{14}	0.000566		190	0.826	0.06316	270	1,270	0.516
4	C_{15}	0.000437	0.002108	206	0.826	0.06325	255	1,304	0.550
	C_{16+}	0.001671		259	0.908	0.0638†	215†	1,467	0.68†

* From Table 2-10 (Chapter 2)
† Calculated

Step 2. Calculate the physical and critical properties of each group by applying Equations 7-14 through 7-19, to give

Group	Z_I	MW_L	γ_L	V_{cL}	p_{cL}	T_{cL}	ω_L
1	0.00822	105.9	0.746	0.0627	424	1,020	0.3076
2	0.002826	139.7	0.787	0.0628	339.7	1,144.5	0.4000
3	0.002246	172.9	0.814	0.0631	288	1,230.6	0.4794
4	0.002108	248	0.892	0.0637	223.3	1,433	0.6531

Behrens and Sandler Lumping Scheme

Behrens and Sandler (1986) used the semi-continuous thermodynamic distribution theory to model the C_{7+} fraction for equation of state calculations. The authors suggested that the heptanes-plus fraction can be fully described with as few as two pseudo-components.

A semi-continuous fluid mixture is defined as one in which the mole fractions of some components, such as C_1 through C_6, have discrete values, while the concentrations of others, (the unidentifiable components) such as C_{7+}, are described by the continuous distribution function $F(I)$. This continuous distribution function $F(I)$ describes the heavy fractions according to the index I, selected to be a property such as boiling point, carbon number, or molecular weight.

For a hydrocarbon system with k discrete components, the following relationship applies:

$$\sum_{i=1}^{k} z_i + z_{7+} = 1.0$$

The mole fraction of C_{7+} in this equation is replaced with the selected distribution function, to give

$$\sum_{i=1}^{k} z_i + \int_{A}^{B} F(I)\, dI = 1.0 \qquad (7\text{-}20)$$

where A = lower limit of integration (beginning of the continuous distribution)

B = upper limit of integration (upper cutoff of the continuous distribution

The above molar distribution behavior is shown schematically in Figure 7-3. This figure shows a semi-log plot of the composition z_i versus the carbon number n of the individual components in a hydrocarbon system. The parameter A can be determined from the plot, or can be defaulted to C_7, i.e.,

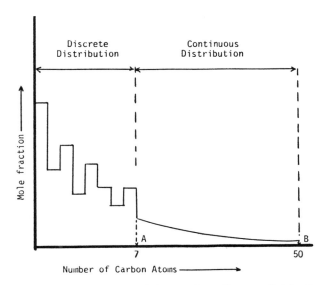

Figure 7-3. Schematic illustration of the semi-continuous distribution model.

A = 7. The value of the second parameter B ranges from 50 to infinity, i.e., $50 \leq B \leq \infty$; however, Behrens and Sandler pointed out that the exact choice of the cutoff is not critical.

Selecting the index I of the distribution function F(I) to be the carbon number n, Behrens and Sandler proposed the following exponential form of F(I).

$$F(n) = D(n)\ e^{-\alpha n}\ dn \qquad (7\text{-}21)$$

with $A \leq n \leq B$

where the parameter α is given by the following expression

$$\frac{1}{\alpha} - \overline{C}_n + A - [A - B]e^{-B\alpha}/[e^{-A\alpha} - e^{-B\alpha}] = 0 \qquad (7\text{-}22)$$

where $\overline{C}_n$ is the average carbon number and defined by the relationship:

$$\overline{C}_n = [MW_{7+} + 4]/14 \qquad (7\text{-}23)$$

Equation 7-22 can be solved for α by successive substitutions or the Newton-Raphson method, with an initial value of $\alpha = 1/\overline{C}_n - A$. Substituting Equation 7-21 into Equation 7-20 yields

$$\sum_{i}^{k} z_i + \int_{A}^{B} D(n) \, e^{-\alpha n} \, dn = 1.0$$

or

$$z_{7+} = \int_{A}^{B} D(n) \, e^{-\alpha n} \, dn$$

By a transformation of variables and changing the range of integration from A and B to 0 and c, the equation becomes

$$z_{7+} = \int_{0}^{c} D(r) \, e^{-r} \, dr \tag{7-24}$$

where $c = (B - A) \, \alpha$ $\qquad\qquad$ (7-25)
$\qquad\quad$ r = dummy variable of integration

The authors applied the "Gaussian quadrature numerical integration method" with a two-point integration to evaluate Equation 7-24, resulting in

$$z_{7+} = \sum_{i=1}^{2} D(r_i) \, w_i \tag{7-26}$$

where $\quad r_i$ = roots for quadrature of integrals after variable transformation
$\qquad\quad w_i$ = weighing factor of Gaussian quadrature at point i

The values of r_i and w_i are given in Table 7-1.

The computational sequences of the proposed method are summarized in the following steps.

Step 1. Find the endpoints A and B of the distribution. Since the endpoints are assumed to start and end at the midpoint between the two carbon numbers, the effective endpoints become

$\qquad$ A = starting carbon number − 1/2 $\qquad\qquad$ (7-27)
$\qquad$ B = ending carbon number + 1/2 $\qquad\qquad$ (7-28)

Step 2. Calculate the value of the parameter α by solving Equation 7-22.

Step 3. Determine the upper limit of integration c by applying Equation 7-25.

Step 4. Find the integration points r_1 and r_2; and the weighting factors w_1 and w_2 from Table 7-1.

Table 7-1
Behrens and Sandler Roots and Weights for Two-point Integration

c	r_1	r_2	w_1	w_2
0.30	0.0615	0.2347	0.5324	0.4676
0.40	0.0795	0.3101	0.5353	0.4647
0.50	0.0977	0.3857	0.5431	0.4569
0.60	0.1155	0.4607	0.5518	0.4482
0.70	0.1326	0.5347	0.5601	0.4399
0.80	0.1492	0.6082	0.5685	0.4315
0.90	0.1652	0.6807	0.5767	0.4233
1.00	0.1808	0.7524	0.5849	0.4151
1.10	0.1959	0.8233	0.5932	0.4068
1.20	0.2104	0.8933	0.6011	0.3989
1.30	0.2245	0.9625	0.6091	0.3909
1.40	0.2381	1.0307	0.6169	0.3831
1.50	0.2512	1.0980	0.6245	0.3755
1.60	0.2639	1.1644	0.6321	0.3679
1.70	0.2763	1.2299	0.6395	0.3605
1.80	0.2881	1.2944	0.6468	0.3532
1.90	0.2996	1.3579	0.6539	0.3461
2.00	0.3107	1.4204	0.6610	0.3390
2.10	0.3215	1.4819	0.6678	0.3322
2.20	0.3318	1.5424	0.6745	0.3255
2.30	0.3418	1.6018	0.6810	0.3190
2.40	0.3515	1.6602	0.6874	0.3126
2.50	0.3608	1.7175	0.6937	0.3063
2.60	0.3699	1.7738	0.6997	0.3003
2.70	0.3786	1.8289	0.7056	0.2944
2.80	0.3870	1.8830	0.7114	0.2886
2.90	0.3951	1.9360	0.7170	0.2830
3.00	0.4029	1.9878	0.7224	0.2776
3.10	0.4104	2.0386	0.7277	0.2723
3.20	0.4177	2.0882	0.7328	0.2672
3.30	0.4247	2.1367	0.7378	0.2622
3.40	0.4315	2.1840	0.7426	0.2574
3.50	0.4380	2.2303	0.7472	0.2528
3.60	0.4443	2.2754	0.7517	0.2483
3.70	0.4504	2.3193	0.7561	0.2439
3.80	0.4562	2.3621	0.7603	0.2397
3.90	0.4618	2.4038	0.7644	0.2356
4.00	0.4672	2.4444	0.7683	0.2317
4.10	0.4724	2.4838	0.7721	0.2279
4.20	0.4775	2.5221	0.7757	0.2243
4.30	0.4823	2.5593	0.7792	0.2208
4.40	0.4869	2.5954	0.7826	0.2174
4.50	0.4914	2.6304	0.7858	0.2142
4.60	0.4957	2.6643	0.7890	0.2110
4.70	0.4998	2.6971	0.7920	0.2080
4.80	0.5038	2.7289	0.7949	0.2051
4.90	0.5076	2.7596	0.7977	0.2023

(table continued on next page)

Table 7-1
Continued

c	r_1	r_2	w_1	w_2
5.00	0.5112	2.7893	0.8003	0.1997
5.10	0.5148	2.8179	0.8029	0.1971
5.20	0.5181	2.8456	0.8054	0.1946
5.30	0.5214	2.8722	0.8077	0.1923
5.40	0.5245	2.8979	0.8100	0.1900
5.50	0.5274	2.9226	0.8121	0.1879
5.60	0.5303	2.9464	0.8142	0.1858
5.70	0.5330	2.9693	0.8162	0.1838
5.80	0.5356	2.9913	0.8181	0.1819
5.90	0.5381	3.0124	0.8199	0.1801
6.00	0.5405	3.0327	0.8216	0.1784
6.20	0.5450	3.0707	0.8248	0.1752
6.40	0.5491	3.1056	0.8278	0.1722
6.60	0.5528	3.1375	0.8305	0.1695
6.80	0.5562	3.1666	0.8329	0.1671
7.00	0.5593	3.1930	0.8351	0.1649
7.20	0.5621	3.2170	0.8371	0.1629
7.40	0.5646	3.2388	0.8389	0.1611
7.70	0.5680	3.2674	0.8413	0.1587
8.10	0.5717	3.2992	0.8439	0.1561
8.50	0.5748	3.3247	0.8460	0.1540
9.00	0.5777	3.3494	0.8480	0.1520
10.00	0.5816	3.3811	0.8507	0.1493
11.00	0.5836	3.3978	0.8521	0.1479
12.00	0.5847	3.4063	0.8529	0.1471
14.00	0.5856	3.4125	0.8534	0.1466
16.00	0.5857	3.4139	0.8535	0.1465
18.00	0.5858	3.4141	0.8536	0.1464
20.00	0.5858	3.4142	0.8536	0.1464
25.00	0.5858	3.4142	0.8536	0.1464
30.00	0.5858	3.4142	0.8536	0.1464
40.00	0.5858	3.4142	0.8536	0.1464
60.00	0.5858	3.4142	0.8536	0.1464
100.00	0.5858	3.4142	0.8536	0.1464
∞	0.5858	3.4142	0.8536	0.1464

(Permission to publish by the Society of Petroleum Engineers of AIME. Copyright SPE-AIME.)

Step 5. Find the pseudo-component carbon numbers n_i and mole fractions z_i from the following expressions:

  • First pseudo-component:

$$n_1 = \frac{r_1}{\alpha} + A$$

$$z_1 = w_1 \, z_{7+}$$

$$(7\text{-}29)$$

• Second pseudo-component:

$$n_2 = \frac{r_2}{\alpha} + A$$

$$z_2 = w_2 \, z_{7+} \tag{7-30}$$

Step 6. Assign the physical and critical properties of the two pseudo-components from Table 2-10.

Example 7-6 (Given by Behrens and Sandler). A heptanes-plus fraction in a crude oil system has a mole fraction of 0.4608 with a molecular weight of 226. Using the Behrens and Sandler lumping scheme, characterize the C_{7+} by two pseudo-components and calculate their mole fractions.

Solution.

Step 1. Assuming the starting and ending carbon numbers to be C_7 and C_{50}, calculate A and B from Equations 7-27 and 7-28.

$$A = 7 - 0.5 = 6.5$$
$$B = 50 + 0.5 = 50.5$$

Step 2. Calculate $\overline{C}_n$ from Equation 7-23.

$$\overline{C}_n = \frac{226 + 4}{14} = 16.43$$

Step 3. Solve Equation 7-22 for α, to give

$$\alpha = 0.0938967$$

Step 4. Calculate range of integration c from Equation 7-25.

$$c = (50.5 - 6.5) \, 0.0938967 = 4.13$$

Step 5. Find integration points r_i and weights w_i from Table 7-1.

$$r_1 = 0.4741$$
$$r_2 = 2.4965$$
$$w_1 = 0.7733$$
$$w_2 = 0.2267$$

Step 6. Find the pseudo-component carbon numbers n_i and mole fractions z_i by applying Equations 7-29 and 7-30.

• First pseudo-component

$$n_1 = \frac{0.4741}{0.0938967} + 6.5 = 11.55$$

$$z_1 = (0.7733)(0.4608) = 0.3563$$

- Second pseudo-component

$$n_2 = \frac{2.4965}{0.0938967} + 6.5 = 33.08$$

$$z_2 = (0.2267)(0.4608) = 0.1045$$

The C_{7+} fraction is represented then by the following *two* pseudo-components:

Pseudo-Component	Carbon Number	Mole Fraction
1	$C_{11.55}$	0.3563
2	$C_{33.08}$	0.1045

Step 7. Assign the physical properties of the two pseudo-components *according to their number of carbon atoms* by using the Katz and Firoozabadi generalized physical properties as given in Table 2-10. The assigned physical properties for the two fractions are shown below:

Pseudo-Component	n	T_b, °R	γ	MW	T_c, °R	p_c, psia	ω
1	11.55	848	0.799	154	1,185	314	0.437
2	33.08	1,341	0.915	426	1,629	134	0.921

Lee's Lumping Scheme

Lee et al. (1979) devised a simple procedure for regrouping the oil fractions into pseudo-components. Lee and co-workers employed the physical reasoning that crude oil fractions having relatively close physico-chemical properties (such as molecular weight, specific gravity, etc.) can be accurately represented by a single fraction. Having observed that the closeness of these properties is reflected by the slopes of curves when the properties are plotted against the weight-averaged boiling point of each fraction, Lee et al. used the weighted sum of slopes of these curves as a criterion for lumping the crude oil fractions. The authors proposed the following computational steps:

Step 1. Plot the available physico-chemical properties of each original fraction versus its weight-averaged boiling point (WABP).

Step 2. Calculate numerically the slope m_{ij} for each fraction at each WABP where m_{ij} = slope of the curve of property versus boiling point

$i = 1,..., n_f$

$j = 1,..., n_p$

n_f = number of original oil fractions

n_p = number of available physico-chemical properties

Step 3. Compute the normalized absolute slope $\overline{m}_{ij}$ as defined

$$\overline{m}_{ij} = \frac{m_{ij}}{\max_{i = 1,..., n_f} m_{ij}} \qquad (7\text{-}31)$$

Step 4. Compute the weighted sum of slopes $\overline{M}_i$ for each fraction, as follows:

$$\overline{M}_i = \frac{\displaystyle\sum_{j=1}^{np} \overline{m}_{ij}}{np} \qquad (7\text{-}32)$$

where $\overline{M}_i$ represents the averaged change of physico-chemical properties of the crude oil fractions along the boiling-point axis.

Step 5. Judging the numerical values of $\overline{M}_i$ for each fraction, group those fractions which have similar $\overline{M}_i$ values.

Step 6. Using the mixing rules given by Equations 7-14 through 7-19, calculate the physical properties of pseudo-components.

Example 7-7. The following data, as given by Hariu and Sage (1969), represent the average boiling point molecular weight, specific gravity, and molar distribution of 15 pseudo-components in a crude oil system.

Pseudo-Component	T_b, °R	MW	γ	z_i
1	600	95	0.680	0.0681
2	654	101	0.710	0.0686
3	698	108	0.732	0.0662
4	732	116	0.750	0.0631
5	770	126	0.767	0.0743
6	808	139	0.781	0.0686
7	851	154	0.793	0.0628
8	895	173	0.800	0.0564
9	938	191	0.826	0.0528
10	983	215	0.836	0.0474
11	1,052	248	0.850	0.0836
12	1,154	322	0.883	0.0669
13	1,257	415	0.910	0.0535
14	1,382	540	0.940	0.0425
15	1,540	700	0.975	0.0340

Using the Lee Lumping Scheme, re-group the given system into an appropriate number of pseudo-component groups and estimate the physical properties of each group.

Solution.

Step 1. Plot the molecular weights and specific gravities versus the boiling points and calculate the slope m_{ij} for each pseudo-component as follows:

Pseudo-Component	T_b	$(\partial MW/\partial T_b)$ m_{i1}	$(\partial\gamma/\partial T_b)$ m_{i2}
1	600	0.1111	0.00056
2	654	0.1327	0.00053
3	698	0.1923	0.00051
4	732	0.2500	0.00049
5	770	0.3026	0.00041
6	808	0.3457	0.00032
7	851	0.3908	0.00022
8	895	0.4253	0.00038
9	938	0.4773	0.00041
10	983	0.5000	0.00021
11	1,052	0.6257	0.00027
12	1,154	0.8146	0.00029
13	1,257	0.9561	0.00025
14	1,382	1.0071	0.00023
15	1,540	1.0127	0.00022

Step 2. Calculate the normalized absolute slope $\underline{m}_{ij}$ and the weighted sum of slopes $\overline{M}_i$ using Equations 7-31 and 7-32, respectively.

Pseudo-Component	$\overline{m}_{i1} = m_{ij}/$ 1.0127	$\overline{m}_{i2} = m_{ij}/$ 0.00056	$\overline{M}_i = (\overline{m}_{i1} + \overline{m}_{i2})/$ 2
1	0.1097	1.0000	0.55485
2	0.1310	0.9464	0.5387
3	0.1899	0.9107	0.5503
4	0.2469	0.8750	0.5610
5	0.2988	0.7321	0.5155
6	0.3414	0.5714	0.4564
7	0.3859	0.3929	0.3894
8	0.4200	0.6786	0.5493
9	0.4713	0.7321	0.6017
10	0.4937	0.3750	0.4344

(*example 7-7 continued on next page*)

Example 7-7. Continued.

Pseudo-Component	$\overline{m}_{i1} = m_{ij}/$ 1.0127	$\overline{m}_{i2} = m_{ij}/$ 0.00056	$\overline{M}_i = (\overline{m}_{i1} + \overline{m}_{i2})/$ 2
11	0.6179	0.4821	0.5500
12	0.8044	0.5179	0.6612
13	0.9441	0.4464	0.6953
14	0.9945	0.4107	0.7026
15	1.0000	0.3929	0.6965

Examining the values of $\overline{M}_i$, the pseudo-components can be lumped into three groups as follows:

Group 1: Combine fractions 1–5 with a *total* mole fraction of 0.3403.

Group 2: Combine fractions 6–10 with a *total* mole fraction of 0.2880.

Group 3: Combine fractions 11–15 with a *total* mole fraction of 0.2805.

Step 3. Calculate the physical properties of each group. This can be achieved by computing MW and γ of each group by applying Equations 7-14 and 7-15 respectively, followed by employing the Riazi-Daubert correlation (Equation 2-1) to characterize each group. Results of the calculations are shown next.

Group I	Mole Fraction Z_i	MW Eq. 7-14	γ Eq. 7-15	T_b Eq. 2-1	T_c Eq. 2-1	p_c Eq. 2-1	V_c Eq. 2-1
1	0.3403	109.4	0.7299	694	1,019	404	0.0637
2	0.2880	171.0	0.8073	891	1,224	287	0.0634
3	0.2805	396.5	0.9115	1,383	1,656	137	0.0656

PROBLEMS

1. A heptanes plus fraction with a molecular weight of 198 and a specific gravity of 0.8135 presents a naturally occurring condensate system. The reported mole fraction of the C_{7+} is 0.1145. Predict the molar distribution of the plus fraction by using:

a. Katz's correlation

b. Ahmed's correlation

Characterize the last fraction in the predicted extended analysis in terms of its physical and critical properties.

2. A naturally occurring crude oil system has a heptanes plus fraction with the following properties:

MW_{7+} = 213
γ_{7+} = 0.8405
x_{7+} = 0.3497

Extend the molar distribution of the plus fraction and determine the critical properties and acentric factor of the last component.

3. A crude oil system has the following composition:

Component	x_i
C_1	0.3100
C_2	0.1042
C_3	0.1187
C_4	0.0732
C_5	0.0441
C_6	0.0255
C_7	0.0571
C_8	0.0472
C_9	0.0246
C_{10}	0.0233
C_{11}	0.0212
C_{12}	0.0169
C_{13+}	0.1340

The molecular weight and specific gravity of C_{13+} are 325 and 0.842. Calculate the appropriate number of pseudo-components necessary to adequately represent the above components. Use

a. Whitson's Lumping Method
b. Behrens-Sandler Method

Characterize the resulting pseudo-components.

REFERENCES

1. Ahmed, T., Cady, G., and Story, A., "A Generalized Correlation for Characterizing the Hydrocarbon Heavy Fractions," Paper SPE 14266, presented at the 60th Annual Technical Conference of the SPE, held in Las Vegas, September 22–25, 1985.

2. Behrens, R. and Sandler, S., "The Use of Semi-continuous Description to Model the C_{7+} Fraction in Equation of State Calculation," Paper SPE/DOE 14925, presented at the 5th Annual Symposium on EOR, held in Tulsa, Oklahoma, April 20–23, 1986.
3. Gonzalez, E., Colonomos, P., and Rusinek, I., "A New Approach for Characterizing Oil Fractions and for Selecting Pseudo-Components of Hydrocarbons," *Canadian JPT*, March–April 1986, pp. 78–84.
4. Hong, K. C., "Lumped-Component Characterization of Crude Oils for Compositional Simulation," Paper SPE/DOE 10691, presented at the 3rd Joint Symposium on EOR, held in Tulsa, Oklahoma April 4–7, 1982.
5. Hariu, O. and Sage, R., "Crude Split Figured by Computer," *Hydrocarbon Proces.*, April 1969, pp. 143–148.
6. Katz, D., "Overview of Phase Behavior of Oil and Gas Production," *JPT*, June 1983, pp. 1205–1214.
7. Lee, S., et al., "Experimental and Theoretical Studies on the Fluid Properties Required for Simulation of Thermal Processes," Paper SPE 8393, presented at the 54th Annual Technical Conference of the SPE, held in Las Vegas, September 23–26, 1979.
8. Maddox, R. N. and Erbar, J. H., *Gas Conditioning and Processing, Vol. 3—Advanced Techniques and Applications*, Campbell Petroleum Series, Norman, Oklahoma, 1982.
9. Maddox, R. N. and Erbar, J. H., "Improve P-V-T Predictions," *Hydrocarbon Processing*, January, 1984, pp. 119–121.
10. Mehra, R., et al., "A Statistical Approach for Combining Reservoir Fluids into Pseudo Components for Compositional Model Studies," Paper SPE 11201, presented at the 57th Annual Meeting of the SPE, New Orleans, September 26–29, 1983.
11. Montel, F. and Gouel, P., "A New Lumping Scheme of Analytical Data for Composition Studies," Paper SPE 13119, presented at the 59th Annual SPE Technical Conference held in Houston, TX, September 16–19, 1984.
12. Pedersen, K., Thomassen, P., and Fredenslund, A., "Phase Equilibria and Separation Processes," Report SEP 8207, Inst. for Kemiteknik, Denmark Tekniske Hojskole (July 1982).
13. Schlijper, A. G., "Simulation of Compositional Process: The Use of Pseudo-Components in Equation of State Calculations," Paper SPE/DOE 12633, presented at the SPE/DOE 4th Symposium on EOR, held in Tulsa, Oklahoma, April 15–18, 1984.
14. Whitson, C., "Characterizing Hydrocarbon Plus Fractions," Paper EUR 183, presented at the European Offshore Petroleum Conference held in London, October 21–24, 1980.

8
Simulation of Laboratory PVT Data by Equations of State

Accurate laboratory studies of PVT and phase-equilibria behavior of reservoir fluids are necessary for characterizing these fluids and evaluating their volumetric performance at various pressure levels. There are many laboratory analyses that can be made on a reservoir fluid sample. The amount of data desired determines the number of tests performed in the laboratory. In general, there are three types of laboratory tests used to measure hydrocarbon reservoir samples.

1. Primary Tests: These routine tests involve the measurements of the specific gravity and the gas-oil ratio of the produced hydrocarbon fluids.
2. Detailed Laboratory Tests: These typical tests are performed to characterize the hydrocarbon fluid; they involve the measurement of the fluid composition, molecular weight, viscosity, compressibility, saturation pressure, formation volume factor, solubility, differential liberation, constant volume depletion, and constant composition expansion characteristics.
3. Swelling Tests: In addition to primary tests and detailed laboratory tests, which are fairly standard, another type of test may be performed for very specific applications. If a reservoir is to be depleted under a gas injection or dry gas cycling scheme, a swelling test should be performed.

Because of the relatively high cost of performing these experimental PVT tests, and the uncertainties in the accuracy of such laboratory measurements, equations of state offer an attractive approach for generating these necessary data. Equations of state, when properly "tuned," are capable of adequately simulating the PVT properties of reservoir fluids, and can consequently save significant time and expense by eliminating the need to perform a complete set of experimental PVT-type tests on each and every new reservoir fluid. In addition, equations of state can be used to check the quality of fluid samples collected, and relative accuracy of many laboratory PVT tests.

The objective of this chapter then is to briefly review some PVT experiments and to show, mathematically, how equations of state can be used to generate these laboratory measurements.

Constant-Volume Depletion Test

A reliable prediction of the pressure depletion performance of a gas-condensate reservoir is necessary in determining reserves and evaluating field separation methods. The predicted performance is also used in planning future operations and studying the economics of projects for increasing liquid recovery by gas cycling. Such predictions can be performed with aid of the experimental data collected from conducting constant volume depletion (CVD) tests on gas condensates. These tests are performed on a reservoir fluid sample in such a manner as to simulate pressure depletion of the actual reservoir, assuming that retrograde liquid appearing during production would remain *immobile* in the reservoir.

The CVD test provides five important laboratory measurements which can be used in a variety of reservoir engineering predictions. These measurements are:

a. Dew-point pressure
b. Composition changes of the gas phase with pressure depletion
c. Compressibility factor at reservoir pressure and temperature
d. Recovery of original in-place hydrocarbons at any pressure
e. Retrograde condensate accumulation, i.e., liquid saturation

The laboratory procedure of the CVD test (with immobile condensate), as shown schematically in Figure 8-1, is summarized in the following steps:

Step 1. A measured amount of a representative sample of the original reservoir fluid with a known overall composition of z_i is charged to a visual PVT cell at the dew-point pressure p_d (Figure 8-1a). The temperature of the PVT cell is maintained at the reservoir temperature T throughout the experiment. The initial volume V_i of the saturated fluid is used as a reference volume.

Step 2. The initial gas compressibility factor is calculated from the real gas equation

$$z_d = \frac{p_d \, V_i}{n_i \, RT} \tag{8-1}$$

where p_d = dew-point pressure, psia
V_i = initial gas volume, ft^3

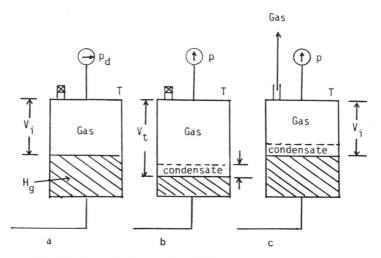

Figure 8-1. A schematic illustration of the constant volume depletion test.

$$n_i = \text{initial number of moles of the gas} = m/MW_a$$
$$MW_a = \text{apparent molecular weight, lb/lb-mole}$$
$$m = \text{mass of the initial gas in-place, lb}$$
$$R = \text{gas constant, 10.731}$$
$$T = \text{temperature, }^\circ R$$
$$z_d = \text{compressibility factor at dew-point pressure}$$

with the Gas Initially In Place as expressed in standard units by

$$GIIP = 379.4 \, n_i \qquad (8\text{-}2)$$

where GIIP = gas initially in place, scf
       n_i = initial number of moles

Step 3. The cell pressure is reduced from the saturation pressure to a pre-determined level P. This can be achieved by withdrawing mercury from the cell, as illustrated in Figure 8-1b. During the process, a second phase (retrograde liquid) is formed. The fluid in the cell is brought to equilibrium and the total volume V_t and volume of the retrograde liquid V_L are visually measured. This retrograde volume is reported as a percent of the initial volume V_i which basically represents the retrograde liquid saturation S_L:

$$S_L = \frac{V_L}{V_i} \, 100$$

Step 4. Mercury is reinjected into the PVT cell at constant pressure P while an equivalent volume of gas is simultaneously removed.

When the initial volume V_i is reached, mercury injection is ceased, as illustrated in Figure 8-1c. The volume of the removed gas is measured at the cell conditions and recorded as $(V_{gp})_{p,T}$. This step simulates a reservoir producing only gas, with retrograde liquid remaining immobile in the reservoir.

Step 5. The removed gas is charged to analytical equipment where its composition y_i is determined, and its volume is measured at standard conditions and recorded as $(V_{gp})_{sc}$. The corresponding moles of gas produced can be calculated from the expression

$$n_p = \frac{(V_{gp})_{sc}}{379.4} \tag{8-3}$$

where n_p = moles of gas produced
            $(V_{gp})_{sc}$ = volume of gas produced measured at standard conditions, scf

Step 6. The compressibility factor of the gas phase at cell pressure and temperature is calculated by invoking the following expression (see Equation 3-16 in Chapter 3)

$$Z = \frac{V_{actual}}{V_{ideal}} = \frac{(V_{gp})_{p,T}}{V_{ideal}} \tag{8-4}$$

where $V_{ideal} = \dfrac{RT\, n_p}{p}$

Combining Equations 8-3 and 8-4 and solving for the compressibility factor Z gives

$$Z = \frac{379.4\, P\, (V_{gp})_{p,T}}{RT\, (V_{gp})_{sc}} \tag{8-5}$$

Another property, the two-phase compressibility factor, is also calculated. The two-phase Z factor represents the total compressibility of all the remaining fluid (gas and retrograde liquid) in the cell and is computed from the real gas law as

$$Z \text{ (two-phase)} = \frac{P\, V_i}{(n_i - n_p)RT} \tag{8-6}$$

where $(n_i - n_p)$ = represents the remaining moles of fluid in the cell
       n_i = initial moles in the cell
       n_p = cumulative moles of gas removed

The two-phase Z factor is a significant property because it is used when the P/Z vs. cumulative-gas produced plot is constructed for evaluating gas-condensate production.

Step 7. The volume of gas produced as a percentage of gas initially in place is calculated by dividing the cumulative volume of the produced gas by the gas initially in place, both at standard conditions

$$\% \, G_p = \left[\frac{\Sigma \, (V_{gp})_{sc}}{GIIP} \right] 100 \tag{8-7}$$

or

$$\% \, G_p = \left[\frac{\Sigma \, n_p}{(n_i)_{original}} \right] 100$$

The above experimental procedure is repeated several times until a minimum test pressure is reached, after which the quantity and composition of the gas and retrograde liquid remaining in the cell are determined.

This test procedure can also be conducted on a volatile oil sample. In this case the PVT cell initially contains liquid, instead of gas, at its bubble-point pressure.

Simulation of the CVD Test

In the absence of the CVD test data on a specific gas-condensate system, predictions of pressure-depletion behavior can be obtained by using any of the well-established equations of state to compute the phase behavior when the composition of the total gas-condensate system is known. The stepwise computational procedure using the Peng-Robinson EOS as a representative equation of state is summarized below in conjunction with the flow diagram shown in Figure 8-2.

Step 1. Assume that the original hydrocarbon system with a total composition z_i occupies an initial volume of one cubic foot at the dew-point pressure p_d and system temperature T:

$$V_i = 1$$

Step 2. Calculate the gas compressibility factor z_d from Equation 6-72.
Step 3. Calculate the initial moles in place by applying the real gas law

$$n_i = \frac{(1)(p_d)}{z_d \, RT} \tag{8-8}$$

where p_d = dew-point pressure, psi
z_d = gas compressibility factor at the dew-point pressure

Step 4. Reduce the pressure to a predetermined value P. At this pressure level, the equilibrium ratios (K-values) are calculated (as outlined previously in Chapter 6) and used in performing flash calculations. The calculated numerical results include:
- Equilibrium ratios (K-values)
- Composition of the liquid phase (i.e., retrograde liquid), x_i

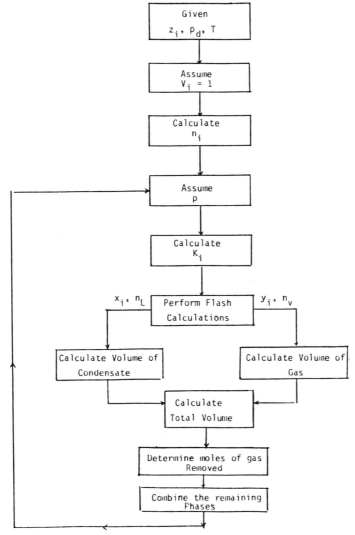

Figure 8-2. Flow diagram for simulating the constant volume depletion test.

- Moles of the liquid phase, n_L
- Composition of the gas phase, y_i
- Moles of the gas phase, n_v
- Compressibility factor of the liquid phase, Z^L
- Compressibility factor of the gas phase, Z^v

The composition of the gas phase y_i should reasonably match the experimental gas composition at pressure P.

Step 5. Because flash calculations are usually performed assuming the total moles are equal to one, calculate the actual moles of the liquid and gas phases from

$$(n_L)_{actual} = n_i \, n_L \tag{8-9}$$

$$(n_v)_{actual} = n_i \, n_v \tag{8-10}$$

where n_L and n_v are moles of liquid and gas, respectively, as determined from flash calculations.

Step 6. Calculate the volume of each hydrocarbon phase by applying the expression

$$V_L = \frac{(n_L)_{actual} \, Z^L \, R \, T}{P} \tag{8-11}$$

$$V_g = \frac{(n_v)_{actual} \, Z^v \, R \, T}{P} \tag{8-12}$$

where V_L = volume of the retrograde liquid, ft^3/ft^3
V_g = volume of the gas phase, ft^3/ft^3
Z^L, Z^v = compressibility factors of the liquid and gas phase
T = cell (reservoir) temperature, °R

Since $V_i = 1$, then

$$S_L = (V_L)100 \tag{8-13}$$

This value should match the experimental value if the equation of state is properly tuned.

Step 7. Calculate the total volume of fluid in the cell

$$V_t = V_L + V_g \tag{8-14}$$

where V_t = total volume of the fluid, ft^3

Step 8. Since the volume of the cell is constant at 1 ft^3, remove the following excess gas volume from the cell

$$(V_{gp})_{P,T} = V_t - 1 \tag{8-15}$$

Step 9. Calculate the number of moles of gas removed

$$n_p = \frac{p \ (V_{gp})_{p,T}}{Z^v \ R \ T} \tag{8-16}$$

Step 10. Calculate cumulative gas produced as a percentage of gas initially in place by dividing cumulative moles of gas removed, $\Sigma \ n_p$, by the original moles in place, or

$$\% \ G_p = \left[\frac{\Sigma \ n_p}{(n_i)_{original}} \right] 100$$

Step 11. Calculate the two-phase gas deviation factor from the relationship

$$Z \ (\text{two-phase}) = \frac{(P)(1)}{(n_i - n_p) \ RT} \tag{8-17}$$

Step 12. Calculate the remaining moles of gas $(n_v)_r$ by subtracting the moles produced n_p from the actual number of moles of the gas phase $(n_v)_{actual}$

$$(n_v)_r = (n_v)_{actual} - n_p \tag{8-18}$$

Step 13. Calculate the new total moles and new composition remaining in the cell by applying the molal and component balances, respectively

$$n_i = (n_L)_{actual} + (n_v)_r \tag{8-19}$$

$$z_i = \frac{x_i \ (n_L)_{actual} + y_i \ (n_v)_r}{n_i} \tag{8-20}$$

Step 14. Consider a new lower pressure and repeat Steps 4 through 13.

Constant Composition Expansion Test

Constant composition expansion experiments, commonly called pressure-volume tests, are performed on gas condensates or crude oil to simulate the pressure-volume relations of these hydrocarbon systems. The test is conducted for the purposes of determining

a. Saturation pressure (bubble-point or dew-point pressure)
b. Isothermal compressibility coefficients of the single-phase fluid in excess of saturation pressure
c. Compressibility factors of the gas phase
d. Total hydrocarbon volume as a function of pressure

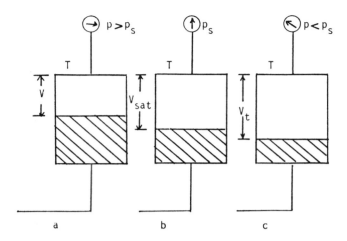

Figure 8-3. A schematic illustration of the constant composition expansion test.

The experimental procedure, as shown schematically in Figure 8-3, involves placing a hydrocarbon fluid sample (oil or gas) in a visual PVT cell at reservoir temperature and at a pressure in excess of the initial reservoir pressure (Figure 8-3a). The pressure is reduced in steps at constant temperature by removing mercury from the cell, and the change in the hydrocarbon volume V is measured for each pressure increment. The saturation pressure (bubble-point or dew-point pressure) and the corresponding volume are observed and recorded (Figure 8-3b). The volume at the saturation pressure is used as a reference volume. At pressure levels higher than the saturation pressure, the volume of the hydrocarbon system is recorded as a ratio of the reference volume. This volume is commonly termed the relative volume and is expressed mathematically by the following equation:

$$V_{rel} = \frac{V}{V_{sat}} \tag{8-21}$$

where V_{rel} = relative volume
V = volume of the hydrocarbon system
V_{sat} = volume at the saturation pressure

Also above the saturation pressure, the isothermal compressibility coefficient of the single phase fluid is usually determined from the expression

$$C = -\frac{1}{V_{rel}} \left[\frac{\partial V_{rel}}{\partial p} \right]_T \tag{8-22}$$

where C = isothermal compressibility coefficient, psi^{-1}

For gas-condensate systems, the gas compressibility factor Z is determined in addition to the above experimentally determined properties.

Below the saturation pressure, the two-phase volume V_t is measured relative to the volume at saturation pressure and expressed as

$$\text{Relative total volume} = \frac{V_t}{V_{sat}} \qquad (8\text{-}23)$$

where V_t = total hydrocarbon volume

It should be noted that no hydrocarbon material is removed from the cell, thus, the composition of the total hydrocarbon mixture in the cell remains fixed at the original composition.

Simulation of the Constant Composition Expansion Test

The simulation procedure utilizing the Peng-Robinson equation of state is illustrated in the following steps.

Step 1. Given the total composition of the hydrocarbon system z_i and saturation pressure (p_b for oil systems; p_d for gas systems), calculate the total volume occupied by one mole of the system. This volume corresponds to the reference volume V_{sat} (volume at the saturation pressure). Mathematically, the volume is calculated from the relationship

$$V_{sat} = \frac{(1)\ Z\ R\ T}{p_{sat}} \qquad (8\text{-}24)$$

where V_{sat} = volume of saturation pressure, ft^3/mole
p_{sat} = saturation pressure (dew-point or bubble-point pressure), psia
T = system temperature, °R
Z = compressibility factor Z^L or Z^v depending on the type of system, calculated by applying Equation 6-72.

Step 2. The pressure is increased in steps above the saturation pressure. At each pressure, the compressibility factor Z^L or Z^v is calculated by solving Equation 6-72, and used to determine the fluid volume

$$V = \frac{(1)\ Z\ R\ T}{p}$$

where V = compressed liquid or gas volume at the pressure level
p, ft^3/mole
Z = compressibility factor of the compressed liquid or gas
p = system pressure, $p > p_{sat}$, psia
R = gas constant

The corresponding relative phase volume V_{rel} is calculated from
the expression

$$V_{rel} = \frac{V}{V_{sat}}$$

Step 3. The pressure is then reduced in steps below the saturation pressure p_{sat}. The equilibrium ratios are calculated and flash calculations are performed at each pressure level. The resulting data include K_i, x_i, y_i, n_L, n_v, Z^v, and Z^L. Since no hydrocarbon material is removed during pressure depletion, the original moles ($n_i = 1$) and composition z_i remain constant. The volumes of the liquid and gas phases can then be calculated from the expressions

$$V_L = \frac{(1) \ (n_L) \ Z^L \ RT}{p} \tag{8-25}$$

$$V_g = \frac{(1) \ (n_v) \ Z^v \ RT}{p} \tag{8-26}$$

and

$$V_t = V_L + V_g$$

where n_L, n_v = moles of liquid and gas as calculated from flash
calculations
Z^L, Z^v = compressibility factors of liquid and gas as calculated by solving Equation 6-73
V_t = total volume of the hydrocarbon system

Step 4. Calculate the relative total volume from the following expression.

$$\text{Relative total volume} = \frac{V_t}{V_{sat}}$$

Differential Liberation Test

The laboratory procedure of performing the differential liberation test is described previously in Chapter 4. The test is carried out on reservoir oil samples and involves charging a visual PVT cell with a liquid sample at the bubble-point pressure and reservoir temperature. The pressure is reduced in steps, usually 10 to 15 pressure levels, and all the liberated gas is removed and its volume G_p is measured at standard conditions. The volume of oil remaining V_L is also measured at each pressure level. It should be noted that the remaining oil is subjected to continual compositional changes as it becomes progressively richer in the heavier components.

The above procedure is continued to atmospheric pressure where the volume of the residual (remaining) oil is measured and converted to a volume at 60°F, V_{sc}. The differential oil formation volume factors B_{od} (commonly called the relative oil volume factors) at all the various pressure levels are calculated by dividing the recorded oil volumes V_L by the volume of residual oil V_{sc}, or

$$B_{od} = \frac{V_L}{V_{sc}} \qquad (8\text{-}27)$$

The differential solution gas-oil ratio R_{sd} is also calculated by dividing the volume of gas in solution by the residual oil volume.

Typical laboratory results of the test are shown in Table 8-1 and Figures 8-4 and 8-5.

Moses (1986), in an excellent discussion of the phase behavior of crude oil and condensate systems, stated that reporting the experimental data relative to the residual oil volume at 60°F (as shown graphically in Figure 8-4) gives

Table 8-1
Differential Vaporization at 200°F

Pressure (psig)	Solution GOR,* R_{sd}	Relative Oil Volume, B_{od}**
2,620	854	1.600
2,350	763	1.554
2,100	684	1.515
1,850	612	1.479
1,600	544	1.445
1,350	479	1.412
1,100	416	1.382
850	354	1.351
600	292	1.320
350	223	1.283
159	157	1.244
0	0	1.075
		at 60°F = 1.000

* Cubic feet of gas at 14.65 psia and 60°F per barrel of residual oil at 60°F.
** Barrels of oil at indicated pressure and temperature per barrel of residual oil at 60°F.
(Permission to publish by the Society of Petroleum Engineers of AIME. Copyright SPE-AIME.)

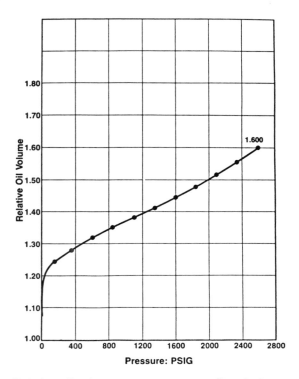

Figure 8-4. Relative oil volume versus pressure. Permission to publish by the Society of Petroleum Engineers of AIME. Copyright SPE-AIME.

the relative-oil volume curve the appearance of the formation volume factor curve, leading to its misuse in reservoir calculations. Moses suggested that a better method of reporting these data is in the form of a shrinkage curve, as shown in Figure 8-6. The relative oil volume data in Figure 8-4 and Table 8-1 can be converted to a shrinkage curve by dividing each relative oil volume factor B_{od} by the relative oil volume factor at the bubble-point B_{odb}, or

$$S_{od} = \frac{B_{od}}{B_{odb}} \qquad (8\text{-}28)$$

where B_{od} = differential relative oil volume factor at pressure p, bbl/STB
  B_{odb} = differential relative oil volume factor at the bubble-point pressure p_b, psia, bbl/STB
  S_{od} = differential oil shrinkage factor, bbl/bbl of bubble-point oil

The shrinkage curve has a value of one at the bubble-point and a value less than one at subsequent pressures below p_b. In this suggested form, the

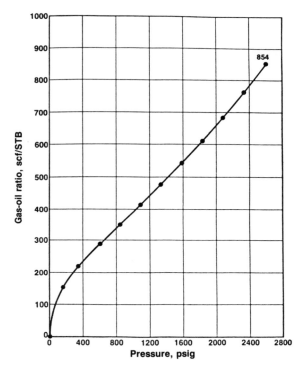

Figure 8-5. Gas-oil ratio versus pressure. Permission to publish by the Society of Petroleum Engineers of AIME. Copyright SPE-AIME.

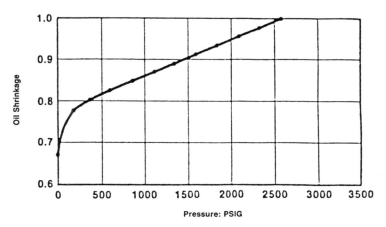

Figure 8-6. Oil shrinkage curve. Permission to publish by the Society of Petroleum Engineers of AIME. Copyright SPE-AIME.

shrinkage curve describes the changes in the volume of the bubble-point oil as reservoir pressure declines.

The results of the differential liberation test when combined with the results of the flash separation test provide a means of calculating the oil formation volume factors and gas solubilities as a function of reservoir pressure. These calculations are outlined in the next section.

Simulation of the Differential Liberation Test

The simulation procedure of the test by the Peng-Robinson EOS is summarized in the following steps and in conjunction with the flow diagram shown in Figure 8-7.

Step 1. Starting with the saturation pressure p_{sat} and reservoir temperature T, calculate the volume occupied by a total of one mole, i.e., $n_i = 1$, of the hydrocarbon system with an overall composition of z_i. This volume, V_{sat}, is calculated by applying Equation 8-24.

Step 2. Reduce the pressure to a predetermined value of p at which the equilibrium ratios are calculated and used in performing flash calculations. The actual number of moles of the liquid phase, with a composition of x_i, and the actual number of moles of the gas phase, with a composition of y_i, are then calculated from the expressions

$$(n_L)_{actual} = n_i \, n_L$$
$$(n_v)_{actual} = n_i \, n_v$$

where $(n_L)_{actual}, (n_v)_{actual}$ = actual number of moles of the liquid and gas phases, respectively

n_L, n_v = number of moles of the liquid and gas phases as computed by performing flash calculations

Step 3. Determine the volume of the liquid and gas phase from

$$V_L = \frac{Z^L \, RT \, (n_L)_{actual}}{p} \tag{8-29}$$

$$V_g = \frac{Z \, RT \, (n_v)_{actual}}{p} \tag{8-30}$$

where V_L, V_g = volumes of the liquid and gas phases, ft³

Z^L, Z^v = compressibility factors of the liquid and gas phases

The volume of the produced gas measured at standard conditions is also determined from the relationship

$$G_p = 379.4 \ (n_v)_{actual} \tag{8-31}$$

where G_p = gas produced during depletion pressure p, scf

The total cumulative gas produced, in scf, should also be calculated from the expression

$$(G_p)_t = \Sigma \ G_p$$

where $(G_p)_t$ is the total cumulative gas produced.

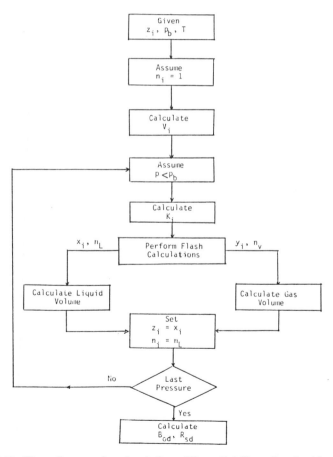

Figure 8-7. Flow diagram for simulating differential liberation test by an equation of state.

Step 4. Assume that all the equilibrium gas is removed from contact with the oil. This can be mathematically achieved by setting the overall composition z_i equal to the composition of the liquid phase x_i and setting the total moles equal to the moles of the liquid phase.

$$z_i = x_i$$
$$n_i = (n_L)_{actual}$$

Step 5. Using the new overall composition and total moles, Steps 2 through 4 are repeated. When the depletion pressure reaches the atmospheric pressure, the temperature is changed to 60°F and the residual oil volume is calculated.

Step 6. The calculated volumes of the oil and removed gas are then divided by the residual oil volume to calculate the relative oil volumes (B_{od}) and the solution GOR, or

$$B_{od} = \frac{V_L}{V_{sat}}$$ (8-32)

$$R_{sd} = \frac{(5.615)(\text{volume of gas in solution})}{V_{sat}}$$

Flash Separation Tests

Flash separation tests, commonly called separator tests, are conducted to determine the changes in the volumetric behavior of the reservoir fluid as the fluid passes through the separator (or separators) and then into the stock-tank. The resulting volumetric behavior is influenced to a large extent by the operating conditions, i.e., pressures and temperatures of the surface separation facilities. The primary objective of conducting separator tests, therefore, is to provide the essential laboratory information necessary for determining the optimum surface separation conditions, which in turn, will maximize the stock-tank oil production. In addition, the results of the test, when appropriately combined with the differential liberation test data, provide a means of obtaining the PVT parameters (B_o, R_s, and B_t) required for petroleum engineering calculations. The laboratory procedure of the flash separation test involves the following steps.

Step 1. The oil sample is charged into a PVT cell at reservoir temperature and its bubble-point pressure.

Step 2. A small amount of the oil sample, with a volume of $(V_o)_{pb}$, is removed from the PVT cell at constant pressure and is flashed through a multi-stage separator system, with each separator at fixed pressure and temperature. The gas liberated from each stage (separator) is removed and its volume and specific gravity are measured. The volume of the remaining oil in the last stage (stock-tank condition) is recorded as $(V_o)_{st}$.

Step 3. The total solution gas-oil ratio and the oil formation volume factor at the bubble-point pressure are then calculated

$$B_{ofb} = (V_o)_{pb}/(V_o)_{st} \tag{8-33}$$
$$R_{sfb} = (V_g)_{sc}/(V_o)_{st} \tag{8-34}$$

where B_{ofb} = bubble-point oil formation volume factor, as measured by flash liberation, bbl of the bubble-point oil/STB

B_{sfb} = bubble-point solution gas-oil ratio as measured by flash liberation, scf/STB

$(V_g)_{sc}$ = total volume of gas removed from separators, scf

Steps 1 through 3 are repeated at series of different separator pressures and at a fixed temperature.

A typical example of a set of separator tests for a two-stage separation system, as reported by Moses (1986), are shown in Table 8-2. By examining the laboratory results reported in Table 8-2, it should be noted that the optimum separator pressure is 100 psia, considered to be the separator pressure that results in the minimum oil formation volume factor.

Table 8-2
Separator Tests

Separator Pressure (psig)	Temperature (°F)	GOR, R_{sfb}*	Stock-Tank Oil Gravity (°API at 60°F)	FVF, B_{ofb}**
50	75	737		
to 0	75	41	40.5	1.481
		778		
100	75	676		
to 0	75	92	40.7	1.474
		768		
200	75	602		
to 0	75	178	40.4	1.483
		780		
300	75	549		
to 0	75	246	40.1	1.495
		795		

* GOR in cubic feet of gas at 14.65 psia and 60°F per barrel of stock-tank oil at 60°F.
** FVF in barrels of saturated oil at 2,620 psig and 220°F per barrel of stock-tank oil at 60°F.
(Permission to publish by the Society of Petroleum Engineers of AIME. Copyright SPE-AIME.)

Amyx et al. (1960) and Dake (1978) proposed a procedure for constructing the oil formation volume factor and gas solubility curves by using the differential liberation data (as shown in Table 8-1) in conjunction with the experimental separator flash data (as shown in Table 8-2) for a given set of separator conditions. The procedure calls for multiplying the flash oil formation factor at the bubble-point B_{ofb} (as defined by Equation 8-33) by the differential oil shrinkage factor S_{od} (as defined by Equation 8-28) at various reservoir pressures. Mathematically, this relationship is expressed as follows

$$B_o = B_{ofb} \, S_{od} \tag{8-35}$$

where B_o = oil formation volume factor, bbl/STB
  B_{ofb} = bubble-point oil formation volume factor, bbl of the bubble-point oil/STB
  S_{od} = differential oil shrinkage factor, bbl/bbl of bubble-point oil

Amyx and Dake proposed the following expression for adjusting the differential gas solubility data R_{sd} to give the required gas solubility factor R_s

$$R_s = R_{sfb} - (R_{sdb} - R_{sd}) \, \frac{B_{ofb}}{B_{odb}} \tag{8-36}$$

where R_s = gas solubility, scf/STB
  R_{sfb} = bubble-point solution gas-oil ratio as defined by Equation 8-34, scf/STB
  R_{sdb} = solution gas-oil at the bubble-point pressure as measured by the differential liberation test, scf/STB
  R_{sd} = solution gas-oil ratio at various pressure levels as measured by the differential liberation test, scf/STB

These adjustments will typically produce lower formation volume factors and gas solubilities than the differential liberation data.

The above adjustment procedure is illustrated graphically in Figure 8-8 and Figure 8-9.

Example 5-7 in Chapter 5 shows how the separator test can be simulated mathematically by applying the concept of equilibrium ratios.

Composite Liberation Test

The laboratory procedures for conducting the differential liberation and flash liberation tests for the purpose of generating B_o and R_s versus p relationships are considered approximations to the actual relationships. Another

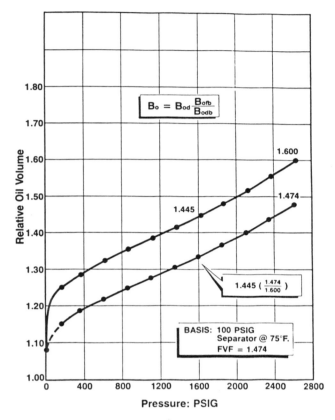

Figure 8-8. Adjustment of oil-volume curve to separator conditions. Permission to publish by the Society of Petroleum Engineers of AIME. Copyright SPE-AIME.

type of test, called a composite liberation test, has been suggested by Dodson et al. (1953) and represents a combination of differential and flash liberation processes. The laboratory test, which is commonly called the Dodson test, provides a better means of describing the PVT relationships. The experimental procedure for this composite liberation, as proposed by Dodson, is summarized in the following steps.

Step 1. A large representative fluid sample is placed into a cell at a pressure higher than the bubble-point pressure of the sample. The temperature of the cell is then raised to reservoir temperature.

Step 2. The pressure is reduced in steps by removing mercury from the cell, and the change in the oil volume is recorded. The process is repeated until the bubble point of the hydrocarbon is reached.

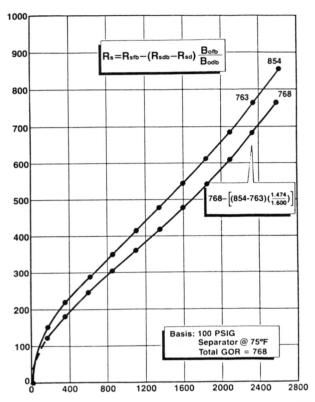

Figure 8-9. Adjustment of gas-in-solution curve to separator conditions. Permission to publish by the Society of Petroleum Engineers of AIME. Copyright SPE-AIME.

Step 3. A carefully measured small volume of oil is removed from the cell at constant pressure and then is flashed at temperatures and pressures equal to those in the surface separators and stock-tank. The liberated gas volume and stock-tank oil volume are measured. The oil formation volume factor B_o and gas solubility R_s are then calculated from the measured volumes.

$$B_o = (V_o)_{p,T}/(V_o)_{st} \qquad (8\text{-}37)$$
$$R_s = (V_g)_{sc}/(V_o)_{st} \qquad (8\text{-}38)$$

where $(V_o)_{p,T}$ = volume of oil removed from the cell at constant pressure p

$(V_g)_{sc}$ = total volume of the liberated gas as measured under standard conditions

$(V_o)_{st}$ = volume of the stock-tank oil

Step 4. The volume of the oil remaining in the cell is allowed to expand through a pressure decrement and the gas evolved is then removed as in the differential liberation.

Step 5. Following the gas removal, Step 3 is then repeated and B_o and R_s are calculated.

Step 6. Steps 3 through 5 are repeated at several progressively lower reservoir pressures to secure a complete PVT relationship.

It should be noted that this type of test, while more accurately representing the PVT behavior of complex hydrocarbon systems, is more difficult and costly to perform than other liberation tests. Consequently, these experiments are not usually included in a routine fluid property analysis.

Simulation of the Composite Liberation Test

The stepwise computational procedure for simulating the composite liberation test by using the Peng-Robinson EOS is summarized below in conjunction with the flow diagram shown in Figure 8-10.

Step 1. Assume a large reservoir fluid sample with an overall composition of z_i is placed in a cell at the bubble-point pressure p_b and reservoir temperature T.

Step 2. Remove one lb-mole, i.e., $n_i = 1$, of the liquid from the cell and calculate the corresponding volume at p_b and T by applying the Peng-Robinson EOS to estimate the liquid compressibility factor Z^L (Equation 6-73).

$$(V_o)_{p,T} = \frac{(1)\ Z^L\ RT}{p_b} \qquad (8\text{-}39)$$

where $(V_o)_{p,T}$ = volume of one mole of the oil, ft^3

Step 3. Flash the resulting volume at temperatures and pressures equal to those in the surface separators and stock-tank. The procedure was outlined previously in Chapter 5 and Example 5-7. Designating the resulting total liberated gas volume as $(V_g)_{sc}$ and the volume of the stock-tank oil as $(V_o)_{st}$, calculate the B_o and R_s by applying Equations 8-37 and 8-38, respectively.

Step 4. Set the cell pressure to a lower pressure level p and using the original composition z_i, generate the K_i-values through the application of the P-R EOS.

Step 5. Perform flash calculations based on z_i and the calculated K_i-values.

Step 6. To simulate the differential liberation step of removing the equilibrium liberated gas from the cell at constant pressure, simply set the overall composition z_i equal to the mole fraction of the equilibrium liquid x_i and the bubble-point pressure p_b equal to the new pressure level p. Mathematically, this step is summarized by the following relationships:

$z_i = x_i$
for all components, and
$p_b = p$

Step 7. Repeat Steps 2 through 6.

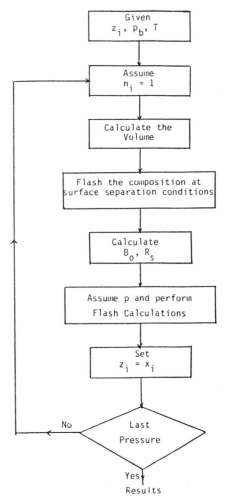

Figure 8-10. Flow diagram for simulating the composite liberation test by an EOS.

Swelling Test

Swelling tests should be performed if a reservoir is to be depleted under gas injection or a dry gas cycling scheme. The swelling test can be conducted on gas-condensate or crude oil samples. The purpose of this laboratory experiment is to determine the degree to which the proposed injection gas will dissolve in the hydrocarbon mixture. The data which can be obtained during a test of this type include

- The relationship of saturation pressure and volume of gas injected
- The volume of the saturated fluid mixture compared to the volume of the original saturated fluid

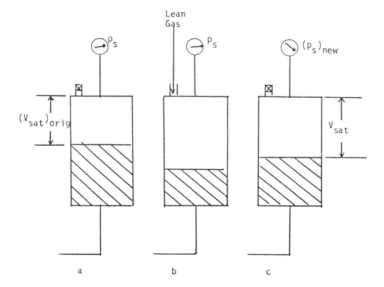

Figure 8-11. A schematic illustration of the swelling test.

The test procedure, as shown schematically in Figure 8-11, is summarized in the following steps:

Step 1. A representative sample of the hydrocarbon mixture with a known overall composition z_i is charged to a visual PVT cell at saturation pressure and reservoir temperature. The volume at the saturation pressure is recorded and designated as $(V_{sat})_{orig.}$. This step is illustrated schematically in Figure 8-11a.

Step 2. A predetermined volume of gas with a composition similar to the proposed injection gas is added to the hydrocarbon mixture. The cell is pressured up until only one phase is present, as shown in Figure 8-11b. The new saturation pressure and volume are recorded and designated as p_s and V_{sat}, respectively. The original saturation volume is used as a reference value and the results are presented as relative total volumes:

$$V_{rel} = V_{sat}/(V_{sat})_{orig.} \qquad (8\text{-}40)$$

where V_{rel} = relative total volume
            $(V_{sat})_{orig.}$ = original saturation volume

Step 3. Repeat Step 2 until the mole percent of the injected gas in the fluid sample reaches a preset value (around 80%).

A typical laboratory result of a gas-condensate swelling test with lean gas of a composition given in Table 8-3 is shown numerically in Table 8-4 and graphically in Figures 8-12 and 8-13. The dew-point pressure behavior, as a function of cumulative lean gas injected, is shown in Table 8-4. It should be noted by examining Table 8-4 that the injection of lean gas into the reservoir

Table 8-3
Composition of the Lean Gas*

Component	y_i
CO_2	0
N_2	0
C_1	0.9468
C_2	0.0527
C_3	0.0005

* Kenyon, D. and Behie, G. "Third SPE Comparative Solution Project: Gas Cycling of Retrograde Condensate Reservoirs," *JPT*, August, 1987, pp. 981–999.

Table 8-4
Solubility and Swelling Test* at 200°F

Cumulative Gas Injected scf/STB	Swollen Volume bbl/bbl	Dew-point Pressure, psig
0	1.0000	3,428
190	1.1224	3,635
572	1.3542	4,015
1,523	1.9248	4,610
2,467	2.5043	4,880

* Kenyon, D. and Behie, G. "Third SPE Comparative Solution Project: Gas Cycling of Retrograde Condensate Reservoirs," *JPT*, August, 1987, pp. 981–999.

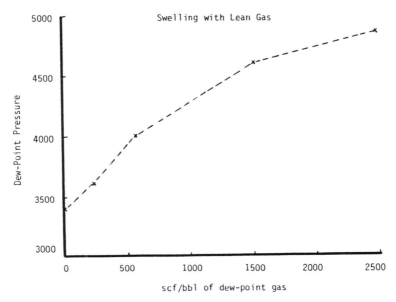

Figure 8-12. Dew-point pressure during swelling of reservoir gas with lean gas at 200°F.

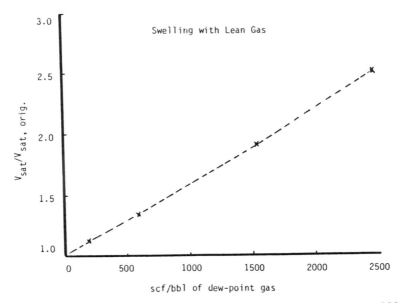

Figure 8-13. Relative volume in swelling of reservoir gas with lean gas at 200°F.

fluid caused the dew-point pressure of the mixture to increase above the original dew-point of the reservoir fluid. Each addition of injection gas caused the dew-point pressure and the relative volume (swollen volume) to increase. The dew-point increased from 3,428 to 4,880 psig and relative total volume from 1 to 2.5043 after the final addition.

Simulation of the Swelling Test

The simulation procedure of the swelling test by the P-R EOS is summarized in the following steps.

Step 1. Assume the hydrocarbon system occupies a total volume of 1 bbl at the saturation pressure p_s and reservoir temperature T. Calculate the initial moles of the hydrocarbon system from the following expression

$$n_i = \frac{5.615\ p_s}{Z\ R\ T} \tag{8-41}$$

where n_i = initial moles of the hydrocarbon system

p_s = saturation pressure, i.e., p_b or p_d, depending on the type of the hydrocarbon system, psia

Z = gas or liquid compressibility factor

Step 2. Given the composition y_{inj} of the proposed injection gas, add a predetermined volume (as measured in scf) of the injection gas to the original hydrocarbon system and calculate the new overall composition by applying the molal and component balances, respectively

$$n_t = n_i + n_{inj} \tag{8-42}$$

$$z_i = \frac{y_{inj}\ n_{inj} + (Z_{sat})_i\ n_i}{n_t} \tag{8-43}$$

with

$$n_{inj} = \frac{V_{inj}}{379.4}$$

where n_{inj} = total moles of the injection gas

V_{inj} = volume of the injection gas, scf

$(Z_{sat})_i$ = mole fraction of component i in the saturated hydrocarbon system

Step 3. Using the new composition z_i, calculate the new saturation pressure, i.e., p_b or p_d, by using the Peng-Robinson EOS as outlined in Chapter 6.

Step 4. Calculate the relative total volume (swollen volume) by applying the relationship

$$V_{rel} = \frac{Z \, n_t \, RT}{5.615 \, p_s}$$

where V_{rel} = swollen volume
       p_s = saturation pressure

Step 5. Steps 2 through 4 are repeated until the last predetermined volume of the lean gas is combined with the original hydrocarbon system.

REFERENCES

1. Amyx, J. M., Bass, D. M., and Whiting, R., *Petroleum Reservoir Engineering-Physical Properties*, New York: McGraw-Hill Book Company, 1960.
2. Dake, L. P., *Fundamentals of Reservoir Engineering*, Amsterdam: Elsevier Scientific Publishing Company, 1978.
3. Dodson, C., Goodwill, D., and Mayer, E., "Application of Laboratory PVT Data to Reservoir Engineering Problems," *JPT*, December 1953, pp. 287–298.
4. Moses, P., "Engineering Application of Phase Behavior of Crude Oil and Condensate Systems," *JPT*, July 1986, pp. 715–723.

APPENDIX
Equilibrium Ratio Curves

PRESSURE, PSIA ⟶

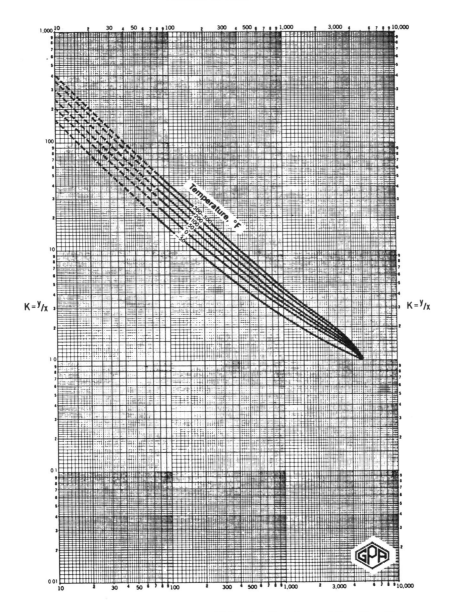

Figure 1. Methane. Conv. press. 5,000 psia. Courtesy of the Gas Processors Suppliers Association. Published in the GPSA Engineering Data Book, Tenth Edition, 1987.

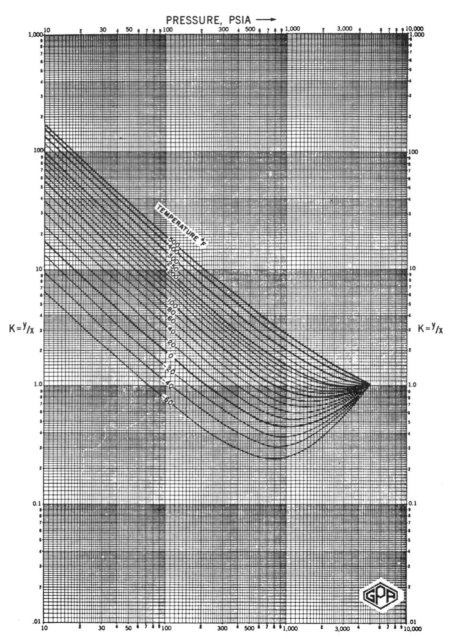

Figure 2. Ethane. Conv. press. 5,000 psia. Courtesy of the Gas Processors Suppliers Association. Published in the GPSA Engineering Data Book, Tenth Edition, 1987.

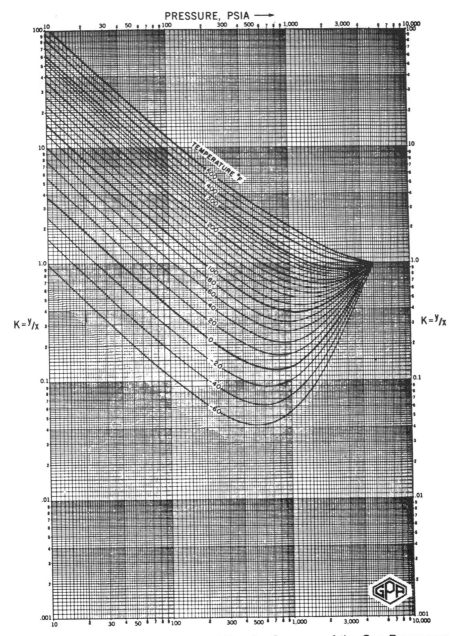

Figure 3. Propane. Conv. press. 5,000 psia. Courtesy of the Gas Processors Suppliers Association. Published in the GPSA Engineering Data Book, Tenth Edition, 1987.

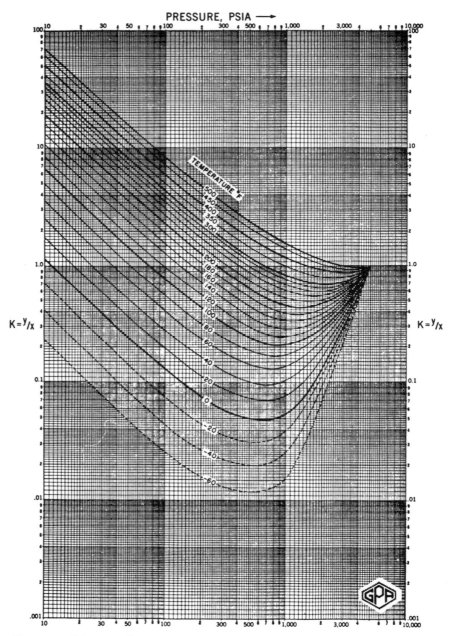

Figure 4. i-Butane. Conv. press. 5,000 psia. Courtesy of the Gas Processors Suppliers Association. Published in the GPSA Engineering Data Book, Tenth Edition, 1987.

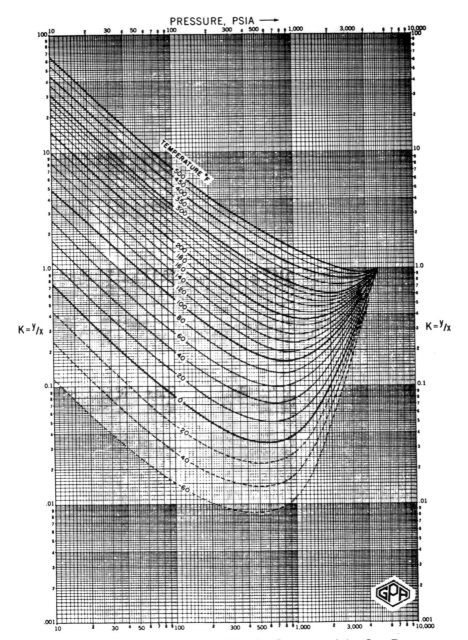

Figure 5. n-Butane. Conv. press. 5,000 psia. Courtesy of the Gas Processors Suppliers Association. Published in the GPSA Engineering Data Book, Tenth Edition, 1987.

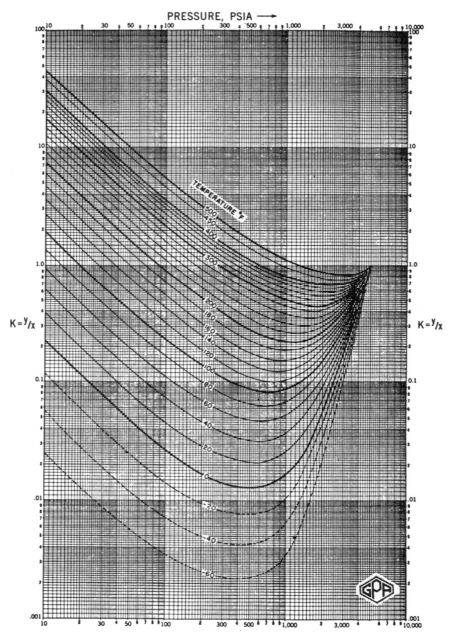

Figure 6. i-Pentane. Conv. press. 5,000 psia. Courtesy of the Gas Processors Suppliers Association. Published in the GPSA Engineering Data Book, Tenth Edition, 1987.

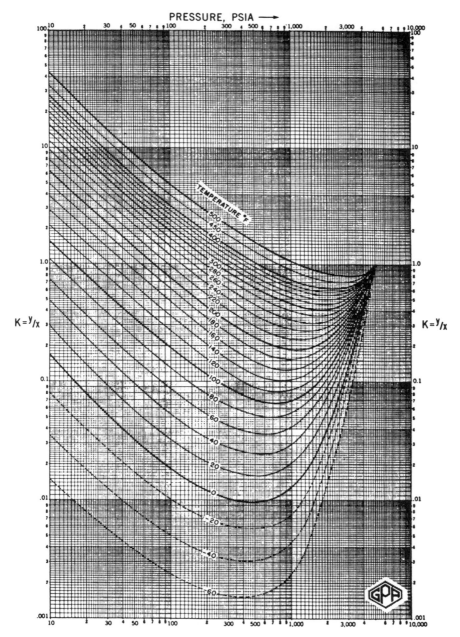

Figure 7. n-Pentane. Conv. press. 5,000 psia. Courtesy of the Gas Processors Suppliers Association. Published in the GPSA Engineering Data Book, Tenth Edition, 1987.

PRESSURE, PSIA ⟶

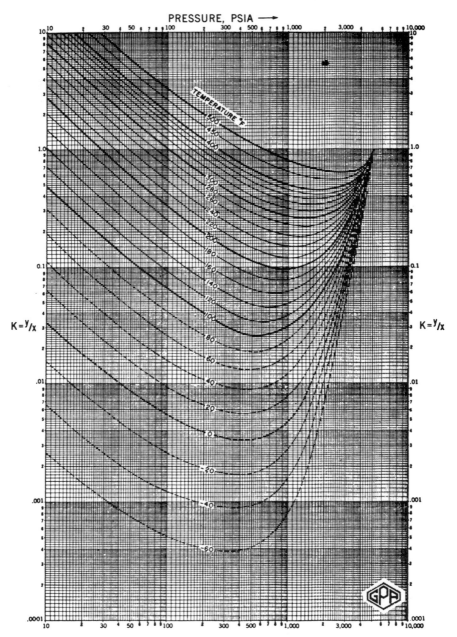

$K = {}^y/x$

$K = {}^y/x$

Figure 8. Hexane. Conv. press. 5,000 psia. Courtesy of the Gas Processors Suppliers Association. Published in the GPSA Engineering Data Book, Tenth Edition, 1987.

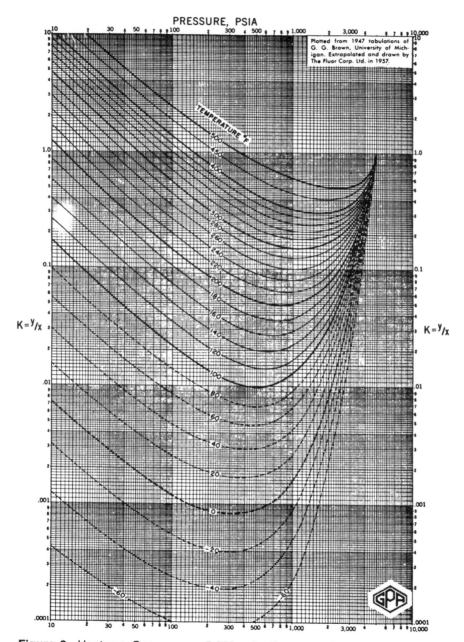

Figure 9. Heptane. Conv. press. 5,000 psia. Courtesy of the Gas Processors Suppliers Association. Published in the GPSA Engineering Data Book, Tenth Edition, 1987.

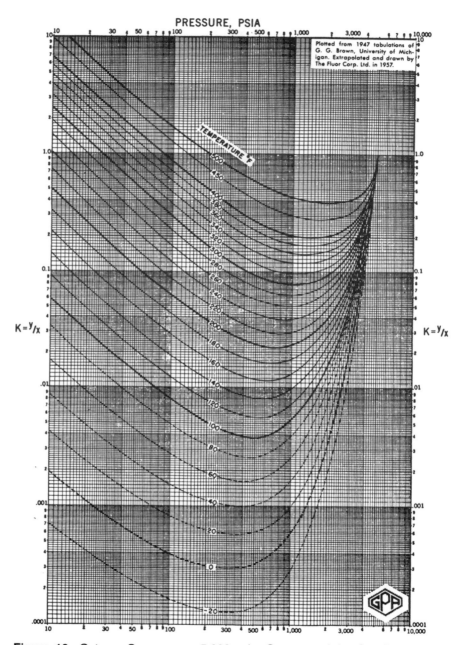

Figure 10. Octane. Conv. press. 5,000 psia. Courtesy of the Gas Processors Suppliers Association. Published in the GPSA Engineering Data Book, Tenth Edition, 1987.

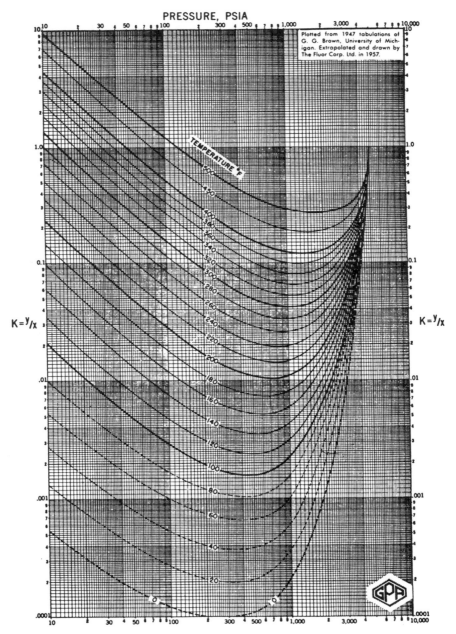

Figure 11. Nonane. Conv. press. 5,000 psia. Courtesy of the Gas Processors Suppliers Association. Published in the GPSA Engineering Data Book, Tenth Edition, 1987.

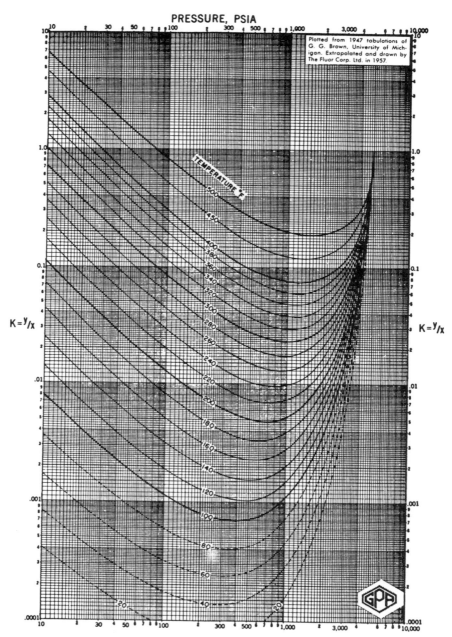

Figure 12. Decane. Conv. press. 5,000 psia. Courtesy of the Gas Processors Suppliers Association. Published in the GPSA Engineering Data Book, Tenth Edition, 1987.

Index

Dodson's (composite) liberation test,
396–400
Dropout, 30, 379–385

E

Equation of state, 287–344
  applications of, 331–343
  definition of, 287
  equilibrium ratio from, 331
  types of
    Heyen, 318
    Kubic, 320
    Patel-Teja, 322
    Peng-Robinson, 310
    Redlich-Kwong, 294
    Schmidt-Wenzel, 325
    Soave-Redlich-Kwong, 298
    Van der Waals, 287
    Yu-Lu, 328
Equilibrium ratio
  charts, 407–418
  definition of, 224
  from equation of state, 306, 331
  methods of determining
    convergence pressure
      approach, 254
    Standing, 251
    Whitson-Torp, 259
    Wilson, 250
    Winn, 360
  of the plus-fraction, 263–270
  three-phase, 339

F

Flash calculations, 247, 275
Flash separation test, 275, 394–396
Formation volume factor
  of gas, 111
  of oil, 183–193
  total, 193–199
  of water, 203

Fugacity, 304
Fugacity coefficient, 305

G

Gas
  compressibility (deviation) factor,
    85, 90–106
  density, 78, 83
  formation volume factor, 111–113
  isothermal compressibility
    coefficient, 106
  specific gravity, 83
  specific volume, 83
  standard volume, 82
Gas solubility, 169–182
  differential liberation, 171,
    174–176
  flash liberation, 171–173
  methods of determining
    Beal, 177
    Glaso, 181
    Lasater, 177
    Marhoun, 181
    Standing, 177
    Vasquez-Beggs, 177
Gas viscosity, 113–123
  methods of determining
    Carr-Kobyashi-Burrows, 114
    Dean-Stiel, 121
    Lee-Gonzalez-Eakin, 120
Gibbs' phase rule, 19

H–K

Heavy (undefined) petroleum
  fractions, 36–74
High shrinkage oil, 24, 27
Ideal gas law, 78
Intensive properties, 4
Isothermal compressibility
  coefficient
    of formation, 236